HRD, OD, and Institution Building

Essays in Memory of Udai Pareek

Edited by

T. V. Rao | Anil K. Khandelwal

ATLANTIC

PUBLISHERS & DISTRIBUTORS (P) LTD

ATLANTIC
PUBLISHERS & DISTRIBUTORS (P) LTD
7/22, Ansari Road, Darya Ganj, New Delhi
Tel.: +91-11-4077 5252, 2327 3880
E-mail: orders@atlanticbooks.com
Web: www.atlanticbooks.com

Published by Atlantic Publishers & Distributors (P) Ltd 2024

Previously Published by SAGE Publications in 2016

Printed & bound in India by Atlantic Print Services

Dedicated to the memory of
Dr Udai Pareek
(1925–2010)

Contents

II. Institution Building

III. Social Development and Nation Building

List of Illustrations

List of Tables

List of Figures

List of Boxes

List of Abbreviations

ADC	assessment development center
AERC	Agro-Economic Research Centre
AHRD	Academy of Human Resources Development
APC	Agricultural Prices Commission
ASK	attitude, skill, and knowledge
BDO	Block Development Officer
BEI	behavioral event interviews
BOB	Bank of Baroda
BSC	balanced scorecard
CBI	Central Bureau of Investigation
CBS	core banking solution
CEO	chief executive officer
CFTs	cross functional teams
CII	Confederation of Indian Industry
CMA	Center for Management in Agriculture
CMD	chairman-cum-managing director
CoE	center of excellence
CUMI	Carborundum Universal
DC	development center
DMRC	Delhi Metro Rail Corporation
ECEL	Escorts Construction Equipment
ED	executive director
EQ	emotional quotient
ER	Employee Relations
ERDC	Employee Relations Development Committee
FDC	faculty development course
FMB	family-managed business
GIGO	Garbage-in, Garbage-out
GM	general manager
GSFC	Gujarat State Fertilizers and Chemicals Limited
HP	Hewlett Packard
HPWP	high-performance work practices

HPWS	high-performance work systems
HR	human resources
HRD	human resource development
HRIDC	Human Resources Initiatives Development Council
IIHMR	Indian Institute of Health Management and Research
IIMA	Indian Institute of Management Ahmedabad
IPS	Indian Police Service
IR	industrial relations
ISABS	Indian Society of Applied Behavioral Sciences
ISRO	Indian Space Research Organisation
JV	joint venture
KCRC	Kutch Comprehensive Rehabilitation Centre
L&T	Larsen and Toubro Ltd
LSIP	large-scale interactive process
MBO	management by objectives
MDP	management development program
MEP	Management Education Program
MNC	multinational company
NES	National Extension Service
NHEH	Navalbhai Hiraba Eye Hospital
NHRDN	National HRD Network
NIBE	National Institute of Basic Education
NTL	National Training Laboratories
NISIET	National Institute of Small Industries Extension Training
OB	organizational behavior
OD	organizational development
OCTAPACE	Openness, Collaboration, Trust, Authenticity, Pro-action, Autonomy, Confrontation, and Experimentation
OD	organization development
PMS	Project Management System
PSC	Plant Steering Committee
PSU	public sector unit
ROI	return on investment
SAFI	Self-Assessment through Feedback on Instruments
SBU	strategic business unit
SEM	structural equation modeling
SGIA	small group improvement activity
SOIL	School of Inspired Leadership
STMP	strategic talent management practices
STMS	strategic talent management system
STS	strategic training systems
TAP	talent appreciation process
TAPMI	T. A. Pai Management Institute
TAT	Thematic Apperception Tests
TCL	Tata Chemicals

TMTC	Tata Management Training Centre
TTS	traditional training systems
TQM	total quality management
VC	Vice Chancellor
VRS	Voluntary Retirement Scheme
WST	whole system transformation
XIMB	Xavier Institute of Management, Bhubaneswar
YEB	young executive board

Preface

This book is in continuation of the great work done and left behind by Dr Udai Pareek and is dedicated to his memory. T. V. Rao was associated with Dr Pareek for more than three decades and partnered many programs and projects with him. Mentored by him, Anil K. Khandelwal who rose from human resource (HR) to chief executive officer (CEO) position effectively used human resource development/organizational development (HRD/OD) concepts in transforming two large public sector banks. Udai Pareek's work had practical significance and eventually gave rise to building HRD movement in India. At a time when India was still an underdeveloped country and the importance of human factor in organizations was not adequately recognized, Pareek's work focussing on human processes, OD, and institution building in some sense raised the level of thinking of emerging management profession and HR professionals. He encouraged hundreds of academic colleagues, scholars, and HR practitioners to think differently, innovate, experiment, and raise the level of personnel job from "welfarism" to its developmental dimensions and institutionalize the human processes in the nervous system of the organization. His pioneering work with T. V. Rao in L&T in the mid-1970s led to an effective and integrated framework of HRD, which was far more comprehensive than the understanding of HRD by the West at that time.

It was felt that any such volume in his memory should be in continuation with his work and should not merely be a collection of essays but filled with lessons we learnt in our lives or a conceptual contribution we like to make in the fields all of us have been working along with Dr Udai Pareek. Hence, we decided to write to all those whom we knew as having been influenced in some way or the other by Dr Pareek. This included a number of CEOs and scholars from all over the world. As Dr Pareek spent a large part of his life in India and worked for Indian organizations and institutions, naturally the impact was much larger in India. As early as in 1967, he wrote an article on a paradigm of development that got published in the most prestigious *Journal of Social Issues*. In that article, Dr Pareek introduced the concept of extension motivation and formulated a theory behind development. His work on human processes or behavioral sciences, human resources development, institution building, and social development or development of various sectors has been seminal. The contributors to this book have been his colleagues or students or both. When we started this effort, we did not have any a priori focal points, though we had listed out areas of work by Dr Pareek. We requested each writer to share the best of his or her thoughts, experiences, or lessons learnt in any of the fields, such as HRD, OD, institution building, and behavioral sciences. What emerged from this effort is this book. The chapters are narrations of the professional growth and excellence of many who emerged as leaders in thought or practice in human processes, HRD, OD, and institution building areas. The human processes relate to any area starting from individual growth to organizational development and nation building. Each chapter provides the context where it can be used. The HRD-related chapters deal with the insights from the growth of professionals who emerged as leaders in their own right: Dennyson Pereira's, Arvind Agarwal's, and Gopal Mahapatra's chapters draw the lesson for the growth of HR professionals. The book starts very rightly with a note from Rolf Lynton who had been his co-author and colleague of for over 50 years.

Insights from HR Leadership Careers: These chapters will hopefully be a source of inspiration to the younger generation to maximize their contributions and use their potential wherever they work.

- Dennyson Pereira, who lives in Canada, traces an academic-turned-HRD manager's career at various places, especially L&T, where Dr Pareek started the first department of HRD.
- Gopal Mahapatra presents various phases in his own life and how different people and settings enabled him to perform, grow, and contribute.
- Arvind Agarwal presents 10 lessons he learnt from his own career as a CEO, a change agent, and an HR leader.
- P. Sethu Madhavan presents an analysis of the tradeoffs HR professionals should be aware of in various "tradeoff" situations in order for them to make effective and conscious choices. He lists some tradeoffs that he experienced and observed during his career as an HR professional in different countries, as well as in different roles, namely, academician, researcher, trainer, consultant, and executive.

The next set of chapters deal with the current state of the HRD function and suggest ways and means to enhance the HR's impact. This section will be of particular significance to all HR professionals, CEOs, and all managers.

- Pradip Khandwalla presents creative ways of changing mindsets and influencing people in the chapter on creative HR.
- S. Y. Siddiqui presents his experience at managing change at Maruti using various HR interventions.

- Anil Sachdev and team present a detailed case study of the role played by leaders in managing change in one large corporation.
- G. P. Rao, drawing from his rich experience of working in India and Malaysia, deals with various types of groups and teams and presents diversity management principles for organization building. He concludes that inclusion is the mantra for diversity.
- Rajeshwari Narendran presents nine futuristic principles for changing the HR and its impact. She concludes that the entire DNA of HR needs to be changed for making a better impact.
- Drawing from his book *Dare to Lead*, Anil K. Khandelwal very convincingly presents the need for focusing on intangibles that drive tangibles implying the significant role HR leaders and CEOs as HR leaders can play in organizational transformation.
- D. M. Pestonjee and Naresh Mehta suggest practicing HRD managers to adopt an appropriate high-performance work systems as part of their people strategy and to carry out research in various formats in their organizations for adding values.
- Visty Banaji, drawing from his rich experience of working with several top-level corporations in India, presents several maladies that the HRD profession is facing and also suggests various approaches to come out of them and make an impact.
- Rajen Gupta and Vikas Bhatnagar explore the philosophical basis of action research as a superior methodology for approaching organizational science and HRD. They present a case for the HRD profession to jump a few orbits in building functional credibility and impact if approaches toward designing and instituting HRD initiatives are based on action research.

Some of the chapters deal with managing change in various settings using OD and other interventions. The settings vary from small groups to large groups, organizations, societies, and nations. These are classified under OD and social development as the conceptual contributions go beyond managing organizations.

- S. Ramnarayan, as a change champion, suggests various insights for change leaders to make an impact. For example, he suggests that change leaders need to have faith in their ability to bring about positive change and they need to communicate this faith through actions and results. They need courage to challenge dysfunctional mindsets but must also be able to offer viable alternatives.
- Rosemary Viswanath and Tejinder Singh Bhogal describe their proposed "social catalysis" as the ninth metaphor. This metaphor helps us to explain and mange social change better. Metaphors are best used for a diagnostic reading to help look at the familiar with freshness and to

open the doors and windows of the mind. Gareth Morgan in his famous book, *Images of Organizations*, compared organizations using metaphors to machines, organisms, brains, cultures, political systems, psychic prisons, flux and transformation, and instruments of domination.

- Zeb Waturuocha presents and illustrates eight competencies required for interventions in the social development sector; these include ownership and identification, understanding and practice values and ethics, authenticity and transparency, diversity and inclusion, empathy and humanness, learning and research orientation, walking the talk, enhancing, enriching and engaging, and evaluating and reforming.
- Paul Siromani discusses "extension motivation," a concept evolved by Udai Pareek and its applications for developing commitment to organizations and society through extension motivation labs.
- Keith D'Souza and team present through an empirical investigation the gap between professed and practiced values in family-owned business organizations. He points out the great need for orienting family-owned business managers.
- Varun Arya outlines the development needs of the country and the role people with passion can play in building the nation.
- Inderjit Khanna presents and argues strongly the need for values and value training in civil services for nation building.
- Aquil Busrai presents very valuable suggestions for bridging the theory–practice gap by making educational institutions and industry play some new roles and adopting new approaches.
- Roland Sullivan and team present from their global experiences how whole system transformation enables and empowers human talent in organizations to accomplish faster, cheaper, and sustainable positive change. It marries leadership and the organization through the use of small and large social groups resulting in enterprise-wide transformation. It involves and engages the entire system, allowing the ownership of the process (people support what they co-create), suggesting and causing change to happen in the moment, and focusing on the alignment rather than planned or segmented change.

The next set of chapters deal with the important theme of institution building. They cover institution building of corporations, government departments, universities, national institutions, central university, registered not-for-profit societies, and management schools.

- Kavil Ramachandran presents experiences of institution building in India and other places and presents insights for institution builders. He concludes that good institution building is an essential step for nation building.
- V. S. Vyas presents his own experiences of contributing to building of Indian Institute of Management Ahmedabad as its Director.

- Abad Ahmad describes his experiences of building the South Campus of the University of Delhi.
- D. Nagabrahmam describes the process of building T. A. Pai Management Institute, Manipal.
- Somnath Chattopadhyay presents in detail, starting from the finding of the National Training Laboratories in the USA, the establishment of the Indian Society for Applied Behavioral Sciences (ISABS) and the value dilemmas and developments. ISABS remains a stable voluntary organization based on certain values and is illustrated in this chapter.
- T. V. Rao presents in some detail conceptual contributions by Udai Pareek to HRD, OD, institution building, and role efficacy and extension motivation. He draws lessons for social development.
- Indira Parikh offers various suggestions and thoughts for academic leadership in universities that can build the future of the nation.
- Lalitha Iyer presents role of values in shaping higher education.
- K. K. Verma presents details of the development of the Blind People's Association in Ahmedabad. Various insights into the development of this institution are provided. Values practiced by the educators in the higher education system strongly determine the results that are desired but remain elusive.
- Ganesh Chella presents his experiences and insights for institution building of consulting firms.
- Anil Gupta, using his experiences of building a Honey Bee Network to discover and disseminate grassroots-level innovations, presents thoughts and approaches to make grassroots innovations globalized phenomena. He expresses his optimism that the Honey Bee Network provides a viable framework for transforming the focus and force of globalization. A recent initiative of bringing community/individual grassroots innovations from China, India, and Brazil together is delivering some results and hopes that more and more young people, start-up companies, investors, and entrepreneurs will explore the untapped potential of grassroots to globalized phenomena.

One underlying theme of all chapters in this book is understanding and enhancing "human processes" and their value to maximize the discovery of talent and enhance the impact of individuals, groups, organizations, and institutions. This is the theme that Udai Pareek has always worked for all through his life and his life stands out as an example to emulate for all of us.

Like Udai, we hope this book will continue to inspire and guide all those interested in growth and development beyond management.

T. V. Rao
Anil K. Khandelwal

Acknowledgement

We thank each one of the authors for their contribution.

We thank Atlantic for bringing out the current edition. The first edition was brought out by Sage in 2015 and the contributors' bios are maintained as in the original edition.

We thank Bank of Baroda Staff College for facilitating our editorial meetings.

We must acknowledge the contribution of our respective families (Jaya Rao and Vandna Khandelwal) for their support while we spent enormous amounts of our time on this book, which belongs to them.

We also thank T. V. Rao Learning Systems Pvt. Ltd for giving us all support facilities for making this work possible.

Memoirs from a Professional Twin Brother

Rolf Lynton

"Twins" is what many of our colleagues in India called Udai and me at Small Industry Extension Training Institute (SIET Institute, now known as National Institute for Small Industries Extension Training [NISIET]) in Hyderabad where we both were working together for the first time half a century ago. Very soon Udai became the director of its Extension Education department and so my "counterpart"—a graceless domineering foreign assistance term—as a consultant for Extension Education to the new staff college for the Government of India's small industries consultants.

We immediately saw eye to eye on just about everything. He too was convinced that the consultants needed to work in interdisciplinary teams. For that, they needed interpersonal skills that small-group and sensitivity training develop the best and that Udai had learned from Max Corey to facilitate and highly value. He easily saw that those skills also had to get rooted in actual practice together with operatives, foremen, and managers on the factory floor and not just talked about with managers in the office.

SIET's director had already agreed to all parts of this sequence, and the initial programs had strongly confirmed. So Udai stepped right into pairing with me to facilitate the small-group sessions and organize and review the fieldwork (with me or Sujit Bhattacharji, who had come with me from Aloka).

With Udai—and/or soon after—also came doctoral students who had worked with him in Delhi; some of them also stayed as faculty members in Hyderabad or worked together with us later—most continuously during our stay of six years in the 1980s—to decentralize decision-making in Indonesia's public health system and to develop India's own professional society for applied behavioral science.

In addition to becoming ever-closer colleagues at SIET Institute, we became long-lasting family friends. In Hyderabad, our family visited Udai's home when his son Anagat was born and the hijras (eunuchs) danced on this occasion. Twenty years later, our family visited his home in Jaipur when 'Anu' (nickname of Anagat) married Shilpa and Udai's mother danced in front of the procession to the bride's house. One of the photographs on our wall shows me with the turban put on my head by Udai for the wedding.

We overlapped two years in Chapel Hill—Udai in the Department of Psychology and I in the new Population Center and in Public Health— and Udai's family broke the coconut in our new house just as they did earlier in Hyderabad.

Next we overlapped in Indonesia in his successive provinces and then in Jakarta where they also had a twin-grandson with them. And after that I, often with Ronnie, visited them many times in Jaipur, and each time Rama prepared extra rough and tough Rajasthani bread for me because that's how I liked it best. We also got to know their wider, expanding family, which we like too.

Only weeks before Udai died, I saw him already frail and sick. Then, the third edition of *Training for Development* was under way; the first edition had already brought us close while working together all through Hyderabad to Chapel Hill.

From our working together so closely and personally Udai's two particular features stick out most prominently for me. One is his openness to what mattered most and the other is his quiet determination to go for it. This became most startlingly clear at our first and quite unplanned meeting and its quick sequel at what others would quite understandably assess at an enormous price.

In 1962, at Erik Erikson's first public meeting in Delhi, Dr Pareek chaired a reputed session to introduce psycho-social history, with Krishna Kripilani introducing his biography of Tagore, *Rabindranatha Tagore: A Biography*. Standing around as it finished, this stranger asked me what I was doing in India, and I told him about the new institute, the three-week interpersonal sessions we had started there and, in response to his asking, I presume, how I had come to value those so highly. "Were any position open there," he asked.

Only a few days later, his application arrived, to work with me and in the hope of becoming the department director. What he was foregoing, it turned out, was his 18 years' retirement benefits in a senior research position in the Ministry of Agriculture.

In later years too, I have seen Udai again and again quietly assessing complex situations and personal choices, and come up with particular action(s)—quietly, avoiding open conflict with others, and seeing no need to explain his determination.

The second feature I think of in his working is his relentless recording of experiences. It stands out extra starkly because even otherwise excellent colleagues—everywhere and certainly in my work in India—seem to have great difficulty developing like habits for full systematic recording (even now that the ever new technologies have made it much easier).

Two examples can best illustrate the recording and the sharing of it with participants. First is the content sheet of our standard report to the participants in a three-week program of "Sensitivity Training in Regular Courses":

First, in three sections, 12 single-spaced pages, about

The Training Design
Sensitivity Training in Practice
Effects of the Training on Participants

Then 12 appendices in four sets:

The five readings handed out during the program
Extension sessions at the SIET Institute
Five data forms about talk frequency, interaction patterns, participants' functions in the group, feelings in the group, and group climate

And lastly, the participants' "rating of sensitivity sessions."

The second example of Udai's (and my) characteristic follow-through is that all the readings we referred to we soon included in a hardback SIET book entitled *Agents of Industrial Development.* Our book *Readings for Trainers, Consultants and Policy-Makers* followed some years later and also *Training for Development,* rewritten separately for policy makers and for trainers (SAGE, 1992, 2000).

PART I

HRD and OD

1 My professional journey in HRD: Some reflections

Dennyson F. Pereira

The beginnings

My initial years of working at the Government of Bombay Vocational Guidance Bureau equipped me with a sound background of validation and reliability processes. I was primarily responsible for the standardization of the differential aptitude tests and the Raven Progressive Matrices on a sample of more than 3,000 participants. I also assisted in the training of school counseling. Soon, a valuable opportunity arose to move into clinical psychology and work with a leading US psychiatrist, Dr Louis English. The position was in the Faculty of Social Work, the MS University of Baroda. At the university, I worked, first, on the Margaret Lowenfield Mosaic Test, Bender Visual Gestalt Test, Raven Progressive Matrices, and Alexander Passalong Test. Then, I looked for appropriate pictures to elicit responses and to measure achievement, affiliation, and power motives through the Thematic Apperception Test. This search led me to contact Dr David McClelland of Department of Psychology and Social Relations, Harvard University. Later when David came to India, he met Udai at the Ford Project in Aloka, Mysore. He was so impressed with Udai that he strongly suggested me to get in touch with him. I was also in close contact with Tavistock and NTL.

Later, on joining the Sarabhai Group in 1959, I was fortunate to be one of the four persons in India chosen for the six-month United States Agency for International Development (USAID) training under a management consulting organization in the United States. I was earlier in touch with NTL and looked for the first opportunity to learn more about sensitivity training. In the second week of my stay in Washington, I met Dr Leeland Bradford, one of the three persons who started NTL, and during our discussions he strongly suggested me to develop a relationship with Udai. Furthermore, he informed me about a forthcoming program being organized in Washington, D.C., in April. At this program, two renowned sensitivity trainers, Weldon Moffit and Gordon Lippit, were present and spoke very highly about Udai.

On my return to India, the USAID office in New Delhi requested me to deliver a presentation at a conference to be held for all persons who had been deputed under various US scholarship programs. At the end of my presentation, two persons approached me and introduced themselves as Dr Udai Pareek and Manohar Nadkarni. This was the start of the most fruitful collaboration and professional growth opportunity for me. Manohar and I since then have been responsible for conducting several programs in achievement motivation.

My professional journey at Larsen and Toubro Ltd

Before I joined Larsen and Toubro Ltd (L&T) in October, 1966, the training was imparted only at the apprentice level in the General Workshop (Group II, Caterpillar Service Station, and the Tool Room). It was Mr Gunnar Hansen, director of the Powai Works, who identified a strong need for the training of foremen and chargemen. Mr Hansen equipped the facilities in Powai to make it one of the best training centers in India. He ensured comfortable seating facilities, moving and rolling blackboards, modern audiovisual equipment, and small rooms for group discussions.

The training department commenced in October 1966 and originally was supposed to cater to the needs of Powai Works. However, we started with a company-wide training needs assessment. After six months, an opportunity arose for us during the annual sales conference in April 1967, wherein we were requested to make a presentation on the value of need-based training programs to enable the Powai Training Center to raise the proficiency of foreman and supervisory levels. At that meeting, the sales division became aware of the absences of training for their division. We stressed a need-based systematic identification of training needs, planned deputation of participants, and active involvement of department heads, production, and sales.

Within a few years, the training unit conducted several excellent supervisory and managerial programs that focused on problem-solving, time management, enhancing pertness, value engineering, written and communication skill building, finance—for nonfinancial personnel—and coping with stress. For the different levels of supervisory and managerial staff, we conducted group and team working programs. We moved away from the traditional T-group method and designed programs modeled on Blake and Mouton Managerial Grid, Bernard Bass Case Situations, and Harvard Business School's excellent case studies. We secured appropriate films on leadership and group dynamics. The success and efficacy of these programs resulted in their increased demands. Quarterly evaluations of the programs were sent to all departmental heads, and to the top management.

During a major lockout, sessions of training programs at the Powai Works were held outside the premises, to assist supervisory and managerial staff to cope with the situation on the resumption of work.

Similarly, a major situation occurred at the Utkal Machinery Limited, Khansbahal. L&T was one of its major partners. Therefore, the chairman, Mr Holck-Larsen requested the training unit to devote major energy of the training programs at Khansbahal. We planned and conducted a one-year company-wide training that covered the senior management, the supervisory level, and the foreman level at Utkal Works. The final evaluation indicated the effectiveness of the training inputs. This was independently stated by officers and employees in the final evaluation.

> Holck Larsen himself in a note dated March 23, 1972, to all the directors of Utkal Machinery acknowledged that the courses contributed notably to the development of team spirit.

From training to Human Resource Development (HRD)

In October 1975, a filing cabinet in the personnel section of the head office in L&T House, Ballard Estate, drew the attention of the chairman, Mr N. M. Desai. It contained reports of performance appraisal forms. The exercise was carried out only at two levels: immediate superior and next superior. No action on the forms was taken beyond the next superior.

During our discussion with the chairman, we suggested that Dr Udai Pareek, a professor at IIMA, be requested to conduct a study of the performance appraisal system and the total process. We placed a proposal to the chairman, and he himself presented the proposal before the board of directors.

The proposal was accepted and assigned to Udai. He proposed the formation of an in-company internal task force of senior management personnel and an external task force consisting of Dr T. V. Rao and himself. Based on his study, he felt that the company needs to expand beyond the performance appraisal system.

Under the new proposal, Udai set the pace for renewal, and enlargement of the following functions beyond training: (a) manpower planning, (b) data storage and processing, (c) critical attributes analysis for each job position, (d) feedback and counseling skills by line or staff managers, (e) feedback and counseling skills, (f) conduct of development-based performance appraisal, (g) inclusion of potential appraisal, (h) harmonious industrial relations at the supervisory level, (i) organizational development unit wise and division wise, and (j) career development and career planning for higher positions.

He cited the following benefits of the new game plan:

1. Employees understand and assess their own abilities, strength, and weaknesses that help or hinder them in contributing to the goals of the organization.
2. They have an opportunity to discover and identify their own strengths in working with people and unlock the potential they possess to evolve a future development of their career.
3. They discover the limitation and identify resources that lead to the development of their potential.

4. They would periodically endeavor to contribute to the organization goals through their own growth and development.
5. They plan their career on the basis of the above insights.
6. Organizations work on creating a new developmental organization culture.

The birth of HRD

To achieve the points discussed above, Udai proposed a comprehensive system approach: (a) development-oriented performance appraisal, (b) a scheme for potential appraisal, (c) supportive employee counseling, (d) career planning and development, (e) training programs that include developing multilevel skills, and (f) organization's development for unit and/or divisional change process.

Udai emphasized that the above-mentioned approach should be implemented and monitored in a phased manner to be able to plug any drawback and to ensure synergetic benefits. He proposed a five-to-seven-year phased implementation period that was to be followed by (a) a six-member senior management committee and (b) by Dr Pareek and Dr T. V. Rao. Further, it was proposed that the phases be monitored by a human resource unit, with strong linkage to the personnel.

L&T executive management examined the painstaking report of Udai and Dr T. V. Rao. The President, Mr N. M. Desai, who had himself followed the plan, provided his whole-hearted support, stating, "[i]n view of the considerable growth of the Company ... an Integrated Human Resources Development system has been examined and accepted by the Corporate Management."

Dr T. V. Rao and Dr D. F. Pereira conducted the first training program for 27 senior line and staff managers complete with visual aids and a training manual. These participants spread out throughout the regional offices and departments and conducted such effective programs that their efforts created a stimulating effect about the implementation of the consultants' recommendations. Employees were impressed to witness their own managers conducting the two-day programs.

With the assistance of the HRD unit, groups were formed within the units to examine critical attributes and potential appraisal factors, and to plan for career development. Employees were now able to visualize the path they could aspire to during their term in L&T. They participated in small groups for contributing to various implementation steps. Udai laid the grand plan for a holistic approach to human resources in the corporate world.

The editor of *International Management*, a McGraw-Hill Publication, located in UK, personally visited L&T and commended the training programs, the top management, and the total management culture. Thus, Udai and Dr T. V. Rao were responsible for the creation of the first total system-based renewal of the human resources in India. Various companies, including some leading national banks and India's top 25 companies, requested us to make presentations of the total HRD process to their senior management.

In this tribute, let me cite Udai's brilliance as a change agent in a hospital care organization that was run by an international religious community of sisters.

The Medical Mission Sisters has several hospitals and rural health centers in the world. They, then, planned a worldwide renewal program. Dr White was the head of this worldwide project and Udai was responsible for the all-India group. He involved me in working with the Bombay Group of Religious Sisters. On the completion of the program, Dr T. White was expected to present a report on the program and its implementation plans in Rome. Thus, Udai, because of his brilliance, was requested to make the final presentation that would be submitted to Vatican.

I learnt from some sisters that Udai performed a superb job.

On the personal level, I had conducted an in-depth study of the organizational culture of the Diocesan Priest in the Archdiocese of Bombay. During Udai's frequent meetings in Bombay, I happened to show the results of the findings to him. He immediately suggested me to visit Ahmedabad. On my arrival, he personally took me to the vice chancellor's office and requested permission to register as a PhD candidate. I, then, witnessed the high esteem that the vice chancellor had for Udai. Within two years of my enrollment, I refined the work with Udai's help; the thesis received an outstanding rating from the external examiners, and recognition from Rome, Poland, and Hungary.

Both Udai and I had a common role model: Colonel Sohan Lal, who directed the Indian Army Psychological Wing at army headquarters.

While working on different projects with Udai, before and during L&T Renewal Project, and on receiving his guidance during my doctoral work, I became aware of Udai's unique gifts. He believed in establishing authentic connections with people and thrived on helping them develop their genuine authentic potential. He could perceive the positive in people and in their future. Connecting with people was more important to him than his rules or plans. He excelled in assisting people clarify issues, ideas, and more than anything else values.

After the loss of my wife, who spurred me to complete my master's and doctorate, my teachers from St. Andrew's High School, and my first superior in the Government of Bombay, Mr Fali Chothia, Udai was an inspirational and a personal loss. He had helped me to clarify issues and values, to develop an identity, and also encouraged me for growth and development.

Recognition

At this stage of my life, I feel proud to see L&T growing by leaps and bounds and making marks globally. It is gratifying to see that the strong foundation of HRD made in the 1970s had a role in several achievements and accolades received by L&T. After all, it is the competence, passion, and professionalism of L&Tites that is reflected in many of its achievements. It also vindicates my belief as an HRD professional that investment of intent, time, and money in developing people always bears rich dividends. Today, L&T stands tall globally, as is evident from the fact that in 2012 *Forbes* ranked L&T as the 9th most-innovative company in the world.

Following is the list highlighting some more achievements of the company.

1. In 1997, the Bangalore Works division was awarded the "Best of all" Rajiv Gandhi National Quality Award.
2. In 2012, L&T was ranked 4th by *Newsweek* in the global list of green companies in the industrial sector.
3. L&T was also featured in the *Forbes* "Asia's Annual Fabulous 50" list in 2010, for the fifth time in a row.
4. According to the Brand Trust Report 2012 published by Trust Research Advisory, a brand analytics company, L&T was positioned 47th among India's most-trusted brands. Subsequently, in Brand Trust Report 2013, it was ranked 127th, while according to the Brand Trust Report 2014, it was elevated to the 38th position.
5. In 2013, L&T Power received the "Golden Peacock National Quality Award, 2012" at the 23rd World Congress on "Leadership & Quality of Governance."
6. In 2014, L&T was ranked 500th on *Forbes* list of 2000 world's largest and most powerful public companies based on revenues, profits, assets, and market value. In total, 54 Indian companies made it to the prestigious list, and L&T was the highest ranked company in the engineering and construction section and 10th among all Indian public and private sectors.

2 Toward creative HR

Pradip N. Khandwalla

I have fond memories of Dr Udai Pareek. He was at Indian Institute of Management Ahmedabad (IIMA) from 1973 to 1985. I joined IIMA in 1975. He was a different person than I am, a much more gentle and refined human being. But we also shared a lot in common: a passion for research and scholarship; a shared commitment to the fields of organizational behavior (OB) and human resource development (HRD); a belief that management education should result in greater commitment of students to human good, and not just to profit and self-interest maximization; a shared interest in creativity and innovation; a shared interest in designing systems and organizations that are not only efficient but also effective, humane, and innovative. I was particularly interested in his concept of extension motive and also his views about the institution-building function in organizations. His concept of extension motive led me to think of the altruistic motive in people and how it manifests itself in a distinctive style of management that I called the altruistic style of management (Khandwalla, 1995). His notions about values, processes, and systems that build humane institutions too appealed to me and led me to formulate notions of institutional excellence and missionary excellence as types of organizational design (Khandwalla, 1992). As we all know, Dr Pareek was a towering figure in India in the areas of OB, human resource management, and HRD, and was also a prolific writer, with some 30 books to his credit and several hundred papers and articles. I feel privileged to have been his friend and colleague, and I feel honored to have been asked to contribute a chapter in a book dedicated to his memory.

In this chapter, I wish to share some ideas on creative human resources (HR). Creative HR is indispensable in our times in India because most Indian companies are far behind the best companies in the world in productivity, quality, customer satisfaction, innovativeness, corporate social responsibility, and various management functions including HR. Creativity in HR offers a way of catching up with the world. If we do not catch up with world standards quickly, our companies will not be able to remain competitive in the global business place. But if we can, then opportunities are of Himalayan proportions. Right now a heavily depreciated rupee and relatively low salary and wage levels are bailing our companies out in the global market place; but how long?

Creative HR

Creativity walks on two legs (Khandwalla, 2004, Ch. 1). One is an analytical, logical, and focused mode of thinking. The other leg is imaginative, even fanciful mode of thinking, which extensively uses analogies, suggestions, fantasies, etc., rather than logic and analysis. Professor Guilford called these two modes convergent thinking and divergent thinking, and he argued that both modes are necessary for creativity to yield results (Guilford, 1960). Creativity is also often a play between the known or familiar and the novel or unknown. That is to say, it often grafts the novel on the known. A good example is 360-degree appraisal. Normally, the boss appraises his or her subordinate. But in 360-degree appraisal not just the superior, but also colleagues and subordinates of a manager, as also the manager himself, also assess the manager. In the history of assessment this is quite novel. Creativity is also the fusing of two previously unrelated frames of reference (Koestler, 1970), as when a manufacturing company benchmarks its innovativeness with the innovativeness of a high-performance information technology (IT) company rather than only with the best companies in its industry.

Let me first outline a major difference between creative HR and textbook HR. Textbook HR systems are standardized and tend to be one size fit all solutions. They represent past innovations in HR that have now got refined and standardized. But given the diversity of organizations in terms of organizational size, cultural setting, type of industry, operating environment, and the organization's internal culture, they give good results when the fit between the HR systems and the context of the organization is strong but give at best mediocre results where the fit is weak. Creative HR involves offbeat improvisations by the organization in response to its operating context, and so, if the improvisation is implemented effectively, it is likely to give very good results. In a dynamic world, the capacity to improvise innovations in HR may be crucial for survival. Most of the examples of innovative HR that I am sharing later were initiatives during company sickness that helped the companies to turnaround and become profitable (Khandwalla, 2001, Chapter 15, especially Appendix 15.1).

There are many examples of creative HR that I have come across as a consultant, researcher, and a student of turnarounds of companies from sickness. There are some areas of HR that are important but as yet the scholarly knowledge base is weak. The examples of creative HR practices that I am going to share later were not drawn from textbooks or handed down by experts. They were relatively novel improvisations, and many of them were initiated by the chief executive officer (CEO) and the top management team, sometimes with the help of consultants but more often without such help. These creative HR practices can be subsumed under five heads:

1. *How to bring about a mindset change in the organization.* This is important in contemporary India because, exceptions apart, most companies have

fairly authoritarian, laidback, and conservative cultures, ill-suited to our times, and the need is to change the mindset to one of flexibility, dynamism, innovation, professionalism, ethics, social responsibility, and workplace equality.

2. *How to identify and harness change agents at lower levels of the organization for bringing about organizational transformation.* This too is important for our organizations because change tends to be perceived as the prerogative of the CEO and his or her immediate subordinates, and tends to be concentrated at the top and at best at senior levels. However, given the increasing size and complexity of many Indian organizations, and the need to jack up their performance swiftly, the need is to bring about decentralized changes also in organizational systems, practices, structures, styles of leadership, etc., through change agents. These change agents need to have both the temperament to be agents of change and the needed competencies.
3. *How to establish a better rapport between the top echelons of the organization and the rank-and-file.* In India, the managerial levels tend to be staffed by social elites while the working class mainly staffs the workers and lower level functionaries such as clerks and first-line supervisors. There is a cultural gap, and also a feeling in the lower echelon people that they are only tools in the hands of the managers. This often breeds alienation or militant unionism that can affect the operations adversely.
4. *How to empower the people at the lower rungs of the organization.* There is much talent at lower rungs of the organization, especially among junior-level managers, but the opportunity to contribute to the organization is minimized by authority differences. If the lower levels are empowered they would feel a much stronger stake in the organization's future and they would be able to make much larger contributions.
5. *How to get the lower level employees to contribute to the organization in areas beyond their remit.* There is a tendency to view employees as contributing to the organization only by playing their formal or assigned roles. But they may well be able to contribute in diverse ways beyond their remit to the profitability and well-being of the organization. However, this calls for creative ways of tapping their skills and enthusiasm beyond what is required to fulfill their contributions as formal role-holders.

Creative ways of bringing about mindset change in the organization

Bureaucratic, conservative, and laid-back mindsets of the staff do not work well in a dynamic, competitive environment. In the 1980s, British Air, a public sector company, was notorious like Air India for its poor customer services and laid-back culture (Goodstein & Burke, 1991). One of the problems was a mindset of caring little for the customer. A vigorous effort was made to change this mindset. About 40,000 staff members, mainly those with direct customer contact, went through a two-day Putting People First program to foster good interpersonal relations, especially with the customers. GM, the American car making leviathan, had a strong manufacturing orientation and low entrepreneurial orientation (Kharbanda & Stallworthy, 1987). In the 1980s, GM temporarily transferred 7000 data-processing personnel to its dynamic IT subsidiary EDS to absorb the entrepreneurial culture of EDS. At Jaguar Motors, the company arranged irate customers to meet the workers and managers to share their frustrations with the quality of Jaguar cars and make the staff much more quality conscious (Chambers, 1988). SAIL, the giant Indian public sector steel company, was ailing in the 1980s. Its market dominance, weak leadership, and strong unions had made the organization complacent and conservative. In the mid-1980s, Mr Krishnamurthy, with a career in BHEL, became the CEO. He improvised a fascinating strategy to bring about a mindset change and greater focus on priorities (Krishnamurthy, 1987). At SAIL, the CEO personally conducted several two-day workshops for general managers in batches of 80 each to discuss the turnaround strategy termed Priorities for Action. The general managers, in turn, conducted similar workshops in their plants. The write-up on priorities for action was mailed to every employee of the company, and shift meetings were organized in each plant to discuss it and come up with local innovations. A vast training program was launched. In total, 19,000 workers and other staff and 10,000 managers were trained in one year, and the numbers in the next year were 37,000 and 14,000. At TVS-Suzuki, India, the top executives punched attendance cards just like workers and also got searched like them on departure in order to turn a hierarchical culture into an egalitarian culture (Ravindranath, 1995). Andhra Pradesh Papers conducted classes twice a month to expose workers to new concepts of industrial relations, economics, and productivity to bring about a more professionalist mindset (Shenoy, 1996).

Creative ways of identifying and using change agents to bring about organizational transformation

In hierarchical organizations the responsibility for change rests mainly with the CEO. But as organizations grow larger and more diversified in terms of both product portfolio and territories, unless changes and innovations are continually initiated at lower levels, the organization may not be able to cope with differences in its operating environments. This requires a contingent of empowered change agents. At British Air, selected staff members were retrained to become internal change agents to assist line and staff managers (Goodstein & Burke, 1991). In Lufthansa, 25 middle-level managers were constituted into the Samurai of Change, and after training, they were charged with communicating the gravity of the company's situation to the rest of the staff and managers (Lehrer, 1995). At National Semiconductor, champions were identified for each financial and transformational goal of the company, and also for each focus market and each strategic market segment (Miles, 1997). At Philips Electronics, 400 line and staff persons were given change agentry training lasting 2 to 3 days to create an internal battalion of change agents (Freedman, 1996). Siemens-Nixdorf, a German IT software and hardware company, identified some 2000 potential change agents through a nomination process, and these were sent to the United States for 13 weeks to visit dynamic companies (Kennedy, 1998). The idea was to create a mindset change from an excessive quality and productivity fixation to innovation and entrepreneurship. Of these, 400 were trained in entrepreneurship. On their return from the United States they were made heads of SBUs and profit centers.

Innovative actions to bring about better rapport between top echelons and much lower echelons

As organizations grow larger, a hiatus tends to develop between the perceptions and priorities at top levels and much lower levels. This often affects the implementation of corporate plans adversely. In American Airlines, a giant US airline, the CEO conducted give-and-take sessions with workers throughout the route system, with rewards of travel passes or merchandise for innovative suggestions (Horn, 1990). In BHEL, the giant Indian public sector company manufacturing electrical equipment, every quarter top-level executives made it a point to visit the plants and meet a cross-section of the staff to discuss problems and issues (Krishnamurthy, 1977). At British Steel, the CEO developed a rescue plan for the ailing, over-sized, bureaucratic company involving a large number of separations and other restructuring, canvassed it with the rank-and-file, and put it up to a company-wide vote—and won it (MacGregor, 1982). At Philips Electronics, some 200,000 employees were involved annually in a one-day exercise to search for better ways to satisfy customers (Freedman, 1996). They also communicated via satellite link with the CEO.

Innovative ways to empower people at lower rungs of the organization

Continental Airlines, United States, instituted an annual survey in which subordinates rated how well their bosses communicated with them (Brenneman, 1998). Eastman Kodak instituted a system of "social contract" in which the workers pledged to understand better Kodak's business and customers, to adapt to change, and give 100 percent of effort, in return for Kodak committing to minimum 40 hours of training for each worker that was designed to meet each worker's specific developmental goals (Chakravarty, 1997). Ford Motor Company, a giant American car manufacturing company, had instituted a Code of Quality. It empowered the worker to stop the assembly line if he detected anything that would endanger the quality of a product (Shook, 1990). At Harley Davidson, an American maker of high-quality motorcycles, a system of statistical control was installed that gave employees the responsibility for checking the quality of their own work and then making the appropriate corrective adjustment (Hartley, 1997, pp. 58–70). In JMEL, India, a Rajasthan state company that produced metals and electrical equipment, the CEO got the workers to accept shares in lieu of bonus (Khandekar, 1985). This way, the workers got seats on the board of directors. In Lufthansa, a German airline, the management set up groups under each of six supervisory board level departments such that each group had three representatives of managers and staff. These groups had to be consulted on all important decisions (Lehrer, 1995). Premier Auto, India, replaced a confidential appraisal system by an open appraisal system that emphasized innovation and initiative taking (Nilakant & Ramnarayan, 1998, pp. 314–318). It also created a system in which senior managers got feedback from their peers and subordinates. Siemens-Nixdorf, a German IT software and hardware company, installed a flexible assessment system in which an employee could choose the criteria on which he or she would be assessed out of a given list (Kennedy, 1998). The company also set up 18 communities of interest, whose members, linked across the company by intranet, could pursue a core competence or area of interest, to present what they had learnt to the management. Everyone's improvement commitment was flashed to the entire organization on intranet.

Innovative ways to get lower level staff to contribute to the organization in areas beyond their remit

Aer Lingus, an Irish airline company, leveraged its HR skills in equipping and managing a hospital in Baghdad and nursing homes in the U.K. (Arbose, 1986). In American Airlines, a giant U.S. airline, the CEO invited workers and their families to lavish feasts at which the invitees gave feedback on how good the dishes were (Horn, 1990). The company wanted to introduce these dishes for passengers. Both Bank of America (Clausen, 1990) and Bank of Baroda (Killawala, 1992) launched campaigns in which the members of the staff were asked to bring potential new customers to the bank. In Bank of Baroda, instead of the targeted 100,000 new savings and loan accounts, as many as 300,000 were opened. Chrysler placed 21 recent retirees into the plants of vendors to make them much more conscious of the quality requirements of Chrysler and thereby to improve the quality of components supplied to the company (Gordon, 1987). The company engaged in extensive benchmarking for everything from accounting to manufacturing by creating teams in which both managers and union members were represented.

The above examples from turnaround cases indicate that creative improvisations in HR are both feasible and necessary for boosting the performance of the organization. Many of the companies such as GM, American Airlines, SAIL,

BHEL, and Bank of America were large and had established HR systems. But they either were not suitable or were badly managed. Hence improvizations had become necessary.

Methods for increasing the creativity of HR

There are ways by which the HR function can be made much more creative and innovative. There are a number of creativity techniques that HR people can learn and disseminate to others in the organization (Khandwalla, 2003, Ch. 9). These include brainstorming, creative benchmarking, exnovation, organizational experiments, stakeholders' councils, and creativity training. Let me briefly describe each as it relates to HR.

Brainstorming: This is a group technique developed by Alex Osborn, a marketing man (Osborn, 1961). One of its major principles is to defer evaluation of ideas until the group has exhausted generating ideas for solving a pressing problem or issue. Surveys indicate that this is the most widely used technique for generating creative options (Torrance, 1987; Westberg, 1996). Brainstorming can be used for generating creative options in many HR functions such as recruitment and selection, induction, reward, promotion, job rotation, job enrichment, and training.

Creative benchmarking: Benchmarking is commonplace in industry and usually involves benchmarking with comparable other companies in the industry that are better performing. In creative benchmarking, the benchmarking is done with outstanding organizations that are highly innovative irrespective of their industry affiliation or institutional genre. As far as HR is concerned, once the major differences are identified through creative benchmarking, the effort is made to understand how to incorporate into HR the practices of the benchmarked organization/institution though with suitable modifications. For instance, an extended Hindu family can be highly cohesive. How is this cohesiveness achieved? Insights into this kind of creative benchmarking can enable an organization with low internal cohesion to become much more cohesive. Similarly, benchmarking with an excellent academic institution can yield insights into how to replace a highly hierarchical culture with much more of a peer culture, and benchmarking with a highly creative ad agency can yield insights into how to bring in a culture of creativity into a staid organization.

Exnovation: Peter Drucker proposed that any aging organization should periodically pose to itself the following question: "If we are to start the organization now, which of the systems, procedures, policies, and structures that we currently have that we would like to adopt?" (Drucker, 1985). This sort of zero-base thinking can help to identify a host of things to dump and replace. Exnovation applied to HR can be a powerful way of rejuvenating a stultified HR function.

Organizational experiments: In organizational experiments, one group or division or department gets one kind of treatment and another gets either no such treatment or a different treatment. Then, results are tabulated after some time, say a year, to see which treatment was the most effective. As an example, designing a powerful motivational system is a major HR function. One treatment could be to provide strong financial incentives in one division; the other could be to empower employees in another division by using higher order needs to motivate superior performance, such as the need for recognition and self-actualization. A year later, the performances of the members of the two divisions could be compared to find out which treatment works better.

Stakeholders' council: Among all the management functions, HR has the most stakeholders in the organization, by level as well as by function. A stakeholders' council, consisting of the representatives of HR's "customers" is a way of bringing the voice and concerns of the stakeholders as well as their expectations from HR to the attention of HR. It would also provide an opportunity for two-way communication because HR can communicate its priorities and concerns to the stakeholders through an interactive process and get suggestions in return. Such an interactive process can lead to many innovations in HR practices.

Creativity training: There is substantial evidence that creativity training "works," that is, it increases the capacity for offbeat, out-of-the-box thinking in the trainees (Torrance, 1987; Westberg, 1996). Creativity training can come in many different formats. The one I use is based on a model of durable creative achievement that I have developed (Khandwalla, 2004, Appendix A). Basically, it involves exercises, readings, sharing of views and experiences, and exposure to role models aimed at reducing fears and blocks, at strengthening the motivation to pioneer and innovate, and at strengthening thinking skills such as the capacity to come up with numerous solutions (fluency), the capacity to look at problems from multiple perspectives (flexibility), the capacity to restructure problems in interesting ways (problem-restructuring capacity), the capacity to come up with unique solutions to problems (originality), and the capacity to sense issues and problems that others do not sense (problem sensitivity). It also involves an attempt to examine one's living and operating environment to see how it could be modified to make it more stimulating and creativity-friendly. From the HR perspective, creativity training, when extensively used, can help the organization to enhance the creativity of employees at all levels, and thereby to enhance the organization's capacity to become much more innovative.

Concluding comments

Creative HR can be a new frontier of HR. If we can collect many examples of how creative HR interventions have been improvised and with what effects, we could train HR personnel in much wider and much more effective uses of creativity in HR. Analyzing these interventions could yield many fruitful models of innovative HR practices that could then get widely disseminated. This can enrich not only HR but the entire field of management. Creativity has been a game changer for human civilization itself (Toynbee, 1946). Creative HR can be a game changer for our organizations in the global sweepstakes.

References

Arbose, J. (1986). The unlikely diversifications helping to keep Aer Lingus aloft. *International Management, 41*(5), 57–63.

Brenneman, G. (1998, September–October). Right away and all at once: How we saved Continental. *Harvard Business Review, 76*(5) 162–179.

Chakravarty, S. (1997). How an outsider's vision saved Kodak. *Forbes*, 45–57.

Chambers, D. (1988, March 10–11). *Consumer orientation and the drive for quality*. Paper presented at the Roundtable on Public Enterprise Management: Strategies for Success, New Delhi.

Clausen, A. N. (1990). Strategic issues in managing change: The turnaround at Bank of America. *California Management Review, 32*(2), 98–105.

Drucker, P. (1985). *Innovation and entrepreneurship: Practice and principles*. London: Heinemann.

Freedman, N. (1996). Operation centurion: Managing transformation at Philips. *Long Range Planning, 29*(3), 607–615.

Goodstein, L., & Burke, W. W. (1991). Creating successful organizational change. *Organizational Dynamics, 19*(4), 5–17.

Gordon, M. (1987). *The Iacocca management technique*. New York: Bantam.

Guilford, J. P. (1960). *The structure of intellect model: Its uses and implications*. Los Angeles, CA: University of Southern California.

Hartley, R. F. (1997). *Management mistakes and success*. New York: John Wiley.

Horn, J. (1990). American Airlines (A): Strategy in the 1990s (Harvard Business School case 9-491-044). Boston, MA: President and Fellows of Harvard College.

Kennedy, C. (1998). The roadmap to success: How Gerhard Schulmeyer changed the culture at Siemens Nixdorf. *Long Range Planning, 31*(2), 262–271.

Khandekar, S. (1985, December 15). JMEL: Dramatic turnaround. *India Today*, 103–104.

Khandwalla, P. N. (1992). *Organizational designs for excellence*. New Delhi: Tata McGraw-Hill.

———. (1995). *Management styles*. New Delhi: Tata McGraw-Hill.

———. (2001). *Turnaround excellence: Insights from 120 Cases*. New Delhi: SAGE Publications.

———. (2003). *Corporate creativity: The winning edge*. New Delhi: Tata McGraw-Hill.

———. (2004). *Lifelong creativity: The unending quest*. New Delhi: Tata McGraw-Hill.

Kharbanda, O. P., & Stallworthy, E. A. (1987). Transformation at General Motors. In O. P. Kharbanda & E. A. Stallworthy, *Company rescue: How to manage a company turnaround* (pp. 115–127). London: Heinemann.

Killawala, A. (1992, June 22). Bank of Baroda: Banking for tomorrow. *Business India*, pp. 73–75.

Koestler, A. (1970). *The act of creation*. London: Pan Books.

Krishnamurthy, V. (1977). Management of organizational change: The BHEL experience. *Vikalpa, 2*(2), 113–119.

———. (1987, November 19). SAIL blazes a new trail. *The Economic Times*.

Lehrer, M. (1995). Lufthansa: The turnaround (Abridged version), Case 396-0190-1 of INSEAD: Fontainebleau.

MacGregor, I. (1982). Recovery at British Steel. *Journal of General Management, 7*(3), 5–16.

Miles, R. H. (1997). *Corporate comeback: The story of renewal and transformation at National Semiconductor*. San Francisco: Jossey Bass.

Nilakant, V., & Ramnarayan, S. (1998). *Managing organizational change*. New Delhi: SAGE.

Osborn, A. (1961). *Applied imagination* (Indian edn). Allahabad: St. Paul's Society.

Ravindranath, S. (1995, February 27). Rising from the ashes. *Business India*, pp. 81–84.
Shenoy, M. (1996, September 9). Turnaround and after. *Business India*, pp. 72–73.
Shook, R. L. (1990). *Turnaround: The new Ford motor company.* New York: Prentice-Hall.
Torrance, P. E. (1987). Teaching for creativity. In S. G. Isaksen (Ed.), *Frontiers of creativity research: Beyond the basics* (pp. 189–215). New York: Bearly.
Toynbee, A. (1946). *A study of history*. New York: Dell.
Westberg, K. L. (1996). The effects of teaching students how to invent. *Journal of Creative Behavior, 30*(4), 249–267.

3 Some tradeoffs and trade secrets from the diary of an HR practitioner

*P. Sethu Madhavan**

Introduction

Most human resource (HR) professionals know the fact that Udai Pareek along with T. V. Rao played an important role in setting up the first HR department in India, at Larsen and Toubro Ltd (L&T), during the 1970s (Pareek & Rao, 2008). However, not many people perhaps know that one of the last professional associations of Udai Pareek was with L&T before he passed away in 2010. On the basis of a special invitation from the corporate HR unit and Management Development Center of L&T, Udai Pareek used to visit L&T corporate HR and Management Development Center, on a monthly basis to support various in-house research programs as well as to provide professional HR advice and coaching services. It was during those days I used to work closely with him and discuss many professional issues in detail! During our discussions, most of the time we concluded that HR profession, today, is far away from the ideal picture that the founders envisaged a few decades ago. We felt that, nowadays, most CEOs are not paying enough attention to HR issues and that HR is nowhere near becoming a business partner in most organizations.

During my daily meetings with Udai Pareek, we used to frequently refer to or discuss the issue of the erosion of credibility of HR function over the years, as well as the various dilemmas and tradeoffs that HR practitioners face today. Some of the frequently discussed dilemmas included whether HR should focus on developing people or controlling them and whether we should use HR tools such as appraisal and assessment for developing people or for other purposes. This chapter is an attempt to list some such tradeoffs that I have experienced and observed during my career as an HR professional in different countries, as well as in different roles namely academician, researcher,

* I thank Dr T. V. Rao for encouraging me to write this chapter and my wife Sugatha Puthalath for all her valuable support and feedback for completing this chapter.

trainer, consultant, and executive. Apart from identifying and listing some of the key dilemmas and tradeoffs that HR practitioners frequently face in today's business world, the chapter links the tradeoffs to erosion of credibility of HR function, where applicable and provides some directions for resolving the tradeoffs in an effective manner to protect and enhance credibility of HR. The chapter attempts to ground the professional experiences and observations in the literature that are relevant to the topics.

The golden period of HR

Gill (2007) noted that decades ago, before HRM and HRD (The term "HR" will be used in the remaining part of this chapter to refer HRM/HRD) approaches evolved, the personnel management function used to focus on administration, welfare, and industrial relations and it was treated as less important than other functions such as finance, marketing, and manufacturing! As quoted by Gill (2007), the work of personnel management in those days was perceived as "partly a file clerk's job, partly a house-keeping job, partly a social worker's job, and partly a fire-fighting job to head off union trouble or to settle it" (Drucker, 1954, pp. 275–276). Many authors have observed that the HR approaches have evolved to fill the gaps that existed in the old personnel management model, which viewed people management as a matter of control, administration, and collective negotiations. The HR approach, on the other hand, promoted the function as a strategic partner for the business who could support the organization to achieve business excellence by leveraging employees as a competitive advantage (Gill, 2007; Pareek & Rao, 1982; Ulrich, 1997).

The HRD system proposed and implemented by Pareek and Rao in L&T, State Bank of India (SBI), and other organizations during the 1970s was innovative and integrated in nature (Pareek & Rao, 1982; Rao, Rao, & Yadav, 2007). HRD professionals during those days were encouraged to have the spirit of missionaries rather than mercenaries (Rao, 1990). During the 1980s and the 1990s, HR professionals introduced a number of innovative methods and tools for driving success and change in organizations. HR used to be a true business partner in many large organizations during that time, supporting large-scale change efforts by introducing various HR systems and tools such as performance appraisal systems, 360-degree feedback, assessment and development centers, leadership development, succession plans, training and development, culture change programs, and so forth. During the 1990s, I have observed as an HR consultant that the HR sections in many leading companies assumed great importance and some of them were physically moved next to the CEO's office or to the nearest floor! However, during the last decade or so, credibility of HR function seems to be on a declining trend and some authors have noted that HR has lost respect in business world and that HR should be downsized or outsourced (Ryan, 2005; Ulrich et al., 2007).

The decline of HR and the call for HR as business partner

Perhaps the realization among HR professionals that their profession is losing credibility led to the famous call for transforming HR into a business or strategic partner raised by Ulrich and others during the 1990s (Ulrich, 1997). However, even after more than a decade, it seems that HR as a profession has not been able to enhance its image. HR functions in many organizations have lost their alignment and linkages with business strategies and often they seem to be operating like the old personnel management functions. The results of a study by Rao, Rao, and Yadav (2007) indicate that today HRD functions in companies are not well structured as envisaged in the mid-1970s and convenience rather than systems drives HR functions in most organizations.

According to Hammonds (1997), HR staff today are only competent to administer pay, benefits, and retirement! He argues that the gulf between capabilities of HR staff and HR job requirements appears to be widening, since educational qualifications of HR professionals have not kept pace with demands of the times. The declining status of HR seems to have affected the reporting structure of HR. Though I have not come across any empirical data on the reporting structure of HR units in India, there are studies that have revealed that globally majority of HR units are not reporting to CEOs and some of them are reporting to finance directors or chief finance officers (CFO Research Services and Mercer Human Resource Consulting, 2003). Hammonds (1997) notes that if a company's HR head is reporting to the CFO, then chances are high that HR is heading in the wrong direction. Human resources when reporting to finance is likely to pursue short-term efficiencies and forget about long-term capability building and value addition.

The problems and challenges faced by HR professionals today are perhaps not limited to any particular country and are applicable across the globe. Even though Ulrich (1997) advocated the need for HR departments to become business partner in the 1990s, even today, it has not happened in most organizations. Kochan (2004) stated that two decades of effort to develop a new "strategic human resource management" role in organizations has failed to realize its promised potential of greater status, influence, and achievement. Kochan (2004) further elaborated that human resource management profession has lost trust and legitimacy in the eyes of its major stakeholders due to this failure. Vosburgh (2007) noted that many HR professionals and companies still follow the old personnel administration and policy police and that they fail to see the connection between HR and business results! As noted by Vosburgh (2007), it is alarming that some authors have been criticizing HR using very harsh words and often with anger and frustration (e.g., Hammonds, 1997; Ryan, 2005). Another study reveals that HR professionals face major challenges such as lack of power, dealing with skepticism of customers, vulnerability, and so forth (Kahnweiler, 2006). In summary, all these writings indicate that a number of trends and changes that are sweeping across the globe are putting immense pressure on HR to demonstrate their contribution and value addition to business.

Line and top managers tend to blame HR function for many problems in today's work places such as poor employee engagement, high personnel turnover, disciplinary issues, late coming, absenteeism, low productivity, customer complaints, lack of return on investment, and so forth. I have come across such complaints about HR irrespective of the roles that I was playing and I must have spent many hours trying to analyze such blames with a hope to identify the root causes, and often to save my career as well. While making an effort, where applicable, to analyze issues related to professional accountability in the context of various complaints about HR, the following sections of this chapter present a set of seven tradeoffs and some directions toward resolving them in an effective manner.

Some tradeoffs and trade secrets

1. Blaming the stage versus learning to dance

> It is the act of an ill-instructed man to blame others for his bad condition. It is the act of one who had begun to be instructed to lay the blame on himself. And of one whose instruction is completed neither to blame another nor himself
>
> (Epictetus)

Blaming the top managers and line managers for the issues related to employees is one of the most preferred pastime activities of HR professionals. However, fixing the blame is not the same as fixing the problem. I have come across many situations related to the issue of blame and accountability, some of them being eye opening, some of them easy, and some of them complex! It ranged from answering to the top management's show-cause memo for delay in renewing a contract with supplier due to legal and other complexities, to convincing an inexperienced trainer that his inabilities are responsible for the negative feedback from participants. We need to consider the fact that years of research and trials to find out who is responsible for the Nazi tragedy did not result in any conclusive answer yet! After all, a good lawyer can save even the worst criminal and hold him unaccountable for his crime or vice versa. Is the issue of responsibility for the status of HR so simple that we can learn to dance perfectly, once we stop blaming the stage? The answer is definitely negative, since there are many other factors that might contribute to the effectiveness of HR. A deeper analysis will help us to understand the complexities and reasons of specific situations and dilemmas better.

All human beings interpret reality and these interpretations may not be accurate. Some of us are capable of interpreting reality in multiple ways, whereas some others tend to stick to single interpretation of situations and reality. Depending on our conscious and perhaps unconscious motivation behind our interpretations, we try, either to search for the truth or to save our skin, beliefs and jobs. When we are trying to search for the truth, we are like a scientist who is considering multiple hypotheses and finally accepts the best explanation that is probably the most accurate. When we are trying to save our skin, beliefs, and jobs by offering convincing self-serving arguments, we are perhaps behaving like lawyers trying to save themselves or their clients. Therefore, the tradeoff is about deciding whether to put the hat of a scientist or that of a lawyer when you are analyzing professional issues at work. Often we are tempted to act like lawyers rather than scientists, but unfortunately, the approach of a lawyer yields short-term victory for the HR professionals involved, but results in loss of credibility in the long term. The scientist style, on the other hand helps us to analyze issues more objectively and either accept the blame and take corrective actions or educate the customers and stakeholders about their roles and responsibilities.

We need to accept responsibility and start listening to our customers and stakeholders rather than just believing that we are not responsible and accountable for the crisis that HR is facing. Therefore, the fundamental dilemma is whether to accept responsibility and try to look within to solve the problem or just believe that we are doing great, and expect the line managers and top managers to change! I believe that, it is important to accept responsibility while acknowledging the role and responsibilities of our customers so that we lead ourselves while leading our customers also, in an interactive manner.

Figure 3.1 presents results arising from the interaction of our beliefs about the responsibilities of HR and our stakeholders. The figure illustrates the fact that to be truly interactive and act objectively like scientists, we need to accept the fact that both HR and our stakeholders are responsible for the current crisis in HR.

2. Centralizing HR versus devolving HR

> Maturity, one discovers, has everything to do with the acceptance of "not knowing."
>
> (Danielewski)

During the last few decades, HR professionals have been making significant efforts to "devolve" HR to line managers. The popularity of HR as a business partner approach perhaps added momentum to the efforts to make line managers a partner in HR (Ulrich, 1997). Hutchinson and Tailby (n.d.) state that the transfer of responsibility for day-to-day HR work to the line manager has been a consequence of the new strategic HR model, which emphasizes the importance of a partnering relationship between HR and the line to deliver HR goals. Because of the trends toward collaborating with line managers and involving them in the HR issues and decisions, today, line managers in most organizations are required to play various HR-related roles such as conducting recruitment interviews, performance appraisal, grievance handling and performance counseling, coaching and mentoring, reviewing salary of their staff, and so forth.

Brewster and Larsen (2000) highlighted the benefits of involving line managers in HR such as cost reduction, faster decision-making, transferring

Figure 3.1. Causal attributions of the HR team regarding complaints about HR performance

HR team members believe that they are responsible for the complaints about HR	**No**	**Yes**
Yes	**Proactive HR** *(HR takes all the blame and makes efforts to anticipate and correct itself without making any efforts to educate the customers and stakeholders)*	**Interactive HR** *(HR makes objective analysis of complaints and makes all the efforts towards continuous improvements, while educating and leading the customers and stakeholders)*
No	**Inactive HR** *(HR blames forces outside the immediate organization such as government, group HR, board, etc., and remains inactive)*	**Reactive HR** *(HR blames the stakeholders and tries to conform to their demands unwillingly while trying to convince them that the root cause of the problem lies outside HR)*

HR team members believe that HR is not responsible for the complaints about HR and that customers are either unreasonable or they lack understanding of HR issues

responsibility to direct line managers, and even as an alternative to outsourcing HR. The concept of business partnership with line managers and idea of devolution of HR roles to the line manager sound logically perfect. However, in practice, they are perhaps not easy to implement or achieve. HR professionals, who have worked in different organizations, and especially in organizations, which are at different levels of maturity, or organizations in different cultures find that it is not easy to implement the concept of HR business partnership. Employees in organizations, which are not familiar with the concept, consider that the HR unit is responsible for all HR issues, decisions and problems. Gilberta, Winnea, and Selsa (n.d.) stated that "HRM" is frequently held synonymous with "the HR department," whereas line managers also have a substantial responsibility in the implementation of HR practices in most contemporary organizations (Purcell & Kinnie, 2007). Similarly, involving line managers and sometimes even top management team members before educating them properly could lead to wrong decisions. Hutchinson and Tailby (n.d.) noted that there is growing theoretical and empirical evidence to show that the way in which line managers deliver HR policies and practices affects the status and credibility of HR functions. In other words, a number of factors are required to be in place to ensure success of the HR efforts for establishing an effective business partnership with line managers. Such critical factors include organizational maturity, maturity level of HR processes and policies, organizational culture, HR skills and competencies of line managers, and top management beliefs regarding HR issues. So the tradeoff is whether to aim at business partnership or not, when the basic enablers are not established in the organization.

CEOs and operating executives are not experts on workplace issues (Cappelli, 2015). HR responsibilities mean additional power and therefore line managers will be eager to accept them, even if they do not have the required experience and competencies. In the absence of proper training and preparation, line managers will grab the "HR power" offered to them, but they might start using it incorrectly and start making a mockery of HR tools and methods. Once the line managers have learned to "misuse" or "incorrectly use" the HR tools and methods, it is very difficult to correct them later and by then, the employees would have developed negative perceptions about HR. To summarize, HR devolution to line managers is highly recommended, but mature HR systems, tools, and processes as well as HR competencies of line managers are critical prerequisites for ensuring its success. If the line managers are not competent, any attempts to transfer HR responsibilities to them will lead to disastrous results and it might take years to recover from the damage it can cause to the credibility of the HR function and professionals involved. On the other hand, if the HR is mature and yet HR is not involving the line managers in HR decisions, it might look like traditional HR function. Figure 3.2 illustrates and elaborates these situations further.

Figure 3.2. Interaction between involvement of line managers and level of maturity of the HR

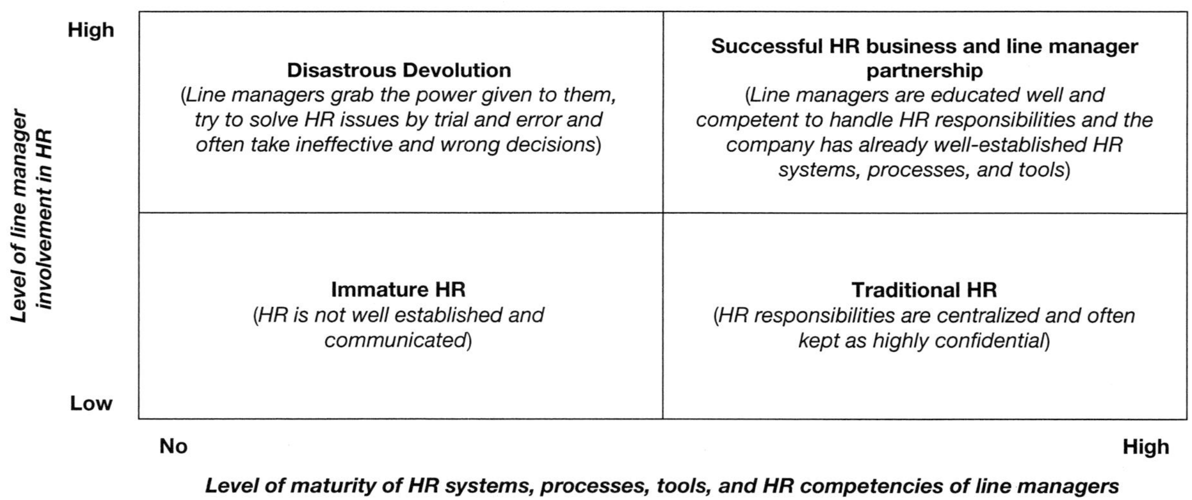

3. Accepting organizational culture versus changing organizational culture

> In matters of style, swim with the current; in matters of principle, stand like a rock
> (Jefferson, n.d.)

Adler and Gundersen (2008) observed that the national culture outweighs organizational culture and therefore, it is important for organizational behavior experts to understand the dynamics of culture. Pareek (2007) had highlighted that some of the dysfunctional aspects of national culture will creep into the organizational culture and it is the role of HR professionals to support the regeneration of culture in organizations to minimize negative impact of such dysfunctional cultural aspects. Giving the example of Indian organizations, Pareek (2007) identified many such negative aspects of Indian culture such as fatalism, nonwork culture, pessimism, and feudalism. He noted that Indian culture does not value the respect for work and workers. Earlier, Pareek (2002) had recommended eight (OCTA) values that will act as the steps (PACE) for creating a positive work ethos in organizations. These values abbreviated as OCTAPACE represent openness, confrontation, trust, authenticity, proactivity, autonomy, collaboration, and experimentation.

Murthy (2009) observed that Indian organizations and communities have a tendency to encourage feudal structure and dictatorship! As noted by Murthy (2009) and Pareek (2007), organizations built around highly dysfunctional aspects of culture are likely to end up as disaster. A knowledge and competitive economy cannot thrive under a feudalistic and fatalistic culture. Murthy (2009) further noted that most of the recent scandals in the Indian corporate world are a result of the tendency to establish feudalistic leadership styles at work. He noted that when nobody could stand up and say to a feudal leader what is happening is wrong, and then disasters happen. Murthy (2009) noted that for building future organizations we need organizations and communities to espouse the cause of democracy and commit to fairness, transparency, and accountability. Katz (n.d.) states that one cannot assume that even a very powerful corporate culture will render the influence of national culture insignificant. This observation makes it clear that if HR and top management of a company fail to drive the organizational culture, then the company culture will drift toward the national culture and often in the more comfortable and dysfunctional directions. In other words, it is important to understand that while we respect the positive aspects of a country's culture, it is also highly important for HR professionals to provide the tools and mechanisms for protecting the organizations from the dysfunctional aspects of national cultures.

Results of a study conducted by Ulrich, Brockbank, Johnson, and Younger (2007) reveal that, during the last decade, "HR's ability to define, create, manage, and change culture has become a unique source of competence." The results of the study indicate that the ability of HR team to change and manage culture is the second-highest-rated factor in predicting performance of both HR professional and HR department effectiveness. Respondents in their study rated the HR competencies related to organization culture management, above the competencies related to strategy, talent management, and organizational design. Unfortunately, senior HR leaders in many organizations are often scared of initiating any cultural transformation exercises as they might go against the leadership styles and values existing in the organization. Who will initiate a series of workshops to encourage empowerment, delegation and participative management in a company where the CEO and his team members follow a dictator style of leadership? Who will dare to highlight the impact of "slave driving style of managers" who distrust their staff, on the employee engagement and motivation, in a company, which has many such senior executives? We need to consider the scary fact that many HR leaders might also live by such dysfunctional values! Consider how the situation might change if the organization is lucky to have a CEO who is futuristic and believes in building a healthy culture. Figure 3.3 illustrates the scenarios that might result from the interaction of these two dimensions, namely, HR beliefs and top management beliefs.

4. Professional versus political approaches

> Politics, it seems to me, for years, or all too long, has been concerned with right or left instead of right or wrong.
>
> (Armour, n.d.)

Organizational politics and undercurrents are likely to influence HR function, more than any other functions in an organization! HR is often used to "lift" and provide growth opportunities for selected few as well as to "size up" or "cut the wings" of some employees by some CEOs and line managers! When HR team consists of well-trained and educated professionals who have the required expertise and professional tools and methods, then it becomes difficult for the top managers, line managers, or anyone else to meddle with HR tools and systems. Though political use or misuse of HR is quite prevalent in many organizations, HR theories, models, and teachings ignore the facts and tend to believe that organizations implement HR tools and methods in an objective and appropriate manner. Some authors have pointed out that HR has been overly optimistic about the possibility of maximizing individual and organizational needs while ignoring the issues of conflict, power, and politics (Ferris & Judge, 1991).

Aronow (2004) argued that human resource professionals must be able to influence highly skilled, fact-based technical experts and their work must demonstrate character. Ferris and King (1991) stated that many of the problems associated with the human resource system relate to ambiguity in the process of matching a person to the work, which involves various processes such as recruitment, job analysis, performance evaluation, training, promotion, and so forth. They noted that when HR standards are clear and uncompromising, the

Figure 3.3. Alignment of HR and top management views about culture change

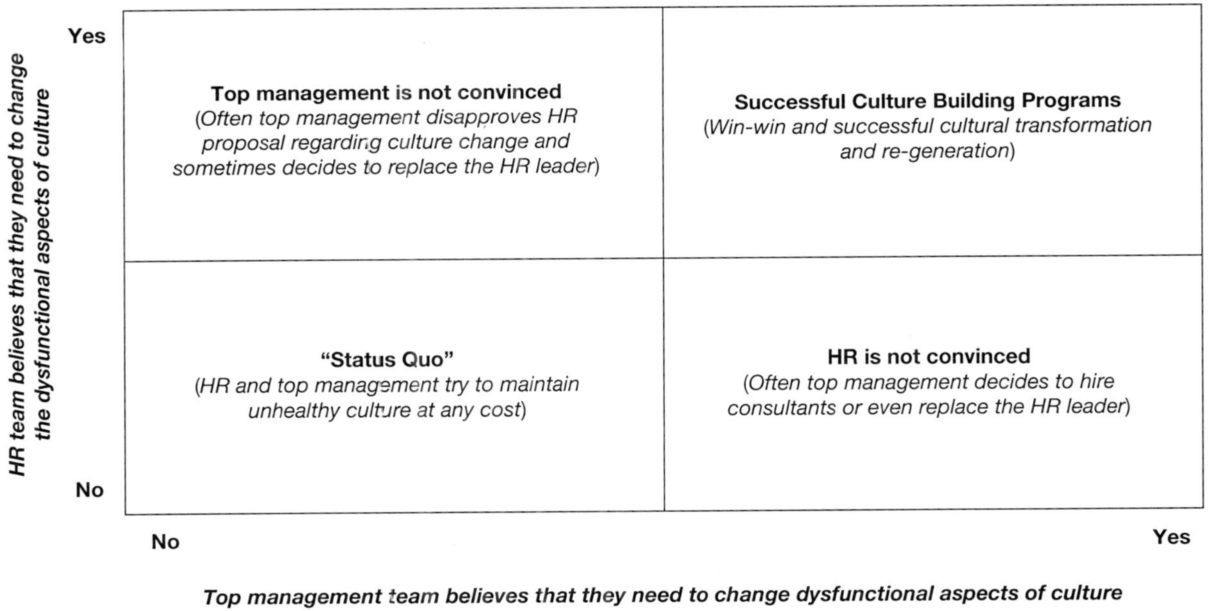

likelihood of organizational politics affecting HR decisions negatively will be very low. However, when standards for staff selection, promotion, or any other change in human roles and tasks are vague, those with greater power and influence will get involved in the processes and try to push decisions based on various personal and unprofessional agenda. Drory and Vigoda-Gadot (2010) identified four types of HRM based on two factors namely the level of organizational politics (high or low) and the actor's type of interest (organizational interest or self-interest). They named these four factors as positive/constructive HRM, negative/destructive HRM, ineffective HRM, and virtual HRM. Though the self-interest of the HR actors is important and will have an impact depending on the level of organizational politics as mentioned by Drory and Vigoda-Gadot (2010), a far more important factor is the level of professionalism of the whole HR team, systems, processes, and tools in an organization. The lower the level of professionalism, the higher the domination by organizational politics on HR decisions. Therefore, the tradeoff is between professionalism and politics! Considering the fact that we cannot wish away the influence of organizational politics in today's corporate life, the choice will be to contain it rather than eradicate it. Figure 3.4 represents this dilemma and indicates the consequences arising out of the interaction of political involvement and skills of HR professionals with the level of professionalism.

5. Serving the employees versus serving the employers

> Cutting wages is not the way to recovery. Raise wages and improve the product.
>
> (Ford, 1934)

Kochan (2004) noted that the tendency of HR professionals to please the top management explains the inability of HR profession to prevent top management excesses that produced the corporate scandals, runaway CEO compensation, and the overall breakdown in trust in corporations. HR professionals seem to be increasingly focusing on pleasing the top management, while ignoring the interests of the employees, line managers and even the company. Many HR professionals today serve only the top management and owner, rather than trying to balance and meet the expectations of all stakeholders. Hammonds (1997) accused HR about using methods such as salary benchmarks to justify low salaries. He further argued that HR has become "henchmen for the chief financial officer" and a "dark bureaucratic force that blindly enforces nonsensical rules, resists creativity, and impedes constructive change"!

I have observed that some organizations are willing to spend huge amount of money on buildings, facilities, and even unwanted luxury, but they do everything possible to avoid providing employees a better salary! Ford (1934), however, realized the importance of employees as a major stakeholder, when he states that "Low wages are the most costly any employer can pay. It is like using low-grade material—the waste makes it very expensive in the end." Aronow (2004) noted that the desire of HR professionals to become a strategic business

Figure 3.4. Professionalism versus political skills of the HR team

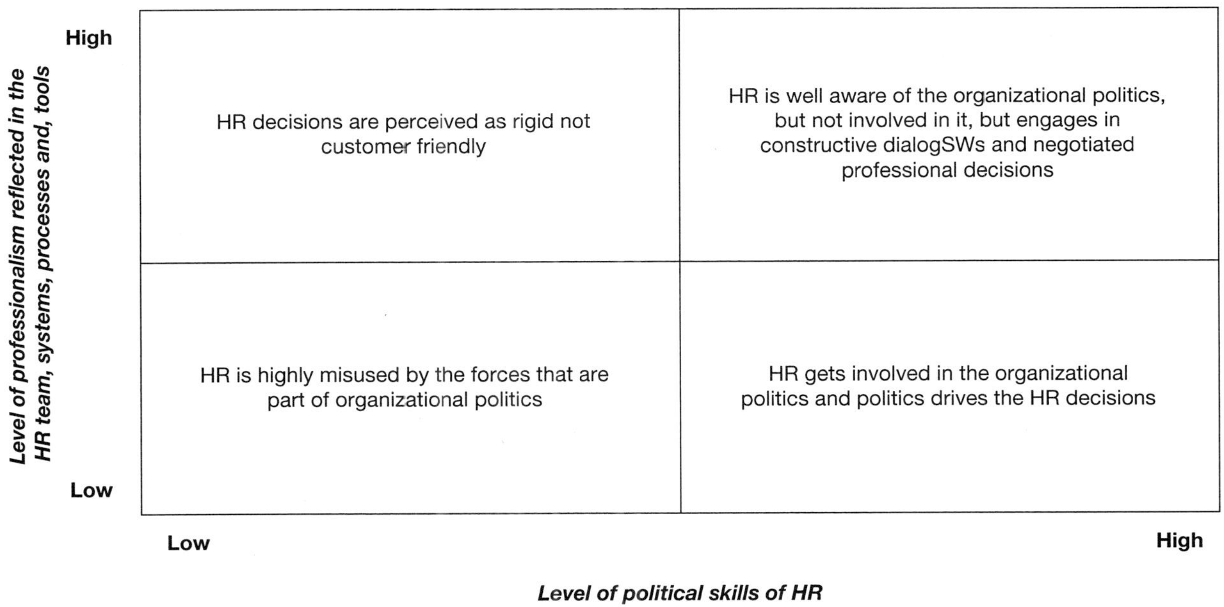

partner at times may force a tradeoff between value-based decision-making and making decisions as a bottom-line driven business partner. Vosburgh (2007) stated that throughout the development of HR profession, there has been a tension between the roles of "employee advocate" and "business leader" and it is not easy to wear these two hats at the same time. However, in the context of the major scandals about the excessive compensation and bonus payments to the top management teams in large organizations, Vosburgh (2007) further emphasized that to be a great business partner does not mean to be co-opted and HR should confront such ethical issues and blow the whistle.

Many organizations are very good at cutting down the benefits during recession times, but not usually eager to raise employee benefits during a boom. Often compensation packages of the top layers in companies are unjustifiable and very high compared to the compensation packages offered to lower level employees, and many HR units seem to accept such unhealthy practices. The policy makers and policy approval authorities are often clever enough to exclude themselves from the highly cost-effective packages offered to the rest of the staff. I have come across some highly profitable companies paying relatively very low salaries to workers while their top layers are earning huge salaries. These practices are highly unjustifiable not only from HR point of view but also from ethical and humanitarian points of views. Low salary in a company, which is highly profitable, is exploitation. Unfortunately, in the absence of any legal obligations or collective negotiations by trade unions or any other checks and balances, such practices have become very rampant nowadays. High profitability with high salary represents progress and real excellence.

The lack of interest in the interests of employees is evident not only in the case of monetary benefits, but also in relation to nonmonetary benefits. HR in many organizations fails to see the importance of meeting the other expectations of staff such as non-monitory recognitions, learning opportunities, training, job rotations, welfare schemes, and so forth. So the question is whom do we represent. It seems that the problem is only when we fail to balance our attention across the interests of different stakeholders. Figure 3.5 illustrates the consequences arising out of the degree of support offered by HR units for employees and the employer.

6. Policing employees versus developing the employees and their careers

Douglas McGregor (1960) differentiated between "Theory X" and "Theory Y" management philosophies. Theory X considers that by their nature, people are lazy and not motivated to work and therefore managers should supervise and control them closely. On the other hand, theory Y assumes that people are self-motivated and they will be more satisfied and engaged at work if they are empowered and given the freedom. These contrasting management approaches are similar to the "control" vs "commitment" approaches (Walton, 1985). The widely discussed "hard vs soft" models of HR reflect

Figure 3.5. Focus on the interests of employees versus the interests of employers

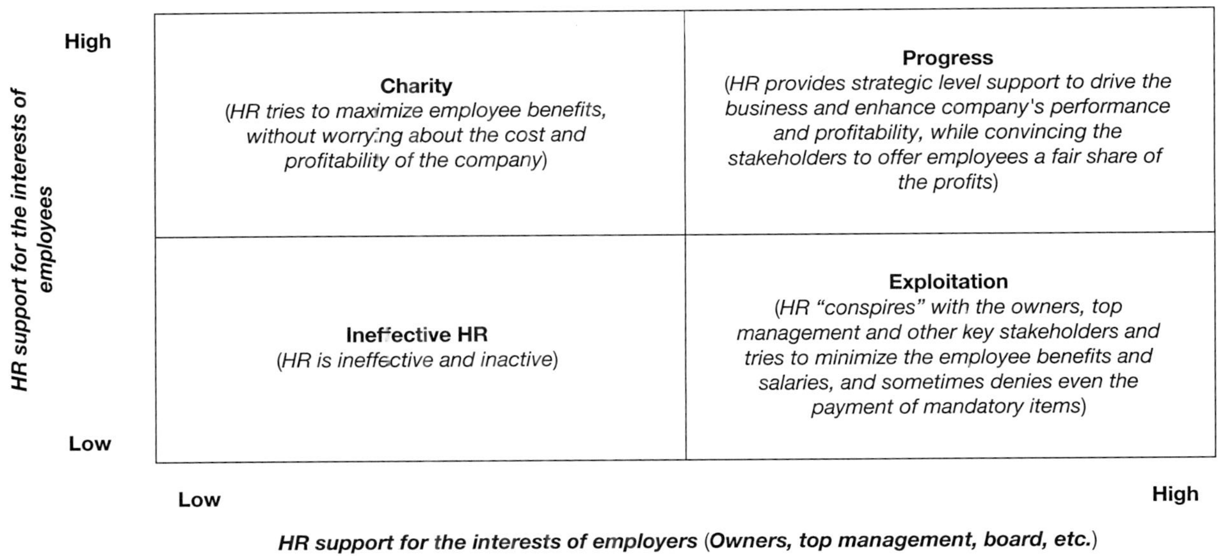

similar philosophical assumptions regarding people and their motivation to work. The "hard" model recommends that organizations should control and mold people whereas the "soft" model emphasizes employee participation and well-being as the key drivers required for motivating them (Beardwell & Claydon , 2007; Guest, 1987; Legge, 1995). If we look at the HR departments today and make an analysis about their focus, often we may find that most of them are following the "hard" approach! They spend most of their time focusing on policing and controlling employees, rather than working toward winning their commitment through robust people development and culture-building activities.

HR departments seem to love the operational and day-to-day administrative activities such as payroll, leave administration, employee welfare, recruitment-related work, routine performance appraisal administration, and so forth. They like to enforce policies and standards. For some reason, they shy away from the strategic part of HR, which might include strategy support, talent management, career development, succession planning, leadership development, change management, culture building, and so forth. Some argue that most HR professionals are not trained or competent to drive such strategic activities and support core business! Hammonds (1997) in his famous and brutal critique of HR stated that HR pursues efficiency in lieu of value because it is easier to do and achieve.

It is important to understand that career and talent development are not the same as having corporate universities and training courses in place! Many HR units take pride in conducting numerous training workshops, which does not align with the needs of employees, job families, and organization. They tend to ignore important activities such as identification of the functional needs and organizational needs arising out of the unit level and organizational level strategies and plans. Madhavan (1997) differentiated strategic training systems (STS) that are aligned with the business needs from traditional training systems (TTS) that conduct such routine and random training. HR units operating in the TTS mode avoid asking employees about their career aspirations and developmental needs! Instead, they tend to keep the career aspirations of employees under control, by enforcing standards and criteria that are not easy to follow or achieve! Some employees who are good at self-learning survive in such places, but most others remain underdeveloped within their roles and positions. When the pressure to perform in such places increases, line managers start requesting HR to recruit "experienced and competent" staff from outside, as developed staff are not available internally, to be promoted to occupy such roles! Ironically, often such HR units run after some short-term trends and fads and end up imposing them on to employees and organizations as a new and "strategic initiative"! HR units often tend to measure such isolated and random training and present the measures and results to stakeholders, as an evidence of HR performance and effectiveness. From a purely logical analysis of return on investment (ROI) of such training, one may conclude that such training amounts to huge loss of company efforts, resources, and money and the ROI

Figure 3.6. Focus on employee development versus focus on policy enforcement

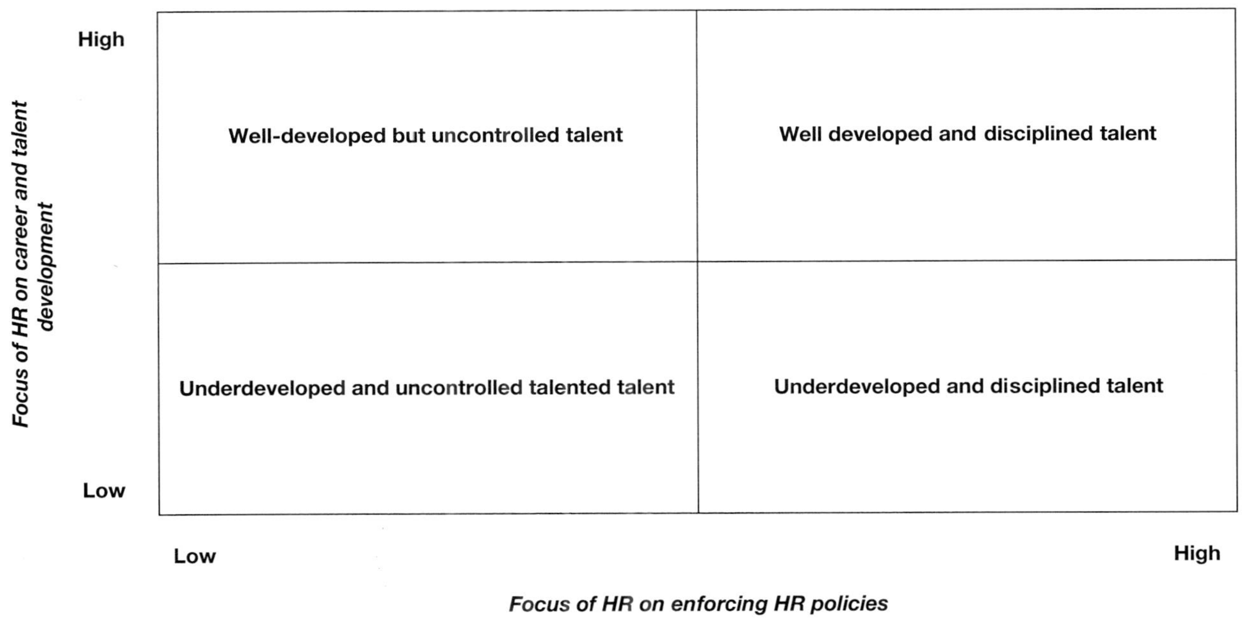

of such training is close to zero. Such training wastes not only the money spent on sourcing such courses but also the time of employees who attend and HR employees who organize such courses.

To conclude, it is important for HR to enforce some professional standards and policies at work and some of them support key strategic initiatives such as culture building, corporate performance management, and so forth. To give an example, the time-punching system, though a "theory X" tool, still helps to develop positive habits and a culture of punctuality over a period of time. However, the existence of HR unit cannot be justified if they are not supporting the staff on their personal development, career aspirations, and upward and lateral career movements. The cost of this mistake again is very high, as it will lead to turnover of staff, high cost of lateral recruitment, low employee engagement and so forth. For achieving excellence, HR should focus on both policy enforcement and employee development. Figure 3.6 highlights the consequences of the manner in which HR units resolve this dilemma.

7. Short-term gimmicks versus long-term capability development

CEOs and line managers of established businesses, deal with predictable business processes related to manufacturing and selling products or services. They monitor the various input–output ratios, which indicate the efficiency, effectiveness, and return on investment (ROI) related to the core business. Often they take calculated business risks, but by virtue of their roles and accountabilities toward shareholders and stakeholders, they want fast and predictable outcomes. Unfortunately, HR is not comparable to a manufacturing or service factory process. To be fair, we should compare HR to an agricultural process that farmers experience when they are planting fruit trees that take years to mature and start returning benefits. However, some CEOs and line managers start putting pressure on HR to demonstrate value and ROI sometimes, on a yearly basis. Many HR professionals succumb to this unreasonable pressure and start planting seasonal flowers and perennials in an attempt to satisfy the top management and line managers. They start chasing short-term deliverables and start using them to convince the top management and customers that HR is doing something and adding some value! As noted by Cappelli (2015), HR often invests heavily in many programs that lack impact and it is important for HR to walk away from the time wasters. Often such short gimmicks include selling some latest fads or flavors of the month to the organization or implementing some of the HR tools such as 360-degree feedback as a "new initiative" that should have been part of the capability development DNA of the organization long time ago! The other methods used by HR professionals include using cleverly chosen performance measures and KPIs for the HR unit, which includes many activity measures, short-term quantitative indicators, and sometimes even cost per employees. So question might be what is the dilemma or tradeoff that HR professionals face in relation to this pressure?

The dilemma arises only when you start planting trees that take years to bear fruit, but fruits that make the organization stand real strong, tall, and sustainable. It is not easy to convince CEOs and line managers who are used to seeing results on their investment on a monthly basis and yearly basis to wait for 2–3 years or even 5–10 years to see the results of investment in HR and people development! HR is expected to contribute to bottom line and business results through a number of methods such as recruitment, development, employee utilization, employee motivation and engagement, organizational culture building, knowledge management, succession planning, leadership development, and so forth. However, when we look at these expectations closely, we realize that we cannot achieve them in a short span of one or two years! For example, to recruit the best talent, the organization should have created a high value brand image in the employment market and it takes years to build! Developing and molding people takes years, and making them feel proud of their organization and engaging them at work takes even longer! Organizational culture building is not easy and it requires years of investment of time and resources by top management, line managers and HR team. Obviously, like trees, it is impossible to grow leaders overnight! Perhaps the trade secret is to balance both short term and long term to ensure some "quick wins" while aiming at "big win" in the long term! Figure 3.7 further illustrates this argument by showing the effect of the focus of HR on short-term and long-term issues.

8. Concluding remarks

This chapter briefly outlined the emergence of HR as a new and promising approach for managing people and talent and the recent frustrations regarding the failure of HR to live up to its promises. The chapter listed and analyzed seven typical dilemmas and tradeoffs that HR practitioners experience in today's world and attempted to link them to the issue of credibility and effectiveness of HR units. The analysis of the tradeoffs indicates that HR professionals should be aware of the "tradeoff" situations in order for them to make effective and conscious choices. If they are not aware or if personal preferences and perceptions drive their decisions, it might lead to loss of image and credibility of HR. The analysis of various tradeoffs reveals that the directions for effective practice are often clear, but we cannot perceive them, until we become aware of the dilemma and tradeoffs involved. The trade secret is therefore to try to unravel the hidden alternatives by identifying the tradeoffs involved before making major decisions about HR issues. This approach offers huge opportunity for HR professionals to deal with the dilemmas at work in a more effective and efficient manner.

Figure 3.7. Short-term gimmicks versus long-term capability development

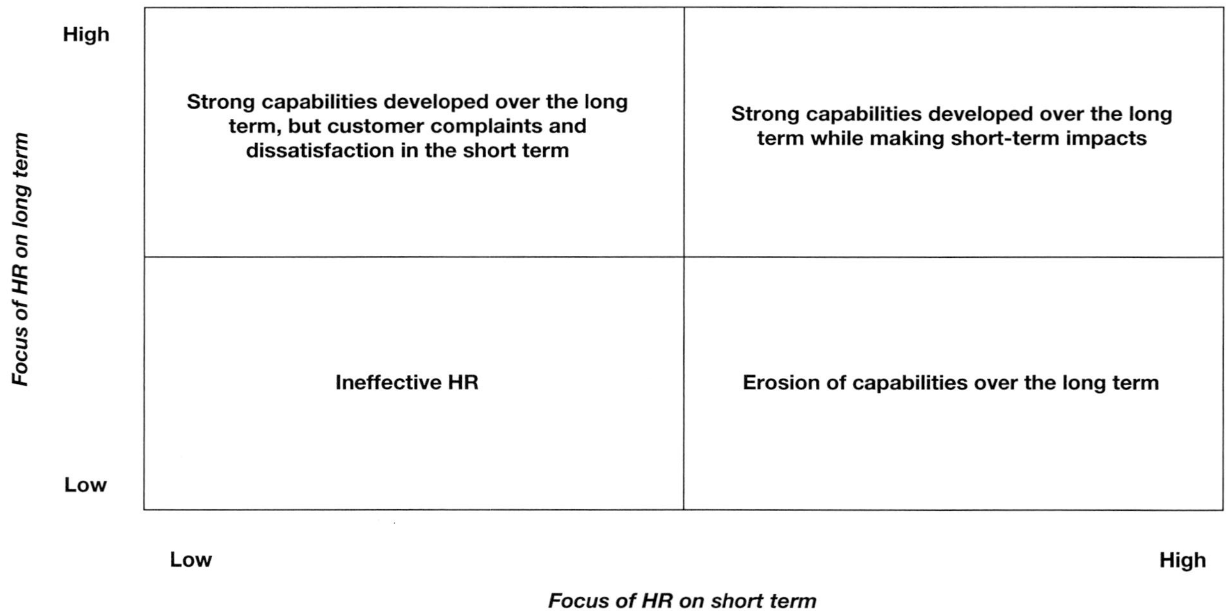

Bibliography

Adler, N. J., & Gundersen, A. (2008). *International dimensions of organizational behavior.* Mason: Thomson/South-Western.

Armour, R. (n.d.). *BrainyQuote.com.* Retrieved June 21, 2015, from http://www.brainyquote.com/quotes/quotes/r/richardarm385684.html

Aronow, J. A. (2004). *The impact of the organizational politics on the work of the internal human resources professional.* Retrieved June 25, 2015, from http://www2.uwstout.edu/content/lib/thesis/2004/2004aronowj.pdf

Beardwell, J., & Claydon , T. (2007). *Human resource management: A contemporary approach* (5th ed.). England: Pearson Education Limited.

Brewster, C., & Larsen, H. (2000). *Human resource management in northern Europe: Trends, dilemmas and strategy.* Oxford: Blackwell.

Cappelli, P. (2015, July–August). Why we love to hate HR...and what HR can do about it. *Harvard Business Review,* 54–61.

CFO Research Services and Mercer Human Resource Consulting. (2003). *Human capital management: The CFO's perspective.* Retrieved June 19, 2015, from http://www.workinfo.com/free/downloads/cfohcstudy.pdf

Drory, A., & Vigoda-Gadot, E. (2010). Organizational politics and human resource management: A typology and the Israeli experience. *Human Resource Management Review, 20*(3), 194–202.

Drucker, P. (1954). *The practice of management.* New York: Harper & Row.

Ferris, G., & Judge, T. (1991). Personnel/Human resources management: A political influence perspective. *Journal of Management, 17*(2), 447–488.

Ferris, G., & King, T. (1991). Politics in human resources decisions: A walk on the dark side. *Organizational Dynamics, 20*(2), 59–71.

Ford, H. (1934). *Henry Ford quotations.* Retrieved June 2015 from https://www.thehenryford.org/research/henryFordQuotes.aspx

Gilberta, C., Winnea, S. D., & Selsa, L. (n.d.). *The influence of line managers and HR department on employees' affective commitment.* Leuven: Catholic University of Leuven. Retrieved June 15, 2015, from https://lirias.kuleuven.be/bitstream/123456789/271894/1/MO_1004.pdf

Gill, C. (2007, January). *A review of the critical perspective on human resource management.* Melbourne: Melbourne Business School. Retrieved June 14, 2015, from http://works.bepress.com/carol_gill/

Guest, D. (1987). Human resource management and industrial relations. *Journal of Management Studies, 24*(5), 503–521.

Hammonds, K. H. (1997, December). *Why we hate HR?* Retrieved June 19, 2016, from http://www.fastcompany.com/magazine/97/open_hr.html

Hutchinson, S., & Tailby, S. (n.d.). HR and the line: How can front line managers manage (CIPD, Producer). Retrieved June 15, 2015, from http://www.cipd.co.uk/NR/rdonlyres/8B514C60-2E18-4488-9DD0-21F1F30BFB84/0/HutchinsonandTailbyresearchpaperF.pdf

Jefferson, T. (n.d.). *BrainyQuote.com.* Retrieved June 15, 2015, from BrainyQuote.com: http://www.brainyquote.com/quotes/quotes/t/thomasjeff121032.html

Kahnweiler, W. M. (2006). Sustaining success in human resources: Some key career self management strategies. *Human Resource Planning, 29*(1).

Katz, L. (n.d.). *Organizational versus national culture.* Retrieved June 12, 2015, from http://www.leadershipcrossroads.com/mat/Organizational%20vs%20National%20Culture.pdf

Kochan, T. A. (2004, August). Restoring trust in the human resource management profession. *Asia Pacific Journal of Human Resources, 42*(2), 132–146.

Legge, K. (1995). *HRM: Rhetorics and realities.* Basingstoke: Macmilla Business.

Madhavan, S. P. (1997). Strategic evaluation and audit of training systems. In U. Pareek (Ed.), *Evaluating HRD in organisations.* Jaipur: National HRD network , Jaipur Chapter.

McGregor, D. (1960). *The human side of enterprise.* New York: McGraw-Hill.

Murthy, N. (2009, April 20). *Feudalism responsible for satyam fiasco.*Retrieved from http://economictimes.indiatimes.com/opinion/interviews/feudalism-responsible-for-satyam-fiasco-n-r-narayana-murthy/articleshow/4422219.cms

Pareek, U. (2002). *Training instruments in HRD and OD.* New Delhi: Tata McGraw-Hill.

———. (2007). *Organisational culture and climate.* Hyderabad: ICFAI Books.

Pareek, U., & Rao, T. (1982). *Designing and managing human resource systems.* New Delhi: Oxford & IBH.

———. (2008, November). From a sapling to the forest: The saga of the development of HRD in India. *Human Resource Development International, 11*(5), 555–564.

Purcell, J., & Kinnie, N. (2007). HRM and business performance. In P. Boxall, J. Purcell, & P. Wright (Eds), *The Oxford handbook of human resource management* (pp. 533–551). Oxford: University Press.

Rao, T. V. (1990). *The HRD missionary.* New Delhi: Oxford and IBH.

Rao, T. V., Rao, R., & Yadav, T. (2007, January–March). A study of HRD concepts, structure of HRD departments, and HRD practices in India. *Vikalpa, 26*(1), 49–63.

Ryan, L. (2005, May 31). *Why HR Gets No Respect.* Retrieved from http://www.bloomberg.com/bw/stories/2005-05-31/why-hr-gets-no-respect (accessed on November 6, 2015).

Ulrich, D. (1997). *Human resource champions.* Boston: Harvard University Press.

Ulrich, D., Brockbank, W., Johnson, D., & Younger, J. (2007). Human resource competencies: Responding to increased expectations. *Employment Relations Today, 34*(3), 1–12.

Vosburgh, R. M. (2007). The evolution of HR: Developing HR as an internal consulting organization. *Human Resource Planning, 30*(3), 11–23.

Walton, R. (1985, March–April). From control to commitment in the workplace. *Harvard Business Review*, 77–84.

4 HR as a catalyst for change management

S. Y. Siddiqui

In my thinking one of the most critical values added from human resources (HR) in the present volatile, dynamic, and complex business environment is leading change. Whether it is the global competition on the business front, changing demography, and high aspirations of a young workforce or the new leadership challenges, flexibility and willingness toward change management will increase the competitive edge of corporate. However, for HR to be a catalyst for change it has to reflect a strong understanding of the overall business environment and a clear resolve and intent to first undergo the change itself. In this context, the role of an HR leader will be the key.

Joining Maruti Udyog Limited in September 2003 turned out to be a very exciting phase of my career. Maruti, known for revolutionizing the auto industry in India, maintaining consistent top performance, customer satisfaction, and growth, was a respected brand in the country. Coming from a European MNC New Holland Tractors, a 100 percent subsidiary of CNH Limited and part of FIAT Group, the initial induction at Maruti was a bit difficult one for me—from the first name culture of an MNC like New Holland to a very formal hierarchy-driven culture at Maruti. It was very different than the work culture I was used to in private sector companies. On the positive side, I could also observe that the scale of business operations was huge with consistent profitability and growth; hence there was an amazing sense of pride in every Marutian I met.

Maruti at that point of time was trying to find its way toward transition to a private sector identity. Therefore, it became a big challenge for me in HR to lead a "critical change initiative" in the company. Maruti started as a public sector company in 1982–1983 and in 1992 it became a 50:50 joint venture between Government of India and Suzuki Japan. Then, in 2003 after the disinvestment by the Government of India, Maruti Udyog Ltd. became a private sector company, part of Suzuki, Japan. In 2006–2007, the name of the company also was changed to Maruti Suzuki India Limited. It was quite a huge change from the original start as a public sector company. However, the bigger challenge was to facilitate the real change in the company—work culture, policies, processes, and most critical being—changing the mindset of people. It was a tough challenge for the top management and HR. At the same time, there was continued business success and an element of pride in people being part of Maruti; hence it was not easy to talk about the "need for change" in such a scenario. Resistance for change in a successful company was likely to be higher as I realized in my initial interactions at Maruti. But the most critical reason for change was to prepare the company for bigger competitive challenges in the future say 2010–11 and to ensure that Maruti not only beats competitor challenges and retains its no. 1 position but also achieves good financial results and growth on a sustained basis. Hence, it was important to first establish the need for change.

The need for change

To set the context clear therefore, it was very important first to define the need for change at Maruti Suzuki. Three important reasons were thus clearly identified during deliberations with the top management team with medium to long-term business strategy:

1. The changing structure and identity of Maruti from a 50:50 JV (between Government and Suzuki Japan) to be a private sector company part of Suzuki Japan from 2003 onward.
2. The changing business environment in India with the globalization setting in almost all sectors including the auto industry and as a consequence the changing competitive landscape in the auto industry in India.
3. The need for Maruti to redefine its "competitive edge" to remain no. 1 in the future too despite increasing global competition in the auto industry in India.

The top management was clear to own and drive the change and they took the most professional and balanced approach of involving the buy-in of all functional heads and then empowering HR to lead the change management initiative in a professional way.

Despite being a government undertaking for many years, Maruti's work culture was still strikingly different from other government organizations because of global exposure to Japanese work culture and management practices coming into the company through Suzuki Japan. The organization culture of Maruti reflected high business focus and commitment, high target and result orientation, and strong business systems and processes. However, it also reflected an inward-looking approach, strong compartmentalization, and functional boundaries, looking up to seniors for direction and decision-making, low risk-taking behavior, bureaucratic and slow decision-making process, and a culture lacking in appreciation and celebration approach.

In my first 6 months at Maruti Suzuki when I had many formal interactions with the senior and top management team, I also consciously met employees at all other levels—middle management, junior level, entry level professionals, supervisors, and union office bearers. During most of these sessions, I just listened to their views and expectations as it gave me tremendous insight into how the people perceived HR as well as the need for change. I could clearly see a very proud past, high target orientation, high-performance culture, and hierarchy-driven work environment. While the older generation was content with past achievements, the younger generation wanted change toward a more progressive and transparent work culture.

Therefore, I could clearly identify the need for reorienting the work culture of Maruti to focus on an inside-out approach, team work and positive inter personal relations, empowerment and responsibility orientation, risk taking and experimentation, and also speedy decision-making. There was also the need to increase the emotional quotient of leadership for better people connect with appreciation and celebration culture to replace the hierarchy-driven work culture.

Another critical observation was with respect to the change of HR functional identity at Maruti. In 2003, the perception about the identity and role of HR function was more oriented to a PSU company. HR was seen as not very progressive and employee oriented. The HR function was IR centric, centralized, administrative, and control oriented. The HR focus was policy and rule driven and not people driven. As I could experience the reflections about HR were as given as follows:

Maruti Suzuki HR Reflections 2003

- Centralized and isolated identity of HR as a function
- Hierarchy-driven decision-making processes
- Lack of transparency in HR processes and decisions
- Basic HR approach reflected conventional policies, systems, and rules
- Predominant industrial relations perspective defining the HR function

The town hall meetings made it relatively easier for me to get a sound understanding of the given situation. That helped me to conceptualize and propose the strategy for the change management initiative to the top management:

The strategy of organization change at Maruti

- To build on the successful past, commitment, and competence of people by launching the change in an appreciation mode
- Emphasize the need for change linked to changing external environment and as a prerequisite to future success of the company
- Create ownership and participation toward the change initiatives through strong and consistent two-way communication and education process
- Focus on the senior pros to take the lead and involve the young population to drive the change
- Organization change to be gradual and linked clearly to business needs

Communication strategy

In any change management initiative, communication plays a very critical role. In fact, the change initiatives that failed in many corporates were largely due to the lack of effective communication as one of the key reasons. The most important aspect is to communicate the "Need for Change." Also to respond to the doubts, anxiety, and various views of the entire employee population, communication strategy becomes a very critical factor.

Therefore, we laid very high focus on internal communications for the ambitious change management initiative. The strategy was to go top down, then bottom up, and across all levels and functions. The strategy also included a structured communication channel, on regular periodicity, at different layers but starting from the senior management level and directly led by the MD and CEO.

Structured communication channels also helped the company not only to manage the somewhat uncertain internal environment and morale but also to evolve an environment of trust, confidence, and mutual cooperation. Strengthening internal communication processes, directly lead by MD and supported by the HR head, resulted in greater employee connect across levels. Maruti Suzuki benefited immensely from these forums by implementing some of its best practices through suggestions coming out of these forums, and by creating buy-in for the change initiative. Top down communication model of yesteryears was replaced by bottom up and horizontal across the functions, thereby encouraging people to voice their opinions and feelings.

"Change in the HR philosophy" as a basic prerequisite

To initiate the organization change at Maruti, the first prerequisite was for the HR function itself to change. Hence, HR team had to prepare itself to change while driving the Organization Change Initiative. It took some real time and effort to first initiate the change in the HR thinking, approach, and the HR credibility at Maruti.

In this regard, I strongly felt that it was important to strengthen the ownership of my business peers toward the change initiative; hence it was important to institutionalize such an ownership process. Therefore, we first created an HR think tank consisting of all my peers from the business functions and titled it HRIDC (Human Resources Initiatives Development Council). The HRIDC became one of the most critical platforms for brainstorming, dialoguing, deliberations, and finalizing the change initiative at Maruti including key HR policies and processes. I was leading the HRIDC meetings, being the head of HR, but focused more on shared leadership perspective. HR conceptualized the new policy concepts and basic proposals for change for HRIDC team to dwell upon and add the business perspective, while we formulated new HR policies and processes for management approval. While it facilitated new HR policies and initiatives, it also built strong ownership of HR processes in the functional heads at Maruti.

Coming back to the starting point of first driving change with the HR team itself was quite a unique challenge. The past background of 21 years for most HR team members was proving to be a tough challenge as it occupied most of the mind-space of the HR team members whether junior or senior. Therefore, the initial effort involved introspection and brainstorming sessions followed by informal and formal training in emerging HR trends and perspectives. One big change that I could also bring in was to connect our HR team with NHRDN, AIMA, and CII to facilitate exposure to our people by meeting and networking with external HR fraternity. Within Maruti I slowly ensured aligning HR to business strategy and plans and encouraging HR measurement indexes. Also I had to bring in some HR professionals through the lateral route from the private sector background to reinforce the HR team with diverse experience.

Just like creating HRIDC through which I involved my business peers in deliberations and formulation of new HR policies, I also started creating cross functional teams on important people processes, e.g., performance management system and career growth policy that ensured feedback and participation from young managers and department heads along with the HR team toward formulating a draft framework of the change of old PMs and career growth policy. The working of CFTs with HR created a high level of actual connect with our employee population. Also in the process change got facilitated gradually without much resistance.

Next was to formally decentralize HR function and also align it closely to business plans of the company. It involved cutting open the HR cross section view to the key stakeholders—people, functional heads, directors, and the board. In the process we created new functional identity for commercial HR (to support sales and marketing business vertical), engineering HR (to support engineering and QA business vertical), and plant HR (to support production). It required a participative, consultative, and flexible approach to define new or redefine existing HR policies, systems, and processes. Some key directions were as follows:

- HR's transition from a standalone identity to integral part of business
- Transition of HR identity and credibility as mere support function to strategic part of business
- From manpower costs to investment in people
- From labor law compliance to strategic IR perspective
- HR administration to people connect
- From technical training to wholesome people development
- From discipline focus to developing and sustaining a positive work culture with ownership

Changing the Maruti work culture

One thing was clear to me that changing the prevailing work culture will be the toughest challenge, especially for a company that was highly successful in business performance on a sustained basis. Therefore, the first prerequisite was to define the need for change in the work culture. The best strategy in my mind was to first influence and involve my business peers to agree for the renewed focus of the Maruti work culture on:

- Global mindset with an inside-out approach and looking at benchmarking with global best practices.
- From compartmentalization to a team-based organization with focus on team work and positive inter personal relations.
- From centralized and hierarchical decision-making to empowerment, delegation, and responsibility orientation in people
- From the tendency for upward delegation to ownership, risk taking, and experimentation and also speedy decision-making.
- From only business focus to higher emotional quotient of leadership for better people connect with appreciation and celebration culture.

HRIDC, the forum consisting of HR team and my business peers, became a very effective platform to discuss, deliberate, and explore the above change through various initiatives along with the change of HR policies and processes over the next five years starting from 2004.

It was also obvious that the work culture of a 20-year-old company cannot be changed overnight; hence I consciously planned for a gradual shift in the work culture and the mindset of people starting from my peers and cascading down. On the reverse I could identify a great opportunity to increase the pace of change through the involvement of younger population of employees in Maruti who not only wanted the change in work culture and people policies and processes but also were excited to become champions of change initiatives.

The success of any organization change initiative depends to a large extent if it is owned by the top management and if it starts from the top. Hence deliberating the organization change perspective with the top management specially the Japanese colleagues at Maruti was critical and cumbersome at times. Sometimes, one realizes that life would have been much simpler if corporate communication were based on a single language—spoken and understood easily by one and all. But in our case it was English and Japanese together where sometimes the communication gaps remained in spite of the interpretation process.

For example, the work culture change toward increasing the emotional quotient in the leadership team was not very much relevant in the context of Japan where the business quotient works fairly well by itself. But in India the emotional quotient is very much essential because of our social construct itself. However, the total empowerment by the top management at Maruti, both

Japanese and Indian, to HR for bringing the change initiatives as we conceptualized was a great ingredient. Initiatives such as quarterly town hall meetings by MD, tea group meetings by HR head, exit interviews, stay interviews, monthly divisional meetings on business and people challenges, annual family day, monthly factory visit of families of associates, long service awards, rewards and recognition policy, higher education policy, OBT initiatives, innovation teams, etc., were instrumental in changing the work culture of Maruti though in a gradual and phased manner.

It was an enormous task to overcome the cultural gaps between SMC, Japan, and the people issues in the local context of our country. At times, we could observe a fairly clear gap between the Indian senior colleagues and our SMC counterparts. Hence, the balancing act required sensitizing them individually for ownership of the organization change initiatives. Above all, it required a tremendously patient, educative, and flexible approach to discuss, deliberate, and convince our SMC colleagues on some of the new concepts, policies, etc.

HR thus had to play the role of a "change catalyst" with a clear intent, high ownership, and high conviction. From 2003 onward, the Organization Change Initiative was further facilitated through OD interventions at levels:

Leadership retreat for top management

The key objective was to strengthen team work at the top leadership level at Maruti Suzuki including the Indian and Japanese directors. Also the effort was to shift gears from an operational focus to strategic business planning. In view of the increasing people challenges and the dynamic business environment, we also aimed at building transformational leadership perspective in the top leadership team. The first initiative in this regard was not easy because of language issues and the cultural subtleties involved between the Japanese culture of SMC Japan and the ground realities in the Indian context. The balancing act was rather tough and required a very patient approach. Therefore, I conceptualized the first Top Management Retreat aimed at bringing together a strategic leadership deliberations comprising of group discussions, outbound activities, and leadership assessment. The Top Management Team of Maruti Suzuki at that time, August 2005, comprised two Indians and four Japanese directors.

The Leadership Retreat involving all directors, over two and a half days residential at Bengaluru, was facilitated by Dr Pritam Singh and Dr Asha Bhandarkar from Management Development Institute (MDI), Gurgaon, with focus on:

- Team building perspective, that is leadership team of Maruti Suzuki
- Self-development focus for each director based on feedback from others
- Develop a strategic perspective of long-term business plan

- Mentoring/coaching role to develop next level people
- Creating a "new curve" for Maruti through transformational leadership—sustained growth and retain no. 1 in future too

The first ever top management retreat was a great success. For most of the Indian and Japanese directors, it was first of its kind formal and structured Leadership Retreat. It reflected very high interest, participation, frank and candid discussions, and an openness to learn and change. The retreat had both class room sessions and outbound exercises and games. We concluded the program with a clear direction for the future including medium to long-term business strategy, strong teaming at the top, as well as Individual self-development plans based on the leadership assessment of each participant. It was also decided that the top management retreat will be an annual event and a must-attend program linked with assessment and evaluation of the progress plan versus actual. One of the most important outcomes that I could observe was the emergence of positive interpersonal relationships in the top group.

Leadership development initiatives for senior management

The focus was on global business perspective, business strategy, people leadership, and leading and managing change. Along with Dr Pritam Singh and Dr Bhandarkar, MDI, Gurgaon, we worked out a specific plan:

- A two and a half-day leadership training at MDI Gurgaon
- External training nominations at premier business schools for each divisional head
- Two-weeks training at Europe—meeting thought leaders from corporates and academia through ESCP EAP Business School (Germany, France, and the UK)

But how this idea of mine regarding the Europe training of divisional heads was received by my training team, my peers at the divisional head level, is another funny reflection of the erstwhile culture that existed in the company.

When I shared the concept of the leadership training with MDI, Gurgaon (Part 1 of the plan), it was received very well with anticipation, excitement, and positivity. But when I talked about the Europe segment—14-day global exposure, it was received with complete disbelief that how a Japanese parent company will digest the idea of global exposure part of leadership development in Europe rather than the obvious given choice of Japan. My concept was more professionally focusing on enabling the "inside-out approach" at Maruti with globalization of business taking strong roots in India and Maruti facing the transition from a monopoly phase to globally competitive phase of organization lifecycle with relatively low global exposure.

But the top leadership, both Japanese directors and the Indian top management, found the idea quite relevant and appreciated the proposal. Rather the two JMDs (Japanese) and MD guided and supported me to cover all 35 divisional heads in three separate groups of 11 to 12 each. They advised me not to attempt the program in one group just to save costs because it would be untenable to continue the company operations if the entire group of divisional heads is away at one go for two weeks.

While for me it was excellent empowerment, free hand, tremendous guidance, support from the top management, for my other colleagues and team members at Maruti, it was a path-breaking, pioneering, and out-of-box opportunity. It also set the right example of top leadership accepting and driving change in the company. Hats off to Suzuki Head Office for the openness, empowerment, and long-term vision for the business in India.

Subsequent to the Europe Training we went for specific additional development actions as given below:

- Leadership assessment of each person and feedback session
- On the job—coaching and mentoring role for young professionals by divisional heads
- 360-degree feedback process
- Developing divisional heads as internal assessors and trainers

Leadership assessment of each divisional head was a first of its type "Development Initiative" at Maruti Suzuki. It was based upon the following:

- Behavioral tools used during the leadership training module at MDI, Gurgaon
- Observation of each individual's participating style, content, and value add during the various sessions
- Performance and involvement during the off-site work groups/projects
- Observation and participation during outbound games/exercises
- Observation and participation during the two-weeks Europe training module including team spirit, social behavior, adaptability, and leadership traits

As part of the leadership development strategy, 360-degree feedback was introduced in Maruti Suzuki in 2006 for the first time for all senior leadership positions starting from department heads to the top management level including MD. The basic objective was the development of their leadership capability in a structured process and as per clearly agreed leadership competencies. We decided to partner with Ernst & Young and initiated the first-ever 360-degree feedback process for about 200 senior leadership positions in the company. But the key focus and stated objective was "development of leadership capability" rather than "assessment for variable pay or career growth decisions."

Even Japanese Directors were covered part of the 360-degree exercise. It was not easy for them. The Japanese are very humble and polite, so it is very difficult for them to say anything negative about others. In spite of this, they agreed to take part in this project. Also language was a major barrier for Japanese colleagues; hence, we translated the same English online questionnaire into Japanese language to make it convenient for them to respond well.

Leadership development of middle management

The focus was on exposure to the external business environment in India, people management and people development, and accepting and facilitating change. The need was to develop them to play an effective leadership role as the first interface with young population. The development was facilitated through intense training interventions over a 3-year period. Some of the key interventions facilitated through the 3-year period are given below:

- Leadership training by GE International
- Nominations to external trainings at leading B-schools
- Leadership training by Dr Pritam Singh and Dr Asha Bhandarkar at MDI, Gurgaon
- 360-degree feedback exercise
- Potential assessment through assessment centers and individual feedback for individual development plans

The entire effort was led by the top management and was linked to their career enhancement process which really brought in the ownership, participation, and willingness to change. The middle management team of about 150 members was considered a very critical link between the senior management and the rest of the company.

The subsequent success of the entire change management initiative gathered faster pace to a great extent due to the total involvement and ownership of the middle management layer after the various interventions covering the entire department head population.

Change initiatives for young managers

One of the most critical groups to facilitate change was identified to be the young engineers and managers part of the junior management layer. While starting the Change Management initiative in early 2004, I did consult my ex-boss in CNH Limited, FIAT Group, who was the Senior Vice President, Global HR based out of USA. During the telecom after he understood the entire Change Management need for Maruti Suzuki, he emphasized that I must concentrate on involving the young engineers and managers as the champions of change.

He supported his advice with sound reasoning that the younger population will be more open, flexible, and energetic about change with no baggage of the past. And his advice turned out to be the most crucial for me in my entire effort over the next three years.

Initially, the big focus was on communication with the younger population across the company through town hall meetings, open sessions, and small group meetings. The basic objective was to communicate the need for change, the fast-changing business environment, and the need to build the competitive edge for the future and retain our no. 1 Leadership position.

Subsequently, specific training initiatives were conceptualized to cover the junior management layer. Some of the highlights of the training interventions are given as follows:

- Leadership training by GE International
- Development and assessment centers with individual feedback
- Performance feedback sessions and action plan for development
- Training series on effective communication skills and also team training (outbound)
- Talent management initiatives as per the feedback received from various communication forums

Additionally, since the Maruti Suzuki Team was being built up further to meet the expanding business needs, a big number of new members were joining the company and we thought to bring them on board with a sound understanding of the change management initiative and the new organizational framework, policies, and processes. At the same time to motivate the junior management layer a need was felt to introduce more progressive HR policies and processes to bring in higher emphasis on career development and career growth perspective. Hence, some new talent management initiatives were introduced which went a long way in creating the total involvement, motivation, and participation of the younger population in the change initiative. Let me share some important ones here:

Talent management initiatives that facilitated the change management process on a sustained basis were as follows:

1. A structured on boarding process was designed with many touch points for the first 12 months of the new employee's settling down process. The induction policy introduced at that time was governed by a policy of 100 percent coverage.
2. Mentoring/coaching for all laterals who were selected from other companies with a career track of 3–7 years. This resulted in "on the job" development of such professionals coming from different background, skill sets, and different company cultures. Senior professionals in the company selected as mentors. Training on mentoring has been provided to all mentors.

3. More critical was to enable the fresh entry level campus joinees (graduate engineer trainees, fresh CAs, MBAs, etc.) to settle in the company. Hence, we conceptualized the "Buddy" system. Buddy was a 3–5 years senior with similar professional education background. Thus, all campus joinees were supported by a Buddy. Each Buddy was given connect with three to four new campus recruits to support, guide, and hand hold on personal, professional, and the job aspects for the first 12 months.
4. With a view to develop all-round business managers aimed at developing the future leadership pipeline, the following initiatives were started:

   a. Job Rotation—good performers who have spent five years in one function have to be necessarily rotated to other functions. The job-rotated individuals can either be considered for a return rotation to their parent function after around 3 years or continue in the new functional area.
   b. Cross Functional Teams (CFTs) on critical projects were formed to give exposure to young managers to foster team working, empowerment, innovation, and experimentation—tolerance for failure but facilitate learning for future.
   c. Stretch Assignments to high potentials to take up roles at one level higher, e.g., manager/senior manager level acting as department managers (general managers) selection of territory sales managers for the senior role as regional managers, etc.

5. Clear career growth paths

   a. Fast track for high performers and high potentials
   b. Normal track for good performers
   c. Extended track growth for solid citizens

6. Stay interviews for higher engagement

Another initiative on stay interview was started as a basic proactive measure to seek views and suggestions of our existing employees for improving and reinforcing the HR policies and processes as well as new ideas on strengthening our business processes. Stay interviews in a way facilitated a "Listening Culture" in the company. The intent of the top leadership to invest time to listen to our people led to high people connect. The objective was to give weightage to the views of people who were adding to retention. Stay Interviews thus became an effective process of hearing out our people. When job-hopping is the rule rather than an exception and when employee retention has become a critical issue, every company wants to know about those factors that can keep an employee engaged and committed; actually, the stay interviews help the organizations in gathering these hard facts through pulse reading to map the expectations of employees. In case of Maruti the policy on higher education (company sponsored) was introduced based on feedback from communication meetings and stay interviews in 2006. Similarly, Family Day Celebration, Service Awards Policy, and Employee Referral for Recruitment, etc., all came through the employee feedback process.

Change initiatives for supervisors and shopfloor technicians

When I started meeting the supervisor population part of the communication sessions, I realized that there were three generations who were forming this important layer for Maruti Suzuki. Old supervisors who rose in their career from shopfloor technician level and diploma engineers who had joined the company at the entry level supervisor in the early years of the company were in the age group of 50 years and more. Then those who joined the company in subsequent years in the age group of around 40 years. And of course young diploma engineers who had joined the company at the asst. supervisor level in recent years below 30 years of age.

It was a mixed population group and also very important from the shopfloor management perspective as they were the most critical link between the company and the shopfloor technicians. Hence, we had to think through this diverse group to conceptualize their training interventions to make the change initiative really cascading to the lower levels. Some of the highlights are as follows:

- Extensive communication sessions led by the senior and top management.
- Training series on change management, business environment, and competitive edge perspective for the older groups.
- Supervisory development training module with focus on people management skills for the young group.
- Information technology skills for the older group of supervisors
- Outbound training to build team work, loyalty, and commitment toward the company for the young group.
- Job rotation and career development policy for all the groups.

The Change Initiatives for Technicians (Workmen) were also given high priority and attention. While Maruti was preparing to sharpen its competitive edge, it decided to undergo a Voluntary Retirement Scheme (VRS) in consultation with the Union in 2003. But it also created a somewhat uncertain and negative environment in the company. Therefore, there was a strong need to bring in better understanding for the need for change, involvement, and ownership in the technician population at the shopfloor. Thus, focused interventions for technicians were launched to bring in a positive culture, participation, and involvement and mindset change. To achieve this, specific training initiatives were conceptualized as a pro-active and positive Industrial relations strategy. The new training interventions were introduced to educate and change the mindset of workmen as per details given below:

"Sankalp," a training program designed with competent external resources focused on the changing business environment and competitive scenario in the auto industry in India. The first program was launched in October 2004 to cover all technicians in two years. The program aimed to achieve the following goals:

- To orient the shopfloor associates to the changing nature of global business environment.
- To help them understand the need for collaborative partnership perspective between the company and all its employees.
- To create involvement, team work excitement, and positive energy at the shopfloor.
- Ownership for a quality mindset and approach among the associates.

The program was designed to be highly participative and interactive so that participants enjoy every moment of the program. Also it was meant to create an excitement so that people should talk about it when they go back to shopfloor encouraging others to come forward and join the program. Furthermore, it was intended to bring involvement of the union leaders in the intervention for them to feel ownership of this change initiative.

The entire initiative was led by Dr Mahavir Jain, a renowned faculty with extensive training background with National Productivity Council. While designing the program objectives, title, content, and methodology we worked with Dr Mahavir Jain to share the background of the company, the larger change initiative at company level, and our intended objectives. When we launched the first training program for the union office bearers the feedback was very positive and encouraging. But the first group of shopfloor associates who came for the Sankalp program were cautious, critiquing the company, even sarcastic with a negative mindset against the company. But after few programs the employees too could understand the company business perspective from a long-term view and the professional approach of the company. Through various participative games, role plays, and work groups they were able to understand and make themselves aware that unless the change starts across levels the ability of the company to fight in market place will suffer badly. A sense of urgency was shared that if change does not take place at the right time and with full ownership, the competition would take advantage. They were also made to understand that the survival of the company may get challenged which may impact their career development and well-being of their families too.

This positive image cascaded down and it created a sense of enthusiasm, positive energy, commitment, and motivation to do more for the company. Our HR intervention was successful as originally conceptualized. The Sankalp Series ran through for two years to cover all the shopfloor associates and we had to put up a quarterly schedule of Sankalp Training Series for the associates to know three months in advance as to when their nomination will take place.

Outbound Team Training program was the second Training Series which was the first ever for the shopfloor associates in Maruti Suzuki. It aimed at installing pride in being a Marutian. The outbound training program was branded with specific and well thought out title "Chunauti" and it went on to become a very successful training intervention creating a high ownership, team work along with an element of pride in being a Marutian.

Subsequently we launched other training series for the associates such as Bulandi and Nischyae to create business focus, build the competitive edge while creating involvement and excitement. The program objectives also laid emphasis on self-development, increasing commitment and motivation, and preparing for future challenges. Outbound training sessions were conducted to improve capabilities, increase team synergy, and create a sense of togetherness among the employees.

Another big impact was created by strengthening the "participative approach" in resolving IR issues with the Union by creating a joint and Empowered Forum—Employee Relations Development Committee (ERDC). A new initiative on Family Connect was introduced through "Parivar Milan" by inviting 40 families of shopfloor technicians on two Saturdays a month for a plant visit during which they interacted with senior management and visited the plant.

The key outcomes of change management

From 2007 to 2008 onward, Maruti entered into the intense global competitive phase in the auto industry in India. But since we had prepared well, we were confident of doing well even in this phase too. The Change Management initiative between 2003 and 2010 triggered a refreshing transformation at Maruti Suzuki at the company level. The company maintained a high level of business success, strong financial performance, and growth while retaining the no. 1 position in the auto industry. The highlights of qualitative outcomes of the change management initiative were as follows:

- Teaming at the top
- Strong concept of One Team-MSIL Team
- Pro-active, positive, and vibrant work environment
- Progressive and people-oriented HR Policies
- HR strategy aligned to business plan
- Winning mindset with a challenging spirit
- Increased competitive edge

Also the impact of the seven-year change management focus on HR and people strategy was quite visible and the Maruti Suzuki HR reflections in 2010 were:

- Proactive and decentralized HR aligned to business plan
- Transparency and people orientation in HR systems and processes
- Culture of appreciation and celebration and focus on team work

- Leadership development and strong teaming at top level
- Competitive salary and benefits but control on labor cost
- Positive and congenial industrial relation scenario
- Maruti HR achieving high credibility in the external environment

I would also like to share the impact of this change initiative driven by HR with the top management direction and ownership, on reinforcing and retaining the competitive edge of Maruti Suzuki and achieve excellent business performance and growth despite globalization and intense completion in the Indian Auto Industry during this phase of 2004 to 2010–2011 and continuing with business success and expansion there on:

S No	Particulars	2004–2005 (Start of Change Mgmt)	2009–2010 (Completion of Change Initiative)	2014–2015 (Last Financial Year Performance)
1	Net sales (₹ in millions)	109,108	289,585	486,055
2	Growth in net sales (%)		42%	14%
3	Vehicles sold (domestic + export)	536,301	1,018,365	1,292,415
4	Market share (%)	45.9	44.6	45.0
5	Net profit (₹ in millions) *	8,536	24,976	37,112
6	Net profit margin (%)	7.8	8.6	7.6
7	EBIDTA (₹ in millions)	16,704	39,509	67,129
8	EBIDTA margin (%)	15.3	13.6	13.8
	* Profit After Tax			

Note: The above performance is despite the intense and increasing competition in the auto industry during this period of 2003–2004 to 2014–2015.

While dealing with business challenges most of us negate or pay less attention to the people aspect in our business, as we are more interested in managing the top line and bottom line. However, in my thinking, the present business environment is pushing organizations to put their belief in "people power" by nurturing and developing them and ensuring a long-term competitive edge for the company. It is high time that we acknowledge the fact that people are the key differentiating factor for any business organization as they "draw, drive, and deliver" the business results. In case of Maruti we could achieve a huge transformation through our people and the business results followed as planned and even better than forecast. The underlying principles of any business model, therefore, should always aim at keeping the employees motivated and engaged that can bring consistency in accomplishment of business objectives. Personally for me, it was a very valuable learning experience but also a highly satisfying one. But this journey of change will go on if we need to keep on evolving and challenging our competitive edge for success in the future.

One more perspective that we need to take into consideration here with respect to the people element as to how much impact the external business environment can have on the internal employee population and people aspirations. I am sharing this more in the context of the younger blue-collar workforce and thus the possibility of larger influence of the external environment on the industrial relations climate of a company. Why even some professionally well-managed companies with excellent corporate image faced serious Industrial Relations conflicts—in recent memory many such reputed and well-managed companies have faced tough IR situations across the country. In some parts of the country the volatile Industrial Relations trend reflects a higher occurrence, maybe a regional bias too. In case of Maruti Suzuki also we faced a rather tough and unexpected situation in 2011 and 2012 which resulted in the loss of life of an extremely good, mature, and well-respected HR professional and serious injury to a large number of supervisors and managers including some Japanese expats.

The incident of violence at Maruti's Manesar plant in July, 2012 was undoubtedly the toughest people-related issue to handle in my entire career. It was rather unfortunate and uncalled for. It happened without any industrial relations dispute or any unresolved long overdue labor issue and therefore all the more unexpected and shocking to say the least. Though we could restore normal operations in three months' time but the bitter memory and scar will remain forever. It was adversity in its toughest form. The responsibility to restore normalcy was the foremost responsibility of HR at that crucial juncture. How did we go about it? We simply followed the professional approach of being "fair but firm." The firm approach in handling industrial relations is tough, but the company decided to take a firm stand and devised countermeasure plans accordingly. Building the morale and confidence of the affected population of management staff was handled as the next priority. It involved frequent communication as well as demonstrative counter measure plan for the future. Rebuilding the team including new shopfloor associates on a time-bound schedule with HR and production team working together was great team work and resulted into speed in action. Building confidence of the employee population through reinforced safety and security measures including state support was also an integral part for us to get back to normal operations at the Plant. But I think the top leadership support and empowerment to HR to handle the entire situation in a fair and firm approach and with openness and flexibility to reinforce our work culture and people processes to ward off any such negative influence from the external environment in the future was the most important aspect of Maruti Suzuki, the way we could handle the adverse situation at that time.

5 HR leaders as change enablers

Anil Sachdev, Surabhi Sharma, and Kritvi Kedia

Human resource development is a noble profession. It serves the highest need of people—to serve and enable others to be happy. When we find work and choose a profession to serve fellow humans, we are discovering the key to our own happiness.

It is essential that human resource (HR) professionals appreciate that their biggest goal is to create "well-being" of all people associated with the organization. This covers the physical, emotional, intellectual, and spiritual dimensions.

HR has a big role to play in creating a sense of family and community in the work place. By emphasizing transparency, openness and timely two-way communication, skip level meetings, open houses, employee portals, etc., the right environment is built. By creating forums for expressing appreciation and gratitude to one another, birthday and anniversary celebrations in unique ways in which people are appreciated for their gifts and contribution, involving families in major landmarks and festivals, HR takes the lead in building emotional well-being. Having counseling sessions for those who need help, mentoring, a buddy system for newcomers, team celebrations, and emphasizing team rewards are other steps that help.

HR also plays a leading role in cocreating vision, values, and strategy in the company. Processes such as large-scale interaction process and open-space methodology for people to have such dialogues are used by HR for this purpose. Organizations with A grade vision and strategy, but B grade involvement of people, often lose to those who have B grade vision and strategy but A grade employee involvement.

By training leaders on developing greater emotional quotient (EQ) and leading initiatives on building high-performance teams, HR builds emotional well-being.

HR also creates intellectual well-being by working with business leaders in improving the strategic decision-making process so that the organization deploys the right people to make the best decisions to create value.

HR is expected to create sense of "higher purpose" in the organization and help people to identify their own "calling" and connect it to the "larger," the organization is serving. It enables leaders to discover the strengths of the organization and dream about leveraging the same to build a visionary institution. HR also encourages people to dream about how they can leverage their own gifts to realize their own vision and find the inspiration within them. This creates an "inspired work place."

HR enables people to learn from each other's faiths, learn from the essence of all major religions, and find ways to make their own lives happier by tapping the divine force within each person.

The above-stated ideas are the core of creating a "happy organization to work for." If HR strives to create well-being at all levels and find newer ways to enable people to be happy, the impact would be felt in improved customer services, better quality, higher innovation, and enhanced reputation.

At the School of Inspired Leadership (SOIL), we have developed a comprehensive approach to how HR can play such a role.

In the remaining part of this chapter, we will share the story of how a significant company has been working with SOIL to lead such a transformation effort to create a happy work place in which all are being enabled to realize their potential. All names have been changed to protect client confidentiality.

The Rama Krishna Group

The Rama Krishna Group was founded by Mr Arun Prakash, the youngest son of a reputed Delhi-based business family. After working for the family business for some years, he decided to be on his own and laid the foundation of the Group. The Group's first venture was in the field of manufacturing. Brick by brick with his hard work, immense passion, and a strong desire to create a name for himself, the Group in a span of 10 years became one of the largest manufacturers of engineering components. The Group quickly became a supplier to many global and Indian original equipment manufacturers and entered long-term partnerships with them. The Group presently has around 3000 employees and operates 12 plants in the NCR area.

Mr Arun Prakash is a visionary leader and believes that people are the most important resource of the organization. He has a strong belief in Indian philosophy and scriptures. He often shares that he would like to leave behind a legacy of happy people in a happy company. He was convinced that "happiness" as a journey was needed in his group for the sake of happiness! At the first meeting itself, he took pains to explain that this was not a strategy to improve quality or productivity and had no hidden agenda. He said, "nothing would give me more happiness than the happiness of my employees and their families."

He invited proposals from the four big consulting firms and finally chose SOIL as his partner as he felt our approach to happiness within individuals and organization based on Indian spiritual wisdom was the one that would work the best in the organization and for its people.

In order to facilitate an organization-wide transformation program, to enable people to realize their true potential, and to discover greater happiness in their professional and personal lives, we planned the entire exercise in four phases.

Phase 1: Preparing the SOIL

During this phase, we conducted interviews based on the principle of appreciative inquiry for a cross section of the company's leadership team using questions such as:

- When were you and your team at your joyful best?
- What happened, who were the people involved, and what was special about the experience?
- What does this teach you about your gifts and that of others in your team?

This served as an important data point to tap into the strengths of the organization which could be leveraged to create happiness in the organization. We also conducted focus group discussions with the middle and junior management to gain their perspective on the state of happiness in the organization.

A survey was designed which looked at the five layers of happiness, namely, physical well-being, communication, emotional well-being, intellectual well-being, and spiritual well-being at the individual level and in the organization. The survey was designed with items that could be answered using a five-point Likert scale. It was created in English and the local language of the region, Hindi, to enable the contractual workforce at the shopfloor to understand and answer the survey. The survey was administered to all the on-roll employees and 33 percent of the blue-collared contractual workforce to be able to understand the level of happiness in the organization.

The one-on-one interviews, focus group discussions, and the survey served to be important data points to lay out the design of the initiative.

We then trained other leaders in the company on conducting similar interviews by practicing "listening with the ears of the child" and cherishing the output with genuine appreciation as if they were "mining for gold." These leaders then started doing these interviews and publishing their stories to all concerned. These stories were shared with all the members of the group through emails, creating spaces in plants to put up the stories and in common meeting forums.

The impact was highly positive and it started creating waves of happiness in the company. We also invited senior leaders from other organizations to address groups of employees within Rama Krishna Group to share what they were doing to bring happiness in their companies. This brought the feeling to the employees that "if they can do it, then why cannot we?" In the first eight weeks of the work, the activity created the right foundation for the transformation exercise.

Phase II: Designing and planning

This phase was the most crucial and set the foundation of the entire program. We set up a Design Team for the initiative that was cross locational, cross functional, cross hierarchical, and cross attitudinal (optimists and cynics were both included). With the help of this team, we did the detailed planning for this journey. The company appointed a senior HR professional and named him as the full-time leader for this exercise. A Steering Team was formed consisting of all senior leaders of the company and the senior HR leader was named as the "Secretary Member" of this team.

The concept of the design team is crucial. What we design in this microcosm is then implemented in the larger organization—the macrocosm, and if we have chosen the design team members effectively, the impact of the initiative is meaningful.

Similarly, the leadership team that meets as the "Steering Team" has to allocate dedicated time to the transformation exercise and the HR function has to do all the background work to make sure that the discussions are productive and meaningful.

Thanks to the encouragement of the chairman who was very passionate about this initiative, it received the deserved attention and right focus. We worked with these forums to design a two-day "offsite" for the leadership team to facilitate a dialogue on creating the first draft of their new vision and values.

After this, true to the concept of "cocreating the vision and values" of the organization, we involved each and every member of the organization in creating the vision and values through a process known as the large-scale interactive process (LSIP). This technology allows effective facilitation and involvement of people across large groups of individuals. There were three LSIPs held, each with approximately 1000 individuals, where inputs were taken from the employees on the perceived strengths of the organization and the vision and values. In this technology, each round table represents a true "microcosm" of the organization in the fact that it represents a cross section of the organization. Employees across levels sat together on each table and visualized the organization of their dream through effective facilitation.

The LSIPs created a lot of positive energy and was the first time that such a large group of the workforce was involved together. The workforce valued the fact that the organization was taking their input in creating something of such great importance to the organization. It helped the organization come closer and bind as a family. It was one of the highlights of the entire journey of creating happiness.

We created teams to help with the facilitation at each table and the logistics of the event. We also trained individuals of the organization in this technology to build sustained internal capability in the organization. Interestingly, the output of the leadership retreat with the senior team and the LSIPs involving the entire workforce of the Ramakrishna Group was similar to a large extent.

The inputs from the leadership retreat and three LSIPs were synthesized together in a one-day confluence workshop, which was attended by the leadership team and a cross section of the representatives from each of the LSIPs to ensure a representative sample.

After the finalization of the vision and values, it was essential that the vision and values are communicated to the entire workforce to close the communication loop with them. The senior management, design team, and the facilitators of the LSIP became our champions and helped us in the dissemination of the vision and values. We enabled the communication cascade through creating small, interactive modules that could be further shared with the workforce in groups of 15–20 led by the identified group of people. Visual stimuli was also created around these and put up in the plants to facilitate recall.

At this point, it was felt by the Steering Committee that there is a need to create a localized structure at the different plants to drive the initiative locally and take up activities that were customized to the plant and its context. Therefore, a Plant Steering Committee (PSC) was created at each plant, which consisted of a sponsor who would enable the resources to drive the initiative, a happiness facilitator who was responsible for understanding the needs of the workforce and designing and implementing the activities and team members who assisted the happiness facilitator in the same.

The PSC members were trained on their roles and responsibilities and given direction on how to implement the initiative at their plants. A plan was also created to cascade the vision and values that consisted of different activities that would help the employees to relate to them and understand the behaviors that are associated with them. Most of the activities initially were dovetailed with the existing activities such as small group improvement activity (SGIA), suggestion meetings, and strategic business unit (SBU) head meetings to make sure that they become a part of the process by integrating them into the culture of Ramakrishna Group.

Processes were created for the new activities as required to ensure standard implementation by SOIL and corporate HR team of Rama Krishna Group, while the local PSCs were given the bandwidth to adapt them to the extent based on the context of their plant. The PSCs were encouraged to document all the processes and activities that they were doing to enable the same to happen.

During this phase of the project, the project was being run locally by the PSCs. To monitor the progress of the initiative, the PSCs were requested to send a monthly status of the work that they had done, with what they thought went well and what could be the areas of improvement. Further, a monthly review was conducted by SOIL with each PSC to understand the progress and give suggestions of the work which could be done in the next month. It was also a platform for the PSCs to get guidance from SOIL and share any concerns that needed to be raised at a higher level.

Slowly, the corporate HR team started to get involved in the reviews by accompanying the SOIL team for the reviews and assessing the progress of the initiative in the different plants. This provided an opportunity for corporate to understand the progress in different plants first hand, and follow up with them on several issues, if needed and build internal capability.

A monthly dipstick survey was also created to understand the pulse of the people. This survey was filled in by the Steering Committee members and the PSCs to get a feel of how they think the initiative is progressing. A report was sent out every month after the analysis of the survey unit wise, for each unit to understand where to focus the activities of the next month.

The creation of the PSCs and the focus on each plant through the same resulted in a deeper focus on happiness and positivity and the initiative slowly and steadily became a part of the DNA of the organization. There were more conversations around this topic and people increasingly looked forward to the different activities that were taken up by the PSCs. There was visible excitement around the initiative in the plants.

The chairman at this point of time felt that it was important for the families of the workforce to be touched by this noble initiative as well and the initiative looked at activities that could involve the families as well. The chairman himself started visiting the homes of the employees in an attempt to build a personal connect with the families and show them that the Ramakrishna Group cares for them.

While the localized structure was implemented for the different activities to be smoothly executed at the different plants, a talent appreciation process (TAP) was conducted to understand the strengths of and develop an individual learning plan for the senior management of the organization. This would help identify the leaders who could take the organization to the next level and develop a succession pipeline as well. The process consisted of a psychometric profile done for each participant, a 360-degree assessment and behavioral event interviews, which is an interview technique that focuses on the past events of an individual's life to assess them on the competencies. An individual report was prepared for each participant and the feedback was shared with the chairman and on a one-on-one basis with the individual.

This exercise served as an important input to understand the uniqueness of each of the senior management team members of the group and to develop the future course of development for each individual. The one-on-one feedback helped the participants become aware of their strengths, development areas, and the perceptions of their seniors, peers, and subordinates. For many of them, this was a real eye opener and helped them become more effective leaders.

Phase III: Expansion

By now, we had successfully completed one year of the transformation initiative and the roadmap and the DNA of the initiative had been laid in place. One of the first steps done now was to reflect on the journey and understand what were the highlights for the organization and where were the areas on which even more focus was needed.

This was done through repeating the annual survey conducted at the start of the initiative and the results demonstrated an increase in the happiness across the different demographics. An informal feedback was also taken from the Steering Team and Design Team on the highlights of the initiative as well. We have traversed this far in our journey with the Rama Krishna Group in this exciting, one of its kind initiative.

Going forward, our plan will include cascading the initiative to touch even more people and actively involve more people. To enable this to happen, it is envisaged to create Happiness Champions at the plants who will be people who are recognized for their great work and support for the initiative. They will then be actively involved in designing and implementing the initiative.

The PSCs will be encouraged to take up some critical initiatives that will address an area of focus for the plant and will be long term in nature. The initiative would be run like a localized project at the plant and the Steering Team and SOIL will help them in project management and impart them the skills required to ensure the success of the same. Over time, more and more individuals in the plant will be deployed in running the initiative ensuring the involvement of individuals.

An important part of this initiative is to realize that the true happiness lies within the individual themselves and hence, the workforce will be encouraged to spread this realization through a custom-designed module. This will be run in a train-the-trainer mode and internal trainers will be trained by us, who will further impart the training to the entire workforce.

Important skills will also be imparted to the workforce in an effort to appeal to the intellectual aspect of their well-being. Train-the-trainer workshops will be held for specific identified skills such as emotional intelligence to help the workforce increase their happiness quotient.

Families of the workforce will also be involved in the initiative through involving them in celebrations and involving them in the successes of the organization. A feeling of an extended family will be created at the group through creating opportunities for people to meet each other, bond with each other, and create a culture of belongingness.

More and more opportunities to celebrate the successes and appreciate and recognize good work would be created. Slowly, the role of SOIL will begin to diminish and the leaders at the Ramakrishna Group will start driving the initiative and creating happiness at their group. While we ensure that we guide the organization in their quest for happiness, we will enable the leaders to take this initiative forward and make it a way of life through the different activities that have been taken up and will continue in the future as well.

Phase IV: Institutionalization

The final phase of transformation will focus most on ensuring that the change created "sticks on." This will be the time when the processes of the organization will be modified to include happiness as an important factor and/or linked to the same. By this time, happiness would have been embraced as a mindset change and this phase would be utilized to ensure that the initiative remains embedded in the organization.

Documentation of all the activities done would be undertaken and organized, the critical learnings would be drawn, and collateral would be created for knowledge management and sharing with the external world.

Some refresher courses/sessions would be held on the important skills/aspects of the initiative to ensure that the organization can sustain the change. The internal change agents will drive the process largely and we would come in only to support them in the process.

By this time, it is envisaged that the organization would have achieved the goals mentioned later, which have been attained to some level over the last one year:

1. The individuals within the organization experience *ananda* and positivity in their daily work, which leads to their personal development.
2. The individuals work toward a shared higher purpose leading to alignment with the vision and values of the organization and greater synergy.
3. The individuals thrive in a culture of growth as happiness is fueled by the space for spontaneous creativity provided by the organization.
4. "Communities of learning" fuel the growth of the organization.
5. People are connected to each other and the organization through a sense of family and belongingness.
6. The individuals evolve emotionally, spiritually, and psychologically by driving out their fears.

It is truly an exciting initiative and journey that we are partnering with Rama Krishna Group on. Our work is helping us to understand the drivers of happiness for people and how we can enable them to realize the true potential that lies within them. The process of creating happiness is not an easy one by any measure and we have faced quite a few challenges, out of which came our greatest learnings.

The biggest challenge was communicating the concept and ensuring its clarity among the people. To explain a "soft" concept such as happiness and what it could mean for people was a difficult task since it was met with inherent skepticism on what the organization meant. Acceptance of the fact that happiness was with an altruistic intention was low. We had to reinforce this fact many times and regular address from the chairman himself was required, and it greatly helped when he started visiting families of the employees to show people that he was truly vested in the program.

There were many "skeptics" and "opponents" of the program we encountered in the process, that is, people who refused to believe in the concept and/or the methodology. We tackled their doubts with patience and answered every question they raised and accommodated their request in the project plan as far as possible. The most critical factor here was to stop them from spreading their doubt to others and for this, we discovered "positive change agents" or people who were bought into the idea and tried to get them to influence them subtly by showing results of change occurring. We also put them in leadership positions so that they could bring about their desired change using the methodologies they thought would be helpful. It was important to not ignore them, but to involve them in the process and make them feel heard.

The leadership of the organization in a change initiative is the people who can make or break the initiative. We found that plants where the senior leadership was vested and believed in the project were the plants that did really well and succeeded in creating happiness. For us, that translated into ensuring the buy-in of every senior leadership team member for the success of the initiative. We did that through creating multiple touch points with every senior leader, and through taking feedback from them on an individual basis as well on what could be done better in the program and suggestions on the way forward. The dissenters were also motivated by seeing the enthusiasm of their own peers and their interactions with them. The learning was that buy-in to change is best created through peer interaction and making people feel that they are heard and valued.

Another challenge was to showcase the results of the initiative and to demonstrate a change taking place in the organization. Since much of the change that we are trying to drive is a mindset and lifestyle change, it was difficult to measure it and show it in tangible terms. After the completion of 9 months of the project, the members of the organization felt that tangible change is needed to be shown to them and for this we adopted the technique of taking feedback from the individuals themselves on what they think is going well and what could be done better in the form of a monthly survey. We actively worked on adopting the recommendations to make sure that the program is a success. Slowly, people themselves started talking about the fact that communication and emotional well-being has improved at the Group.

The implementation of the approach at the Group showed that stronger communication and the feeling of being appreciated, togetherness with the organization were the first outcomes to be felt by people. The implementation has proven to be effective over the last year and has most importantly demonstrated that people can coexist in multiple layers (physical well-being, communication, emotional well-being, intellectual well-being, and spiritual well-being) and that it is important for HR to focus on all the five levels to create happiness within the organization.

One of the most joyous and satisfying elements of working on this project was the realization that organizations and people have an inherent sense of goodness within them. Sometimes, it may get pushed down or hidden due to certain conditions, but when people overcome their fears and insecurities and embrace this goodness within them, great work and change can occur.

The work done at the Rama Krishna Group till now serves as a testimony to the fact that HR leaders can be the true change enabler for organizations and can create positive and long-lasting change at the workplace. The success of this initiative will strengthen our belief in our nation's rich spiritual and cultural heritage and its relevance to the organizations of today and tomorrow!

6 HR professionals as institution builders: Some reflections from experience

Arvind N. Agarwal

There is old fable that I had read in my childhood about Baba Kharak Sing, a very generous old man who was always very helpful to everyone. Once upon a time, Baba was riding his horse to pass through a jungle. On the way, he saw a young man who was somewhat limping. Baba could not bear that this young man is walking with limp in this dense jungle while he himself is comfortably riding the horse. Baba, true to his habit, gave his horse to the young man who bowed in gratitude, thanked Baba, and got on the horse. Few steps ahead, Baba suddenly noticed that the young man yanked the horse and galloped ahead. Young man was not limping anymore. Baba realized instantly that he was cheated by this young man. He thought for a moment and then loudly called the young man to stop right away. As the young man slowed down, Baba came close to the young man and whispered in his ear. Baba urged the young man that he is welcome to go ahead with the horse but he must not narrate this story to anyone, or else people will lose faith in helping others. Baba was more concerned about the larger cause of sustaining the institution of "helping others," even at own personal discomfort.

Human resource professionals have several opportunities to build institution, provided there is conscious focus toward doing so. In my long career spanning over nearly four decades, I have had privilege and honor of serving in leadership positions in human resources (HR), total quality management (TQM), marketing, strategy, and as a CEO. Serving in a variety of these roles, I have learnt several important lessons relevant to institution building.

Lesson 1: Job rotation

Early in my career, I got exposed to TQM, which helped me learn power of collaborative working, process orientation, benchmarking, customer orientation, and systematic problem-solving. I was in numerous cross functional teams (CFTs) as a facilitator and as a team member was involved in solving problems on wide-ranging subjects such as "cash-to-cash cycle," improving market share, increasing sales force productivity, improving employee engagement, etc. Through these CFTs, I got much closer to the business.

I remember as the HR head in a company that had just begun its commercial operations, I found there was suddenly high level of attrition among the leadership team members. Very people who had worked so hard through the difficult phase of its early start-up were now leaving from the very company they had toiled to build. Each of these departures was heart wrenching and it used to adversely impact the energy level in the company. As the HR head of the company, I was at a loss to understand this phenomenon. There were easy fixes such as giving better title and salaries. These were suggested by many in the company. All of these appeared likely options till I could sense that people were leaving as they considered that "job is done" and there is no further challenge in the company. It became very clear that the company needed a compelling vision that was challenging and attractive so that people find it worthwhile staying further. I discussed this with the chairman and other key business leaders in the company. Consequently, the leadership team embarked upon developing vision and strategy for the next phase of the company. The leadership team of the company clearly took the institution-building option rather than increasing salaries and giving higher title which would have been a transactional option.

Consequent to this exercise, I was given added responsibility of strategy planning. There were many contenders who were perhaps more competent but I seem to have displayed more passion for strategy and future growth of the business. And later from strategy, I was moved as marketing director. In due course, I was invited by the company I had served as the HR person to join them as CEO of two of the businesses. In 20 years, I had traveled from young HR trainee to CEO and this journey was through TQM, strategy, and marketing roles. From what I recall now, each of those roles I was doing with passion and excitement. These roles, each time I took them, were an end in themselves. I was not looking at what these roles will lead to in terms of career. I was only focused on how I can contribute toward building the organization through doing honest job of what I had in hand. This honesty of purpose and passion brought the best of me. I was always quick to experimenting with new ideas such as implementation of self-managed team in shop floors, launch of new products, and forging new alliances for new business lines. Slew of these job rotations kept me on the learning curve all the time and kept me passionate about what I was doing. In this way, I could contribute my best toward the institution-building process in each of my roles. It was my approach toward doing the role in its totality rather than treating the role as a means to career advancement which helped me contribute toward institution building. A person who is treating the role as means is perhaps too pre-occupied with self and is more likely to miss the opportunity to excel and serve the larger cause of institution building. I seem to have truly lived the popular saying: "pursue excellence, success follows."

Lesson 2: Work with your CEO

I had great opportunity of working for some great entrepreneurs and business leaders over the years. I have learnt a lot from each of them. For institution building, it was important to earn trust from them before they believed in my ideas. I discovered senior business leaders I reported to, had own ways of testing if one can be trusted and quite often it was through some very mundane transactional activities. If there are many surprises, time-wasting conversations, self-serving recommendations, politicking, mis-information, then one is not likely to earn the trust. I remember once in a joint venture (JV) company where I was working, I had challenge to earn the trust from both the partners. At the outset, the foreign partner believed that as an Indian I am likely to be more aligned to Indian partner and may not take an objective decision. The reason was I was handling recruitment of CFO that is a very sensitive position particularly in the JV situation. I had produced a short list of three candidates for final interviews. I was taken aback when instead of progressing with interviews, the foreign partner wanted to see all the resumes to double check if I was objective in short listing. Only after finding that it was indeed done objectively, he agreed to progress with final interviews. Only after passing through numerous such mundane transactions, I earned trust of the foreign partner. I found this to be a fundamental requirement for any institutional-building initiative that one can do. Having earned the trust, I found I had enormous opportunity to influence the system by leveraging bosses' power and position. For instance, to institutionalize TQM culture in Modi Xerox, I found numerous ways to engage the chairman, Dr B. K. Modi. He attended the five-days training program on TQM and that too at the factory site along with others, led CFT, and spoke about these initiatives in industry and academic forums. In this process, his own conviction on TQM enormously grew and that served as a great role model for the rest of the company to get on board with this new discipline. To make the company customer oriented, Xerox had embarked upon several initiatives across the globe. There was huge challenge as to how one aligns the employees in India. It was clear to me that this would be possible only if Dr Modi personally demonstrated that he is aligned to these initiatives. When I talked to Dr Modi, he readily agreed to lead all the discussion meetings with leadership team on implementation of all the initiatives. And once people found he was fully on board, every one followed the suit. It went a long way toward making Modi Xerox truly customer-oriented organization despite its near monopolistic position in the Indian market.

In Escorts, as a business leader of construction equipment businesses, I found we were all focused on meeting the year-to-year financial plan. Long-term thinking with concrete vision and strategy for the future was needed to be brought in. After talking to the chairman, Mr. Rajan Nanda, when I brought up this long-term plan in the Board, everyone got excited seeing the great future these businesses held. It led to several initiatives such as process re-engineering to make manufacturing more productive, building of new factory, slew of collaborations to bring in new-generation construction equipments such as vibratory compactors and hydraulic excavators, technology agreement with global design house to modernize the mobile cranes, to name a few. This led to investment and subsequent growth of construction equipment businesses in Escorts.

In RPG, Chairman Mr. Harsh Goenka had long cherished vision of making RPG "most exciting place for professionals." In fact, I found this vision compelling enough for me to join him in 1999. Over the last 16 years, I have relentlessly worked with him to help him realize this dream with slew of initiatives. We started with setting the vision and values for the Group and then began to operationalize them through introduction of relevant HR policies and processes. Some of these are balanced business score card based objective performance management process, higher proportion of performance bonuses, employee engagement surveys, 360-degree survey process to align leadership style across the group, organization and management review process for identification and development of high caliber talents, development centers, self-managed teams in all the new factories and coaching culture. Introduction and sustaining of these initiatives needed constant dialogue with the chairman, other senior business leaders, as well as senior HR professionals who were running the HR function in various RPG companies. Chairman's style is truly federal. So this requires everyone to get on board before any policy is implemented. It needed me to understand the business context, contemporary practices and then bring the relevant practices fully customizing them to "RPG way." It required listening to all with openness to learn and then embark upon introduction of the right policies, however mundane these might be. I remember, I was once talking to the business head of retail business and he talked of a new term life insurance he has introduced for all his people, nearly 2000 in number, just by spending ₹15 lakhs in annual premium. Logically, I could come heavily on him as he had no business introducing a policy without discussion with group HR. But, instead, I appreciated his thoughtfulness and initiative. It also occurred to me that this was a great policy and should be done for everybody right across the group. I collected the details of the policy and began to talk to all the business leaders and HR heads of all the companies in the group and ensured this policy got introduced. In large group, there are instances of deaths and we had our share of them. Everyone felt good that we had a right policy to take care of employee's family in such situations. On my own, it would have never occurred to me to do this. I could do this purely because I was not lost in the bureaucracy of the organization. I was listening and was mindful of my responsibilities toward the larger purpose all the time.

Lesson 3: Be authentic

I learnt the importance of authenticity for institution building from Dr Udai Pareek himself.

I was fortunate to first meet Udai when I was participant in the personal growth lab organized by the Indian Society of Applied Behavioral Sciences (ISABS). Udai was the facilitator. This was way back in the early 1980s when I was in my twenties—less than 10 years into corporate world after graduation from the Indian Institute of Management Ahmedabad (IIMA). Midway into the lab, it was probably day 2 or 3, I was taken aback when he said he did not feel very engaged with us and the lab. We were quite startled to hear these words from the facilitator himself. Later we learnt that he was being authentic. And I learnt in those early formative years of my growing as a professional importance of authenticity that some time will mean sharing own vulnerability.

My learning was tested some five years down in my career when I was the HR head in a company where I was handling downsizing for the first time in my career. I had thought through all the details and worked out how the entire exercise is to be managed. I called for the meeting of all the department heads and briefed them about the detailed plan of action. I made sure everyone was on the same page and they all understood what each one needs to do. Every one appeared quite on board. At this moment, I shared with them that this is the first time I am handling this kind of task and I will be happy to hear from them in case they had doubts about any aspect of the "plan." I felt instant support from all of them. There was tremendous sense of camaraderie and bonding all around inside the conference room. We all started talking about what may go wrong and began to fool-proof the "plan." I was no longer alone on this. This positive experience reinforced my belief on authenticity even at the expense of own vulnerability—a lesson I had learned earlier in my career from Udai. I found the value of authenticity in building trust and collaboration.

Lesson 4: Collaborate and integrate

I have worked in numerous project teams, as a member as well as a team leader, right through my career. I learnt power of collaboration through my involvement in TQM early in my career when I was in Modi Xerox. Since then it has become my natural working style. This got reinforced by RPG Chairman Mr. Harsh Goenka who is always encouraging collaborative behavior across management teams. The ultimate in promoting collaborative behavior was introduction of "self-managed team" concept in running all the new factories in RPG. In this concept, workers work in teams. They plan, organize, and monitor work by themselves with no supervisors on top of them. In RPG, there are also other collaborative forums such as "young executive board"

(YEB) and "center of excellence" (CoE). YEB consists of young high caliber professionals who are appointed for a period of two years to function as shadow board reporting to the CEO of the company. CoE is forum consisting of heads of a particular function from every company of the Group. They come together every two months to learn best practices from each other and also do collective problem-solving on issues of common interests. There are now CoEs for manufacturing, materials, finance, legal, and R&D functions. One person from the team is appointed as the chairman for a period of two years. These forums are proving to be great energizer and they are building fellowship among the functional people cutting across the company boundaries. I was not an originator of these concepts. They were proposed by HR people down the line. I was quick in recognizing the power of these new concepts. I then moved on to evangelizing them and got endorsement from all concerned including chairman to implement them. In collaborative working, it does not matter from whom the idea comes. If the idea is useful, one puts full weight behind them. This then leads to institutionalization of the idea.

Lesson 5: Be an institution builder and problem solver than a problem creator

As an institution builder, one has to be a problem solver not a problem creator. I remember, when I joined the RPG Group, many of the companies in the Group were making considerable losses. I could have been easily disheartened and walked away to another company. That would have only compounded the problem. Instead, I stayed on and joined shoulder to shoulder with the leadership team to turn around the businesses. In many cases, it needed hiring right people to strengthen the leadership team, downsizing, and making interventions to making them more performance oriented. It took relentless working for nearly five years to get all our businesses to start posting profits: not very high-end HR interventions but all what the businesses needed at the time.

Lesson 6: Give yourself to the profession. It is a great opportunity to develop

I was pulled into the National HRD Network (NHRDN) by Dr Udai Pareek and Dr T. V. Rao in the mid-1980s.

Big lesson I learnt from Udai toward institution building is importance of engaging people in building and sustaining culture of trust, collaboration, and problem-solving. I remember in the mid-1980s when I first got associated with NHRDN, whenever Udai visited Delhi, there were several short, impromptu meetings. In most of these Dr T. V. Rao was present as well. In these meetings, we used to talk about several issues related to HR profession and NHRDN. Udai and Dr Rao were perfectly capable of thinking through all aspects of running NHRDN and taking appropriate steps for further developing the HR profession in the country. But instead of doing it by themselves, they were always reaching out to many of us to brainstorm new ideas, discuss them threadbare, and then work toward implementation. The process built great sense of belonging and ownership in several young HR professionals including me. This has had deep impact in my making of a leader. I tried to practice it both in my role in NHRDN as well as in companies that I worked for in leadership positions.

In the year 2000 when I became National President, I found there were several senior HR professionals who needed to be involved in NHRDN. I personally reached out to many and I was overwhelmed that they all enthusiastically responded in a very positive manner. Some of them were invited to join the Advisory Board of NHRDN. Also, the board meetings began to be held on a regular basis involving all the Chapter Presidents. All these actions of including the senior HR professionals through advisory board, holding of regular board meeting involving board members and Chapter Presidents re-energized the HR community across the country. This wide-reaching inclusion further strengthened the spirit of trust and collaboration in the working of NHRDN. Toward building NHRDN this was an important institutional landmark.

By working in NHRDN over the years, I built strong and lasting relations with a large number of senior HR professionals. These relationships have been great learning experience and they have greatly influenced my development as a HR professional.

Lesson 7: Maintain and mange professional teams: The way you interact and network with HR professionals

I have always looked for high-caliber HR professionals in my team even if it means they may not be long-term stayer. In Modi Xerox for instance, most of my HR team members moved on to sales function and progressed well in their career. I followed the same philosophy in RPG. I have actively encouraged HR professionals to take cross functional career moves for their overall long-term growth. Many of the HR professionals from the RPG HR team have now moved on to other functions as well and many have moved on to other companies in significant leadership positions. I have this strong conviction that one should be surrounded by good people. That produces the best from every one. My approach is very collegiate, treating each of them as equal and giving them autonomy and respect. Sometimes, I have been let down and regretted having been so trusting. But, on the whole, such incidences have been far and few and I continue to believe that it is well worth taking chances with people.

Lesson 8: Be prepared to learn from juniors

In Escorts Construction Equipment (ECEL), as a CEO, I have seen the power of listening to the juniors in solving complex problems and revitalizing the enterprise. This was in the 1990s when ECEL was an ailing mid-size construction equipment company. Mandate from the chairman was to grow and make the business profitable. ECEL was long-standing business and a leader in its category. Nevertheless due to low volume and high cost the business was not profitable. Workforce, at all level, was very experienced, long serving, but somewhat alienated from the business. Frequent work stoppages on ground of very flimsy and petty reasons were not uncommon. I began to meet cross section of employees at all levels to understand their perception of ECEL problems and also their suggestions to deal with them. I established a small and empowered team to rapidly implement these solutions, which were coming from people. I also formed several cross-functional teams (CFT) to deliberate on problems in which ready solutions were not available. CFTs were given three months' time to have their regular meetings and come up with recommendations. It still took lot of time to solve all the problems and move the business toward growth and profit. But, the energy level in the organization was palpable. Work stoppages were the thing of the past. Friday afternoons, leaving their regular work, the CFTs used to meet to solve the problems taken up by them. I used to go around the factory and office to see what the CFTs were doing. They were reengineering the business processes to cut costs. There were others who were developing newer models of the product which were lower cost but more efficient. Some teams came up with the idea of introducing new product line leveraging the manufacturing and distribution infrastructure which ECEL already had. I used to be quite excited to see how invigorating the place had become. Inclusion of people, reposing of trust, and engaging them into collaborative problem-solving CFTs were beginning to build a profitably growing vibrant ECEL.

Lesson 9: Continuously experiment

For building an institution, one has to constantly experiment. In RPG we experimented with concept of self-managed teams, first in our new factory located in north. After initial struggle, it became very successful and the concept then got implemented in other new plants in Gujarat. Today, SMT is the way of managing in all new plants in RPG. Early in my career in Modi Xerox, I got opportunity to learn and lead TQM initiative in the company. I then got opportunity to take job rotation in marketing and strategy function. away from the main stream HR function. I was exposed to experimenting with myself. Underlying trust, confidence, and resilience are important determinants of success with any experimentation that is so vital to institution building.

Lesson 10: Be value driven. Never give up integrity

Among all these, if I have to pick one single most important factor for success in institution-building journey, it will be integrity. One has to be honest to oneself and all others one interacts with. Honesty of purpose, authentic, and transparent behavior are various aspects of integrity. One gets constantly tested by others in the organization.

There are many theories and frameworks of institution building and the nature of leadership required for institution building. Academic literature is replete with them. These are well researched and undoubtedly credible theories. I have but learnt some from Udai which I have found powerful in building institution both in the context of professional body such as NHRDN where I had privilege of working closely with him and the business organizations where I worked as the HR head and the CEO. These lessons helped me to discover power of authenticity even at the cost of personal inconvenience, role of leadership in going beyond transaction to address larger causes, engaging people, reposing trust, and fostering collaborative working toward problem-solving.

I have narrated several incidences in this chapter which corroborate that an institution builder needs to display:

- Authenticity
- Openness
- Confrontation
- Trust
- Authenticity
- Proactively
- Autonomy
- Experimenting

In conclusion, it is worth sharing a short poem that a colleague passed on to me, just a few days ago. This well captures the spirit of institution building:

The Bridge Builder
by Will Allen Dromgoole (1860–1934)

An old man going a lone highway,
Came, at the evening cold and gray,
To a chasm vast and deep and wide.
Through which was flowing a sullen tide
The old man crossed in the twilight dim,
The sullen stream had no fear for him;
But he turned when safe on the other side
And built a bridge to span the tide.

"Old man," said a fellow pilgrim near,
"You are wasting your strength with building here;
Your journey will end with the ending day,
You never again will pass this way;
You've crossed the chasm, deep and wide,
Why build this bridge at evening tide?"

The builder lifted his old gray head;
"Good friend, in the path I have come," he said,
"There followed after me to-day
A youth whose feet must pass this way.
This chasm that has been as naught to me
To that fair-haired youth may a pitfall be;
He, too, must cross in the twilight dim;
Good friend, I am building this bridge for him!"

7 Multisource learning as a critical variable in professional growth: Reflections on my HRD journey*

Gopal P. Mahapatra

In today's competitive global environment, sustainability of organizations and society has gained great significance. The last three decades has seen the significance of people in general and human resource development (HRD) area in particular, and how it can play an important enabling role in the process for individuals, organizations, and beyond. In this chapter, the author, as a practicing HRD professional, mentee, and well-wisher of Dr Udai Pareek, shares a few critical incidents and interventions in his career, which might have had significance beyond the day-to-day functioning of the HRD in organizations. Keeping Dr Udai Pareek's contributions to various fields and institutions as a benchmark, the author tries to create a framework on what are the cross-roads he experienced and the various steps he took for championing, facilitating, leading, or assisting in the change initiatives and the interventions, for example, total quality management (TQM), restructuring, employee engagement surveys, customer satisfaction surveys, assessment and development centers (DCs), 360-degree feedback, coaching, and other initiatives for leadership development and institution building. The author reviews and reflects on the linking it to the development of individuals, teams, organizations, profession, and professional bodies from an institution-building perspective. The author has learnt a great deal from various people, situations, challenges, and mistakes. The path to development is continuous learning, initiative, and action. The author suggests that for any young professional every organization is a learning opportunity, every person you come across, every role you perform,

* The views expressed in this chapter are of the author and not necessarily of the organizations where author has worked or is working currently.

every difficulty you face, and every accomplishment or mistake you make is a building block in one's own development. The author hopes, other human resource (HR) professionals and leaders would review, reflect, learn, critique, develop further on these perspectives in their professional journey, as well as contribute to this field and beyond through institution-building initiatives.

The beginnings

Growing up in the native of Orissa from an agrarian family background, after having studied SSC in a village school, I had the great opportunity of graduation in English literature blended with postgraduation in personnel management and international relations from Xavier Institute of Social Service, Ranchi, a Jesuit institution with an amazing Belgian Fr. Bogaert as director as well as my doctoral education (Fellow Program in Management) from Indian Institute of Management Bangalore (IIMB). Every institution I studied and the teachers I had and the friends and colleagues with whom I interacted have remained with me until now and have been sources of learning and growth. I remain in touch with several of them especially from the institutions of my professional education and interact to learn and occasionally try to pass on my learning's to the friends, colleagues, and the younger generation. Having been blessed with a rich and diverse career of about three decades in HR (personnel management, IR, HRD, organizational development [OD]) given are a few critical initiatives and threshold moments in my professional life.

Bharat Electronics Ltd; role of chairman and managing director Capt. Prabhala (Retd Indian Navy) in institution building

I took up the first job as a personnel officer at Corporate Office of Bharat Electronics Ltd. (BEL), an India Government-Defence Ministry Company, as my higher studies were nearing completion. I was privileged to work during the tenure of Capt. Prabhala (Retd Indian Navy), a very progressive and dynamic chairman and managing director (CMD). As a very young professional, I experienced visionary (or transformational) leadership by him in a few initiatives highlighted as follows: The CMD initiated "organization restructuring in BEL" with the help of a strategic consulting firm. It was facilitated with the help of a compact in-house team of change agents, where I was a member. He wanted to prepare BEL for becoming a more customer and quality-focussed organization; moving away from a functional organization to business oriented (strategic business unit [SBU] set-up), making BEL competitive and sustainable in the future.

While fostering the CMD-initiated rearticulation of the mission at BEL, I coordinated the event generating few thousand suggestions and received the Special Award for the Company Mission "Quality, Technology & Innovation" Evolved in the 1980s, the mission still stands as relevant today as ever.

With focus on leadership and talent pipeline for growth and sustainability, CMD initiated broadbanding and fast-track promotion and development of internal leaders. This enabled accelerated top talent development, including internal leaders becoming the CMD, directors, and other top management positions in BEL in the next two decades.

In summary, my early career in BEL experienced strong leadership of Capt. Prabhala, CMD, as a leader of change and institution building (transforming ahead of times) and satisfaction in my role as an able associate and coordinator of change (OE) interventions. It was a very strong socialization in my early career. Capt. Prabhala is remembered even today by most in BEL and ex-defence officials. Working with him has been a great learning experience. I kept in touch and continued to learn from him and kept him informed occasionally of my learning's and development.

Great leaders provide great opportunities to learn. Often the theories of leadership focus on teachers and not learners. The lesson from my early career is to find mentors, appreciate, and learn. My advice to young professionals is to constantly look for such mentors around. Some of them may not be great at this point of time. It is not their greatness but the extent of learning they facilitate that matter. You should work on multiple projects, take initiative, and make difference irrespective of your level and the distance between you and your seniors. It paves a path for growth. When I was in BEL, I never knew that I will one day be going back there as a consultant to develop its future leaders. A few years of my experiences there provided me with solid foundations for my own building as an accomplished professional.

Young faculty member at Xavier Institute of Management, Bhubaneswar

After completion of my doctorate (Fellowship from IIMB), I dreamt of contributing as a management teacher in my home state Orissa. I know Fr R. D'Souza, the founder director of Xavier Institute of Management, Bhubaneswar (XIMB), and the then director of Xavier School of Management (XLRI), accepted the invite of the then chief minister of Orissa in strengthening the qualified managerial talent of Orissa by building an institute of management for accelerating future development of the state (This was publicly acknowledged by him 25 years later during the silver jubilee ceremony of XIMB.) I was motivated by Fr D'Souza and then Dean Fr Abraham (who is the current director of XLRI) to not only join, but also motivate many of my friends and batch-mate fellows to join the faculty team. Teaching MBA students, full-time and part-time, managers of SAIL, Rourkela Steel Plant, in their modernization along with the IAS & Group A Officers on HRD areas and management development programs (MDPs) and consulting projects were quite thrilling and demanding too at a very young age. XIMB gave the opportunity of conducting a pilot research on "Collaboration for Stimulating R&D" jointly with my colleague that got presented in the National HRD Network (NHRDN) International Conference in 1991 and published in a Tata McGraw-Hill book.

In a nutshell, a power-packed three years' journey in academics with empowerment and trust by director, dean, and the faculty colleagues in early stages helped in strengthening the foundation of XIMB in institution building bricks by bricks in its early stages.

The three years of stay at XIMB was another foundation stone for my professional growth. I had the opportunity to teach what I have learnt, internalized now many things. I learnt as a student by teaching some of them, and learnt how to influence the young minds. Most of all, working in a Jesuit institution in my own state and observing the commitment to "institution building and people development" they have is a great learning experience. It is these observations of the things around me, the way the institutions are managed and built that gave me a lot of insights into my later development to facilitate institution building. Little did I realize at that time that one day I will be called as a consultant and member of Strategic Advisory Council by the very Institution I worked with to help manage transition from a small institute to a large university.

Academic institutions provide great opportunities to learn. Get the best of all your association with people who build institutions from day one. One day you might be called on to be an institution builder. Academic institutions are great laboratories to experiment and try out a lot of new initiatives. Variety, depth, and focus are great possibilities in an academic institution; more so, the young ones. Do not underestimate the contributions of such institutions to nation building as well as significant personal (and competency) development. Those who get an opportunity to spend time in academic institutions are lucky. Many great people are teachers and they continue to learn by teaching at academic institutions.

Working with a multinational:The INDAL experience

Going back to industry after a stint in academics by joining INDAL, the Indian associate of Alcan Canada, was another great learning experience. When I was joining INDAL, one of my mentors from industry said, "Gopal, it's easier to talk from a higher pedestal (as a faculty at XIMB). You now have to prove yourself by your contributions and hard work with professional colleagues; influencing without power." I was playing an active facilitator role as the corporate HRD manager supporting the transformation initiatives led by the then managing director (MD) and chief executive officer (CEO) through "Cost, Quality, Customer Satisfaction and Exports (CQCX)." In the early liberalization years of India, INDAL was globalizing and retaining its existing competitiveness in the domestic market through TQM, ISO 9000, and other change initiatives. As a trained TQM facilitator I anchored and supported Employee Satisfaction and Organization Health Surveys; getting it translated into regional languages, and facilitating quality improvement projects (QIP) teams. It was a valuable learning experience for a young HRD professional contributing professionally to vital area of employee involvement through employee satisfaction surveys and post follow-up action plans. I also worked closely with and prepared the senior line managers through "Train the Trainers" at corporate office and various manufacturing units. These senior line managers played the role as resource persons for the culture change through TQM in their respective units and divisions. I had done quality check (QC) tools training in Bengali/Oriya and Hindi along with the line managers of many INDAL plants (first time in my life, a unique experience). The cascading of TQM in units for cultural transformation was quite visible and got reflected in the exports and domestic market of INDAL too during those years, creating an excellent work environment in the units and the corporate office.

Working at grassroots provides great opportunities. Visiting all the plants, interacting with people in units at multiple levels, and preparing and supporting them collaboratively than sitting at a corporate office was another professional enrichment. Learn the hard way the nuts and bolts of change and how to collaborate with various stake holders in undertaking and managing change through QIPs. Observe how the top management drive change and the need for change agents in line managers and TQM and HRD for organization transformation. Inclusiveness, collaboration, joint problem-solving, and integrating and closing with local unit and corporate management is very enriching. One learnt practically, how employee involvement can be a very powerful tool for not only building a positive culture but also solving a lot of organizational issues including high level of employee engagement. Overall, learning how to actively be a change agent with the people and critical stakeholders than advising from outside was very powerful and helped me later as a consultant and other companies as an internal change agent too.

Gujarat Gas Company Ltd; a value-driven promoter, Mr Mafatlal, and a progressive CEO, Mr Virani

The other opportunity I got to add significant value was at Gujarat Gas Company Ltd (GGCL), a rapidly growing energy and infrastructure company in Ahmedabad. The value-driven entrepreneur Mr H. A. Mafatlal had a very foresighted MD, Mr F. B. Virani, as the leader. I joined them as an outcome of HRD audit. The pioneering efforts by the initial team in natural gas distribution (domestic and commercial), solar power, financial services, and focused explorations toward LPG and LNG in the early 1990s were noteworthy! As the head of HR, and Corp Communication, I got to be part of a very active senior management team for this very profitable and fast-growing organization in the country. The dynamic MD orchestrated change efforts leading from the front with interventions like: formulating vision, mission, and quality roadmap. The business growth was further fuelled at the group level with facilitation by a strategic management legendary consultant. This got blended with the Customer Satisfaction Survey we conducted with Gallup and Employee Satisfaction Survey with another external agency we initiated as the HR team. Survey findings were very encouraging and many cross-functional QIP project teams on improvement areas followed. Introduction of Management Trainee scheme from B Schools and recruitment graduate engineer trainees from engineering colleges created the base for young talent development that led to the talent pipeline for the company. Setting up of joint ventures (with BPCL) and growing and diversifying into financial services (GFSL), LPG, LNG, and power were the offshoot of this ambitious growth. Gujarat Gas Day and Family Day facilitated by my HR teams at units were epitomes of this people orientation and employee involvement. Through all these initiatives stated earlier, Gujarat Gas taught me how positive thinking with focused efforts and coordinated team work at the senior management level and below can bring miracles. Further, at the growth stage introduction of progressive systems and processes, for example, vision/mission, business plan and strategies, performance review discussions and Perf/Potential Review Councils, Customer and Employee Satisfaction Surveys (CSS& ESS), MT/GET schemes, corporate manual, and the like can accelerate and sustain the growth trajectory of a company keeping people focus alive. I learnt entrepreneurship in Gujarat Gas, typical of Gujarat culture. In addition, I also got encouraged by the company to be associated with Academy of HRD as an advisory board member and as a faculty member—directed a program on "TQM & HRD Facilitators" for 50 odd senior HR professionals.

As you grow and reach senior levels your focus should shift to learn on higher level concepts. Learn about visionary leadership, entrepreneurship, and developing leadership within. Introducing systems, implementing, and institutionalizing them to stabilize and grow the organization and various units with empowerment and talent pipeline was a great learning in retrospect. In all these efforts, how one can continue to facilitate fast growth maintaining the focus on systems and not diluting individual and organizational values was another great lesson.

BPL Ltd: The opportunity to work with promoters and top management professionals

Taking a journey away from West to Bengaluru and joining BPL, a very fast growing family-managed firm in the consumer durables industry (with strong Japanese connections of Sanyo), was crossing another threshold for me. With the sponsorship of ambitious and professionalizing second-generation entrepreneurs as my sponsors (Vice Chairman and MD Mr Ajit Nambiar and Director HR/Chairman's Office Mrs Anju Nambiar), I saw many opportunities to contribute in organization building.

Starting with capability building of sales, marketing, and customer service professionals in thousands, with able guidance of my manager and mentor, an ex-air force engineering veteran empowering me, I got the opportunity to strengthen the performance management and training and development system across 12 group companies, 20 factories, and 40 branches in collaboration with the heads of business units and HR.

We took the initiative of building up a strong talent pipeline by inducting bright young management trainees in middle-management cadre from premium B-schools, IIMA/B/C, XLRI, and Faculty of Management Studies (FMS), Delhi, for the fast-growing consumer durables conglomerate.

For strengthening talent management, we in HR developed and institutionalized a mechanism for assessing and developing talent at critical managerial bands. With HRD Audit and In-house Assessment Development Centers (ADCs) introduced with the help of renowned HRD consultants and sustained with internal assessors. This gave the opportunity to enable the group companies to reach greater heights with diverse talent. Post ADC collaborations with national and local management institutions, such as IIMB, T. A. Pai Management Institute (TAPMI), SDM Institute for Management Development, and IFIM (see HRD Newsletter, August 2001) for further capability building of key talent was a great talent development initiative linked to the ongoing business growth and development in these companies.

During this period, I worked with a leading HRD academician and OD consultant on change management; in restructuring exercise; making the central marketing organization closer to the market through diversified regional offices working closer to the grass root level markets; this changed the market dynamics progressively. I was fortunate to have enjoyed the trust and empowerment of the promoter CEO and directors as well as the various business leaders at 37 years (rather young age) to help build up the organization and people processes of the parent company and associate companies in the group. The top management accelerated my global thinking and broadening perspectives by sponsoring me to global strategic HR program at London Business School followed by visits to International Institute for Management Development (IMD), Switzerland, and International Labor Organization (ILO), Geneva.

Currently, BPL, the parent company, appears to be going through challenging times. However, as the pioneer and leader of quality and diversified consumer electronics and durables group, in the pre-liberalization and early liberalization era, the group taught me the values of dedication and commitment by professionals, the role of leadership in leading change and leading by example, building culture, capability building, and talent management. There were many examples of management of flux in this complex and fast growth oriented organization operating in a highly competitive market in a globalizing India in the 1990s.

Learn from continuous audit and self-renewal exercises. Consultants can help you learn and can be used as instruments of change and in the process you help the organization transform itself. Be prepared to face frustrations as you are still not in-charge? Keep continuing to learn by observing the growth of the organizations you left and interventions you made. Along with the strong alignment with the promoters and top management for leading the change, collaboration with critical stakeholders, for example, business leaders unit heads and functional managers along with strong HR involvement are essential for the success of many business initiatives, more so the HR interventions.

T. V. Rao Learning Systems Pvt. Ltd

My decision to leave a large group as the head of HR and join Dr T. V. Rao, the Indian father of HRD movement, with the vision of creating a great HRD and OD consulting company was a life time opportunity of adding value to HR profession in the country while growing professionally and personally simultaneously. Other than simplifying (or demystifying) assessment centers I remember my involvement with many HR Gurus in few assignments and many more associate consultants in many large public sector units (PSUs) as an ADC TTT (assessor training programs) (e.g., NTPC and ONGC), HRD audit, visioning, performance and leadership development at many corporations, including huge OD and team building interventions for many medium-sized and large corporates.

I also had significant experience of facilitating, and coaching 360-degree feedback with more than 1200 top and senior leaders from various large multinational companies (MNCs), public, and private sectors. This was an excellent opportunity of showing the leaders a mirror for their current strengths and how they needed to do mid-course correction in their leadership style for organizational profitability, growth, and sustainability. While occasionally I was getting involved and obsessed at times with the execution of the 360 coaching and follow-up, Dr T. V. Rao always looked in the OD and institution-building framework. Given the compressed time of 36 months working with Dr T. V. Rao closely and intensely on multiple assignments reinforced those values quality HRD, collaboration, and customer centricity with OD focus.

Working closely with Dr T. V. Rao and other thought leaders and my consulting colleagues from Dr T. V. Rao Learning Systems Pvt. Ltd (TVRLS) was a unique experience. This strengthened my learning multifold in terms of conceptual clarity, applications, as well as sharpening one's capability as a change agent. I also remember the ongoing internal challenge of balancing revenue generation with quality contributions on an ongoing basis. The variety and depth of assignments strengthened mine and my team members' learnings significantly. It enabled me and others not only to be part of some major changes and leadership development, but also to sustain development and continuous value addition to education, health, PSUs, young managers, and HRD professionals contributing to larger parts of society.

Be in-charge when you get an opportunity. Initiative and close work with colleagues and top management at top levels is a great opportunity to grow. In consulting, one is always caught in the conflict between revenue generation/ organization building versus high-quality assignment with customer centricity. Opportunity to learn experiment and collaborate for problem diagnosis, resolution, and reinforcement appears to be plenty though not explicitly stated. Building up the consulting organization of young knowledge workers, motivating them while grooming and preparing them for the client-centric assignments is how learning and growth are simultaneous and correlated.

Leading organization and talent development at Oracle India; a leading American multinational product company

After working in consulting with HRD thought leaders and quite a few leaders of diverse organizations in the private sector, MNC, and PSUs in India, I took a break from the organization and talent development role in Oracle, a very aggressive and fast growing Global IT Product Development American MNC, based in Bengaluru.

In addition to building up the organization and talent development activities, I got a great opportunity for strengthening institutionalized leadership development activities at various lines of business in Oracle India, including use of globally developed 360-degree tools, Cross-cultural management strengthening competency-based assessment and development centers for institutionalizing leadership development (the case was published in Vikalpa 2011 as a case study). With leadership support and my team's continued efforts, "Prima," a young professionals program became an institutionalized leadership/technical pipeline effort that has seen grooming young professionals with 2–3 years of experience for top IT leaders/techies/architects from India Development Center getting recognition in the global context.

When inorganic growth through mergers and acquisitions became a methodology to enhance greater product offerings leading product companies in the sector (60 odd companies) were acquired by Oracle during my tenure. Being the OD leader in India and part of global organization and talent development team, HR leaders, I and my team proactively played the integration of the acquired employees with people focus; leading change without tears or reduced pangs.

At the growth stage of the Corporation in India, I had opportunity to facilitate and strengthen institution-building exercises for many business divisions of the MNC in India, through visioning, enhanced team synergy, and preparing the lines of business/divisions for future through virtualization. Working as an OD and talent development leader in this MNC in India, thousands of miles away from USA headquarters, one learnt implementing the basic value of institution building (influencing without power); understanding the core values and purpose of the global organization and making ongoing interventions for aligning the OD initiatives in line with the ongoing activities of the organization successfully as well as for helping it for the sustained effectiveness of the organization.

For HR professionals change in sector of your work in the initial stages may be unnerving, but provides great opportunities to learn, grow, and make an impact. Continuously building skills, aligning with global and local business realities, and introducing innovative OD/talent development initiatives aligned with business realities get credibility and recognition at global and local levels too. It builds your self-confidence and changes your perspectives.

NHRDN Bangalore (2006–2008); taking the leadership role

Having been associated as a Life Member of National HRD Network from 1989 to 1990, I became more active as Honorary Secretary of NHRDN Calcutta Chapter, Member of NHRDN Bangalore Chapter, and was co-editor of the book on "Emerging Asia" published by Tata McGraw-Hill (which included paper presented at the International Conference). This was followed by the tenure as Honorary President, NHRDN Bangalore Chapter. When I was hesitating, Dr T. V. Rao, the founder President of NHRDN, reminded me: "You have a duty and obligation to the profession, hence, please proactively accept the role." The opportunity of serving NHRDN Bangalore Chapter helped in creating a vision, getting the Chapter registered under Societies Registration Act, strengthening corporate governance of election (consensus) of office bearers, enhancing inclusion and diversity in NHRDN Bangalore by enhancing involvement of public sector and manufacturing organizations and nongovernmental organizations (NGOs), creating an unique professional capability building refresher program PHPD for HR professionals who did not have HR qualification from reputed B schools. I also did editing as guest editor of the special issue of the journal on OD of NHRDN. There were various challenges; managing national and local balance, growth and sustainability of the chapter, initiating newer offerings for Gen Y were quite an opportunity in addition to the regular organizational responsibilities I carried. This taught me the lesion; we owe a lot to our professional body for giving back. NGOs and professional bodies provide great opportunities to learn. The real learning of becoming powerful and significant through expert power is when you have no power. In NGOs you get power by virtue of the work you do and contributions you make. They are places for imitative and leadership. Never miss an opportunity to play leadership roles in NGOs and professional bodies. They are excellent avenues for developing leadership talent.

Two incidents with Dr Udai Pareek highlighting authenticity and openness for learning

My chapter on the reflections in my HRD journey will not be complete without highlighting two memorable incidents I had with Dr Udai Pareek.

An incident involving Udai I would ever remember was my visit to Jaipur for Group Relations Conference during April end to early May 2007. I knew Udai had lost his wife just few months before. As I wanted to personally visit him and pay my condolence, I rang up Udai one evening to ask him when I could visit him that week as I was in Jaipur. He was very warm and told me he was happy knowing I was in Jaipur. He suggested, instead of coming by taxi/auto in the summer and visiting him, he would personally come over to Clark Amer. He insisted that Clark Amer was on his way to work and he will meet me at breakfast, have tea with me before going to work. I wanted to meet Udai at his residence taking little pain, than asking him at 82 years of his age to visit me. He persisted. Lo and behold; Udai came over the next day morning to my hotel at breakfast and straight joined me. Though I adored him and respected him a lot having closely worked with him at NHRDN and TVRLS assignments, I was embarrassed and hesitant to take exclusive time with him. He continued chatting with me; he asked about my family and kids, my role in Oracle, and my learning at GGCL. This was similar to Udai visiting my home in Bengaluru once, he autographed his book and moved on after a quiet dinner. That was the authenticity and humility demonstrated by Udai as a mentor and well-wisher.

Another incident was in 2002 or so. We in TVRLS were conducting an ADC/DC for the Top Management (Executive Director/Chief General Managers) of ONGC at Taj Lands end, Mumbai. I was in DC panel of assessors where Udai was a co-assessor. After the multiple assessments, we discussed for consensus and had to do moderation. I clearly remember telling Udai that though his assessments and insights on the candidates were very powerful, his DC observations were so succinct and powerful, I was not sure, if the EDS/CGMs would adequately appreciate the views. He was like an intern and asked me: "Gopal, please help me elaborate my comments so that they'll benefit by it." I was touched; but worked with him to elaborate the qualitative feedback; that was crucial for the further development of the ONGC top professionals.

The above incidents taught me a lot about authenticity of Udai as a person, humility and willingness to learn in his areas of strengths; much beyond what I would have imagined and whatever I experienced before (more so, with many of my mentors and well-wishers)! To quote Dr T. V. Rao "Great Gurus like Dr Udai Pareek have spent their entire lifetime professing certain things by self-example, and living in certain ways that spread desirable values for the good of the present and future societies."

Mentor's, enablers, and facilitators of my learning and growth

In addition to my multiple and rich on-the-job experiences, my journey as a HRD professional, HRD, change management/OD Learning and teaching have been supplemented and complemented by a great number of luminaries as my gurus /thought leaders whom I have not named here.

Coming from not necessarily an urban sophisticated background many of the powerful growth and development insights surely have come through the powerful association and collaborations. I actively sought and pursed collaborations vigorously with many academic bodies along with my regular industry roles, including XIMB, IISC in Department of Management Studies, Bangalore University, Canbank School of Management IBA, RV Institute of Management, PES Institute of Management, as well as dialogue with many ex-military officers attending the Armed Forces Officers Program (AFOP) at IIMB and many IIMB AFOP/MDP.

Content getting increased by the subtleties of process work is another accelerator for my learning and growth.

In my journey of being a certified executive coach, to enable and enrich great leaders as coach, I have been blessed to be taught by great coaching gurus and many other ICF master coaches. Along with teaching valuable coaching skills, they taught me the major difference between "doing coaching versus being a coach" and its use implication in terms of developing self in the journey of coaching. I have learnt and grown in the last 14 years in the field of mentoring and coaching by informal and formal mentoring, coaching, training, and certification by trainers, certified coaches, and thought leaders.

Figure 7.1. Leadership for institution building

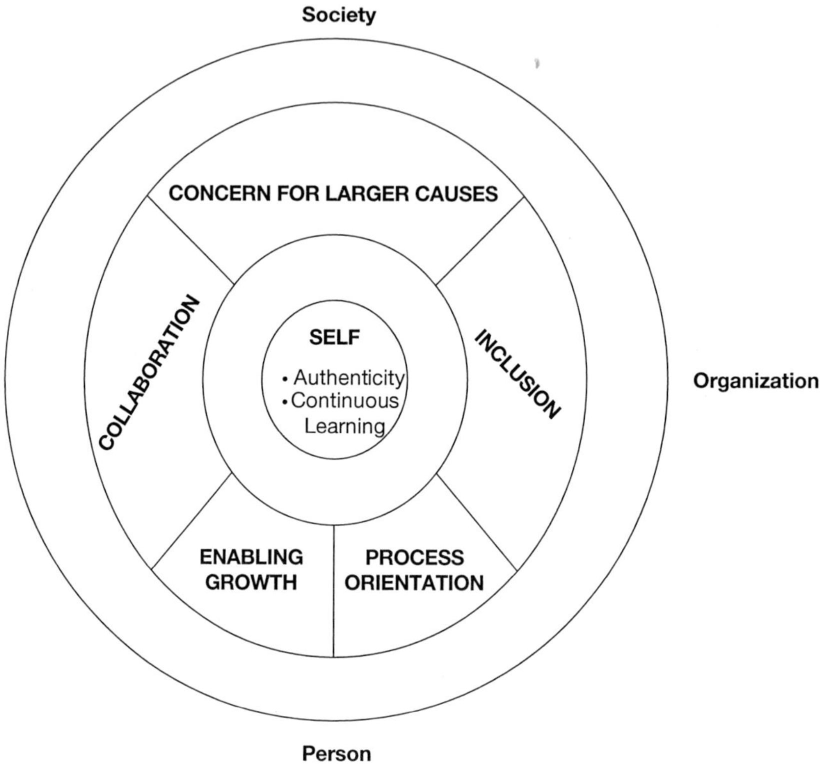

This has been supplemented and accelerated by periodic, formal and informal clients who benefitted from seemingly powerful insights which enabled them as a client (coachee/mentee). Coaching is a journey; many minute accept the growth publicly, but commitment practice and values for excellence in this field are really enablers.

Overall, looking back at the last 29 years of professional life, more so in HR, it appears, being sincere and committed to continuous learning and growth through all opportunities and challenges coming in the way is worthwhile. Also following role models like Dr Udai Pareek and Dr T. V. Rao, working with the philosophy of HRD and change management, one has sincerely attempted associating with and practising change management and institution building with multiple organizations: commercial, professional, and societal alike. I have illustrated in Figures 7.1 and 7.2 my conceptualization of institution building process and any individual's contribution to the same. It also appears the journey ahead is full of opportunities to make the difference in many spheres too.

Figure 7.2. Self and stakeholders in institution building

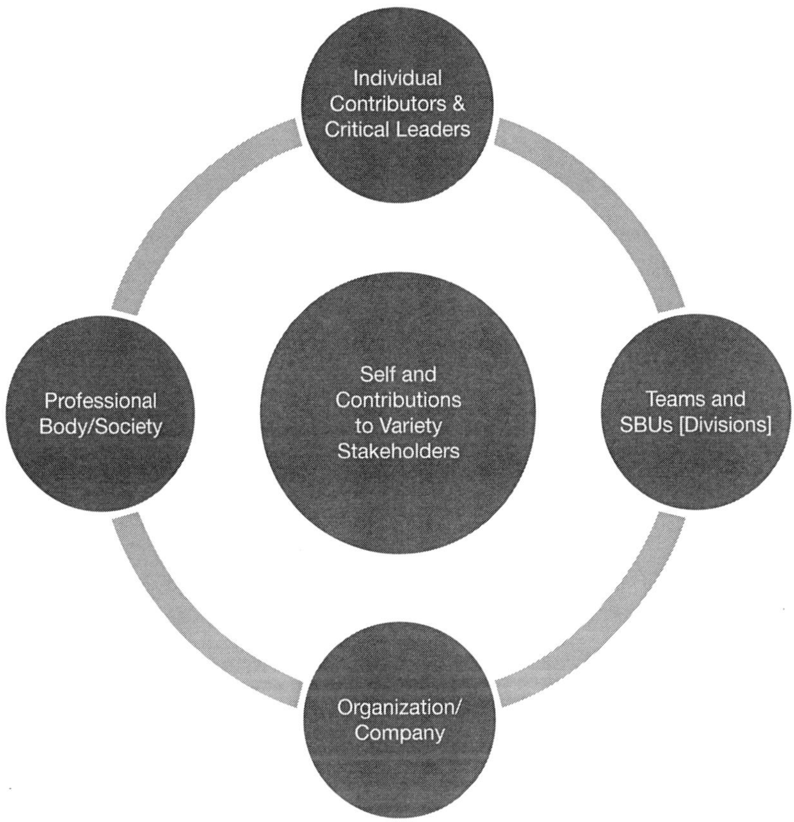

Note: I am grateful to ex-manager, Dr Arvind Agrawal, for articulating the model for me

To conclude "We learn all the time from different sources. We have to find people and incidents from where you learn. If you can't you have to create them. Knowledge keeps getting built and we matter in building our own knowledge. We have to appreciate the setting in which we are and get the best out of the setting and people therein."

The author would like to end with the famous quote from the poem "Stopping by Woods on a Snowy Evening" by Robert Frost:

I have promises to keep
And miles to go before I sleep
And miles to go before I sleep.

8 Cross-cultural diversity: Inclusion is the mantra

G. P. Rao

Preface

Doyen of human resource development (HRD) movement Dr Udai Pareek often used to talk about diversity and inclusion. He also referred to the learnings on culture he got from his stay in Malaysia. Although I lived in Malaysia till recently, for six years I faced the challenge of cultural diversity at the work place and in the society particularly at Rotary and Golf. At the work place, we had a mix of people from eight nationalities and several races. I followed the advice of Udai and greatly sailed through. A noble deed of creating an Udai Pareek memorial book has been done. I thought it appropriate to contribute by sharing my experience of dealing with cross-cultural diversity in the form of a chapter. A few anecdotes have been incorporated which are the basis for the thoughts and beliefs.

Introduction

The human resource (HR) agenda is under transformation in response to the shifts and twists in the business olympiad. While the core of human resource management revolves around building of culture, competence, and creativity in the organization, the priorities and approaches need to change suiting to the context. All leaders and managers are faced with the growing challenges of managing the "3 A's, that is, adversity, ambiguity, and alignment." On the people front, Alignment is more about creating an inclusive work culture taking along a diverse work force. Diversity among the work force is no more an exception. It has become common. One of the priorities for HR is therefore to deal with diversity.

Dimensions of diversity

Over the years, with the ever-changing composition of employee profile and the globalization, the dimensions of diversity have become manifold.

At the society level, involuntarily, all of us regularly face and deal with the diversity of age/generation, gender, physical/mental ability, physical characteristics, life style, language, income, education, marital status, religious beliefs, parental status, personality type, etc. Work place is not immune to the impact of such diversity. During the last two decades, managers/leaders have taken cognizance of the same. They are also proactively finding ways and means to deal with the challenges and leverage the benefits of the same.

A higher level of diversity flows from cultural differences. Cultural behavior depends upon where one is born, nationality and heritage, ethnicity, where one is raised and one's family life, religious preferences, language, traditions, belief system, etc. The differences are seen in the form of body language, verbal and written communication, presentation, dress and food, time management, relationship, values, norms, work habits, respect for hierarchy, social orientation, thinking styles, team spirit, triggers of motivation, straightforwardness, eye contact, etc.

Anecdote 1

We were doing a training program for the local employees (mostly Malaysian Malays), funded by the Government. Four of us (Indian expats) joined the group at the first tea break at 10.45 am. We were surprised to see mutton curry, fish fry, chicken curry, rice, etc., at the table. We asked the hotel staff whether it was the lunch being laid out. The repeated answer was that it is the tea arrangement. It takes time to accept the food habits of others!!

Managers and leaders should have the ability to recognize the differences and deal with the same.

Organizational choices

We normally have four different choices combining the elements of blending and acceptance. Figure 8.1 depicts the choices.

Need for diverse teams

There is a conscious attempt seen among many organizations, to bring in diversity in the composition of teams for various reasons. Gone are the days when diversity was seen as thrust upon or considered an inevitable evil. The following rationale is extended to support the HR strategy of embracing diversity.

Figure 8.1. Organizational choices

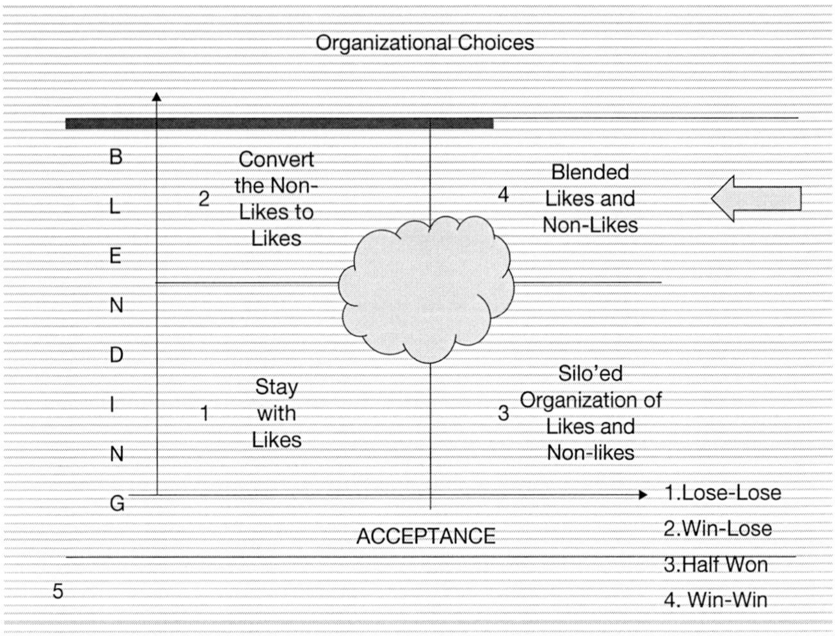

a. The belief that different types of individuals will bring in unique competencies to enhance collective capability.
b. There is a need for understanding new horizons and technologies requiring varied approaches, orientation, and viewpoints.
c. There is a requirement of resources with local understanding at various geographical locations in the global market.
d. The need for flexibility in deployment of resources.
e. Need to inculcate internal competition for creativity and innovation.
f. To convert groups into teams.[1]

Anecdote 2

We had an industrial relations problem with the foreign workers refusing to work, on account of differences between Bangladeshi and Vietnamese workers. As a part of resolution, we were trying out various methods to reach out to and counsel the two communities.

[1] A high-performance organization requires a diverse team of dreamers, thinkers, planners, organizers, controllers, motivators, and doers, characterized by working for a common business agenda through a multicompetent focus.

We found that the Indian expats made a breakthrough with the Bangladeshi workers and our Malaysian Chinese managers succeeded with the Vietnamese workers. That gave us learning that the diversity in our management team is useful in such difficult situations.

Diversity poses challenges to leadership

It is a human tendency to look for the comfort of working with or working for a same or similar type of people. Diversity impairs that comfort and brings in new challenges to the leadership, as follows.

a. The groups within the team will lead to certain pulls, pushes, and undercurrents.
b. Mix of background, language, beliefs, etc., creates serious barriers to smooth flow of communication in the organization.
c. Diverse interest and biases create hurdles in the process of blending.
d. The existence of divergent orientation and interests culminate into softer issues that are harder to solve.
e. The challenge of creating and implementing employee policies that are acceptable to diverse groups.
f. The divergent lifestyles require a variety of facilities making it difficult to provide for.
g. The clash of individual cultures coming in the way of creating a common culture.
h. There is a death of diversity leadership leaving the challenge unmet and the agenda unfinished.

Anecdote 3

We wanted to improve bonding among the management staff (about 800) by organizing an annual dinner at a 5-star hotel. The event included distribution of door gifts, cultural program, complimentary lucky draw with attractive prizes, lavish dinner, etc. The event is generally organized around the Chinese New Year. Incidentally, about 75 percent of the staff belonged to the Chinese race and the event was branded as Chinese New Year Dinner (CNY Dinner). Little we realized that the other races were aggrieved over the special treatment extended to the Chinese race. We recognized the issue and renamed the event as Annual Staff Dinner to ensure better feelings among all races.

Leadership focus to get the best from the diverse

Diversity leadership is not about mere mitigation of the challenges; rather it is about a proactive strategy to leverage the best from the diverse. It is more about driving a strategy with a practical orientation. The approach is based on the following principles.

a. The overall business strategies and plan must recognize the need for creating and leveraging a diverse work force.
b. Creating and nurturing the climate of seeking mutual help, with emphasis on collaboration, empathy, and networking.
c. An all-inclusive approach in the process of planning and implementation at all levels.
d. Encouraging and supporting formal and informal efforts for fellowship and socialization, leading to team spirit.
e. Enable processes for understanding strengths of "self and others" to build synergy and collective capability.
f. To run diversity workshops at all levels, for dealing with the challenges of diversity and of deriving the benefits of diversity.
g. A clear-cut and unbiased policy for intake of diverse talent.

Anecdote 4

At the initial stages of deployment of Indian expats, we received feedback about their strange practices in Malaysia, that is, presenting a painting with any human being in the same is considered a bad omen by a Muslim, numbers 4 and 14 are considered inauspicious numbers by a Chinese, informally enquiring about family considered an intrusion, offering a hug of love to a Muslim female considered a sin, giving a feedback with a raised voice considered an offence, etc. Similarly, there was a confusing indication of silence of the participants at a meeting, not knowing whether it was an acceptance or not. People generally take one instruction at a time and do not assume any instruction that was not spelt out. While the Vietnamese, Bangladeshi, Indonesian, Sri Lanka, and Nepalese workers were eager to work on overtime, the local Malaysian were not that keen. Some opted for hard jobs, some preferred lighter jobs, and some wanted supervisory jobs. Chinese were not seen among the workers and there were very few Indians among executive jobs. Generally, the locals preferred monetary rewards than non-financial rewards or promotions. Keeping in view all such learnings, the expats were guided from time to time to modulate their dealings and adapt to the realities.

Barriers to the success of diversity management framework

Despite the best intentions, often the diversity management framework meets with failures for the following reasons that are both within and beyond the control of the leaders.

a. The legacy of the leadership styles creating rigidity and inflexibility.
b. The lack of willingness and ability to accept others as they are.
c. Excessive knowledge and thinking bringing in multiple lenses of filtering.
d. Political factors at a local or a national level also derailing the efforts of integration.
e. Lack or inadequacy of comprehensive HRD plans and practices to provide for diversity management skills.

Anecdote 5

The Government of Malaysia has a policy of notifying a list of countries from where foreigners can be recruited. To cite a few, in the manufacturing sector, no new Bangladeshis are allowed, Indonesian male, Pilipino female, and Indians are prohibited. There were restrictions on intake of expats in terms of numbers and categories and job titles. Such restrictions required reorganization of deployment mix which was an inconvenience for the managers who were used to a particular pattern. The resistance used to reflect in the form of lower productivity. It took a conservable effort to convince the managers to accept the realities.

Diversity leadership competencies

The diversity leadership competency framework has a mix of technical and soft management areas as enumerated later. Depending upon the context and the complexity of the diversity challenge, the extent of proficiency requirement under each area may vary.

a. Global mindset
b. Inclusive orientation
c. Ambiguity and adversity management
d. Influencing and Inspiring
e. Empathy and empowerment
f. Learnability
g. Diversity development

Anecdote 6

Recognizing a few setbacks, at the time of deployment and induction of Indian expats, there was an emphasis on the specific soft skills required to deal with diversity. Similarly, while the local management staff were considered for responsible positions, the competency of inclusion was evaluated. There was a conscious effort to walk the talk on aspects such as global mind-set, inclusive orientation, ambiguity and adversity management, influencing and inspiring, empathy, and empowerment "learnability and diversity development."

In lieu of conclusion

Instead of attempting to conclude the experiences and recommendations, it may be appropriate to deal with the following commonly asked questions on the subject of inclusive work culture which is the heart and soul of the HR agenda of effectively embracing diversity. The suggestive answers are open to further debate and fine tuning as we grow, evolve, and excel.

Q1. How should organizations align their business/operational systems to be in line with the forming of an inclusive culture in the company?
Ans: Organizations willing to embrace and leverage diversity with an inclusive culture must have a clear business strategy on diversity enumerating policies on places of business, choice of business partners, recruitment, placement, rewards, work ethics, etc. The systems in the organization must possess both positive frames for promoting inclusiveness and negative frames for discouraging discrimination.

Q2. How important is an inclusive organizational culture to the success of a business?
Ans: It is critical and important for the organizational success. In a global market scenario, the winning slots will go to inclusive organizations that are built on the foundation of an inclusive work culture bringing in benefits, viz., varied viewpoints, multiple competencies, flexibility in operations, competitiveness, innovation, and team working.

Q3. How can HR help a company better manage diversity and build an inclusive work culture?
Ans: Inclusiveness is more about people and culture, than systems and processes. HR can play a leading role in establishing, nurturing, and leveraging inclusiveness, through developing diversity mindset and skills, practising inclusive-oriented people policies, and enabling line managers in getting the best from diverse.

Q4. How can organizations leverage on its diverse talents as a form of competitive advantage?
Ans: Diverse mix of talent will provide an edge to the organization both in growth and in downturn. While growth requires talent to shoulder larger, wider, newer, and diverse responsibilities, downturn requires talent to manage adversity and ambiguity. A diverse talent pool with a variety of skill-sets will provide that extra edge for creative thinking and innovative execution in a competitive market.

Q5. What are good ways to bridge communication gaps in a diverse workplace?
Ans: The best way to break communication gap is by opening the channels of communication. The specific tools used to manage communication blocks in a diverse workplace are training on languages and cross-culture and the ability to leverage the uniqueness of the individuals in the team. Audio-visuals, pictures, posters, multilanguage formats, etc., are more effective in such situations. People with listening skills, open-mindedness, tolerance to frustration, learn-ability, etc., will prove very useful communication champions to break the barriers and bring in oneness,

Q6. From your experience in human resources so far, which are some companies that have experienced success in building such an inclusive culture?
Ans: There are many organizations that have excelled in establishing an inclusive culture. For tactical reasons, it may not be appropriate to name any. We need to take benefit of such rich experiences through both formal and informal channels. In Malaysia, we took full benefit of the diverse composition of the 7,500 workforce with eight nationalities, many languages, mix of gender, age, qualification, background, experience, management style, religion, life style, etc. We developed leaders, managers, and supervisors to deal with and leverage the benefits of the diversity in the workforce. We embraced cross cultural diversity for better results through the mantra of inclusion

9 HR today: Looking beyond the faded glass of truism

Rajeshwari Narendran

The present millennium is witnessing human resources (HR) at its biggest crossroad where world-renowned HR professionals are clearly indicating the warning bell "It's time to split HR" (Charan, 2014).Why we love to hate HR (Cappelli, 2015) and cover feature of *Harvard Business Review*, July–August 2015, speaking loud "It's time to blow up HR and build something new." The focus of these articles and cover feature tells something and compels us to think "Is all well in the world of HR?"

While Ram Charan proposes to split the HR function into two parts administration and leadership, his apprehension is echoed by many business leaders expressing fear that there is a clear dissonance in business performance and role of HR in creating a value proposition at the table. It is sad but true that research indicates many HR departments have a long way to go. In Deloitte's recent global human resources trends research study, the reskilling of HR was identified as the third most urgent issue for 2014 (Global Human Capital Trends, 2014, p. 8) though it could be argued that it is the most important trend because creating and executing a strategy capable of affecting change requires the right human resource leader with a right team, a right attitude, and a constructive demonstration of capabilities of moving mountains for business development plans.

The last century has seen HR evolving from the era of industrial and labor relations, disciplinary compliances, blue-color workers, hire and fire policy (1920–1930), social security, unionization, grievance handling, human relations movement (1930–1950), emergence of white-color workers, systemization of personnel function, onset of behavioral training, participatory management techniques, employees suggestion scheme, merit-based pay program (1950–1970), emergence of HR, job enrichment, team building, quality circles, organizational restructuring, balancing work life, IT boom and technology-driven organizations, paradigm shift in service sector, performance management system, assessment centers, 360-degree feedback, competency mapping, organizational health (1970–1990), downsizing, knowledge-based economy, workforce diversity, flexi time, dotcom burst, job hopping, poaching, war of talent, big ethical dilemmas, outsourcing, employee engagement, innovative practices, virtual offices, emergence of new corporate citizen-Gen Y (1990–2010).

The role of human resource development (HRD) founded on the basic roots in administration has had a paradigm shift from service feeder function to being a strategic business partner and "sitting at the table." It is important that while marketing and finance find a strong hold in CEOs charter the HR has to find a niche in chalking out road map of competency and capabilities to deliver the best through human capital. They need not only to be aligned with the business but become a strategic partner that looks outside of the business to help it grow, compete, and win in the market. HR leaders along with all HR executives in the department must become attune to changes in the business and talent landscapes and help organizations to navigate the changes in pursuit of business goals.

The areas of concern churned out from the Global Human Capital Trends report of Deloitte are:

- Leadership, retention, HR skills, and talent acquisition are the top global trends needing urgent attention.
- Companies report generally low levels of readiness to respond to these challenges and business leaders have less confidence in their organization's readiness to deal with future trends than HR leaders themselves.
- The largest capability gaps concerning HR are reported in leadership, analytics, reskilling HR, talent acquisition and access, and the overwhelmed employees.
- Value of HR in the eyes of HR is low and they realize their inability to pioneer winds of changes due to less readiness at their end.
- Leadership crisis is evident in all businesses irrespective of nature of industry region or economy.

HR truism: Challenges to work on

Many research studies have strongly proven that the most haunting challenges in front of HR today in the 21st century start with leadership, retention, and active engagement of employees. The leadership pipeline from within the organization has to be the role of HR by tracking the employee right from the day one, identifying potential and capabilities, identifying the star performers, and separating thinkers and doers from rest of the crowd. Such an exercise can be done only by close connect and an eye for identifying the talent. However, only good HR with a clear strategic vision can retain these star performers with a strong and challenging work environment and best of engagement tactics as a roadmap to corporate success.

Many of the chief executive officers (CEOs) voice that HR leaders are more like "yes man" of CEOs and do not bring forward value "on the table."

They normally do not voice any constructive confrontation, rather they would prefer to be the followers of the instructions. This attitude of HR leaders puts them in a comfort zone, but at the same time their value in the eyes of CEO is greatly reduced. The brand HR must be leading from front like a triumvirate along with CEO and chief financial officer (CFO) (Charan, Barton, & Carey, 2015), which means that HR has to find its voice, if not more, at least equivalent to the CFO in strategic planning of business, where the assessment of human capital requirement, sourcing of leader, development of team, and pool of talent should be the tall order from the HR.

HR cannot work in silos with the business development plan. There has been a felt dissonance by many business leaders as they value marketing as the core or the propeller of the business. Similarly, finance also takes a center stage with lot of value, worth, and data in real and tangible terms. HR has to deeply know the cross functional requirements with clear alignment of functional goals leading to achievement of business goals and success. None of the functions of business can actually work without the support of HR as best brains are hired through them only and it is HR that helps create employer branding.

Let us not forget that the inculcation of organizational culture, ethics, and values is the core function of HR nurturing like a mother. HR cannot be treated as a surrogate mother to any employee where the corporate life begins with HR and HR would leave them after on-boarding and say all is over now and you are at the mercy of your function; rather it has to nurture, support, and inspire the employees continuously toward the value chain of function and make him or her perform the best through pragmatic and developmental interventions.

While multinational corporate culture has brought huge difference through diversity and inclusiveness challenges to HR, at the same time balancing the disparities in expectation, behavior, mannerisms within the organization cannot be let unattended. The role of HR to handle this challenge becomes rather more crucial as it requires a lot of preparedness in the HR executives to understand the differences and to bridge them so that the local and regional needs are taken care of and the team is ready to perform globally. This also calls for value and image of HR in the eyes of HR itself as they need to create ownership in whatever they do to make this happen.

The routine of human resource function is largely attributed and taken over by technology. The automation takes care of all these requirements in such a way that over a period of time the employees feel that role of HR is ritualistic and centered around sermons, forms, and formalities. There is more to this deep-rooted feelings and HR need to respond to it.

There is a clear void and divide in HR's ability to listen to people and feel the pulse of the most valued resource. Many a times, when the organization is preparing for transformation the first doubts are raised by HR stating that this cannot succeed whereas it is them who should be letting the human potential flourish and take the organization up in the ladder of success. The sensitization toward the employee, their needs, and their voices is completely missing and that compels the employees to term HR as a least valued function or a most hated function.

The unpreparedness of HR in creating a brand for themselves is often attributed to the Cobblers barefoot children syndrome as HR is least concerned about training and updating their own knowledge and skill to the demand of time, place, and situation. In one of the research surveys conducted by author on 60 corporate houses and their HR, one simple question was asked as to how many training or learning interventions you yourself have attended in the last five years which has enhanced your proficiency in HR, and only 17 percent could respond in affirmation. This is a clear indication that HR itself is not ready to reskill itself and does not mold it according to changing paradigms of business today.

Being one of the people-focused functions most of the attention of HR goes to executives and their career management issues. In a recent survey conducted on 200 young HR executives the author had asked them to define the percentage of time they spend on shopfloor and to the surprise, it was found that most of the time, say as high as 82 percent, was spent on executives and sitting in their well-decorated AC offices and the 18 percent time only was spent on shopfloor and too mostly only for inspection or any other routine formal visits with dignitaries or festivities. The fallout of such a survey is a clear indication to a nonresponsive HR behavior toward exploring and updating themselves to manufacturing processes, sensitization toward labor and their hardships, and also ignoring the bottom of pyramid of the organization. Let us not forget that when the grass root level employees stir and explode their frustrations then only some very scary incidents may take the corporate world by shock. We all have witnessed some of the dreaded incidents in Indian organizations in the last decade. HR needs to be sensitized in handling industrial relations and labor management and in sensibly tackling the issues by being one of them and one for them without compromising on performance and standards.

While missing out on huge strategic decisions, HR is usually considered to be risk avoidance types. The time has now come where HR has to be in the shoes of risk takers and to create an environment of decision-makers for themselves and to be equal partner in alignment of core of business.

While measuring the agility in the organization based on various functions, often HR comes last in the list as most of the HR's functions are qualitative in nature and do not produce quantifiable or measurable data, except for very miniscule proportion of the work. On the other hand, HR is considered to be very active in organizing various events and functions creating all the more wrong impression of being a country club function. Though we cannot generalize but at the same time this fact cannot be ignored.

Learning and development and performance management system takes almost 60 percent of HR's role and responsibilities. However, these efforts and their measurability in terms of return on investment are often not recorded where a simple tool like hurconomics proposed by the doyen of HR and guru Dr T. V. Rao can be applied for measuring the effectiveness and return on investment (ROI) of every human endeavor but seldom explored and paid attention to.

The diagnostic role of HR is often not finding its footing in corporate agenda, whereby the goldmine of data on which HR sits could be explored for the good of the organization and for making people happier and connected at the workplace. For example, if the HR department knows that most of the employees who leave the organization fall in the age group of 35–38, then a pro-active step can be taken like counseling for the employees aged 33–35. Enhancing the flexi work culture and also multi-job culture, expecting commitment and loyalty on a sustainable basis particularly from the outsourced employees, is completely pointless. It is very crucial and challenging for HR to bridge the gap between outsourced and on-roll employees, their expectations and performances in such a way that answerability and responsibility quotients are equivalent in both types of employees. There cannot be a rule of escapism for the outsourced employees.

Workplace citizenship behavior, widening gap between Gen X and Gen Y, errant and aggressive outbursts, bullying, excessive use of social media, mobile, and other tech gadgets also bring a lot of challenges to HR and the organizations bear the brunt of these undesired human actions. While the mobile apps have made a huge communication connect, at the same time unexpected hyperactive employees on such gadgets pose a huge challenge in terms of vigilance, transparency, ethics, performance, and responsibility. Though many organizations have a non-disclosure policy but apps such as WhatsApp can leak any important information unchecked to any competitive organization. It is next to impossible to infuse and inculcate an environment of 100 percent trust and transparency with these new-age gadgets unless there is some self-control mechanisms developed to track, trace, and tackle.

Another challenge for HR is to understand that employees do not come for job only. They also look for an opportunity to learn and grow in their career path. They see HR as an enabler function to their journey of growth. Many a times, it is felt that the rules, regulations, and policies crafted by HR create a block. Let us not forget that the small and intangible things govern the tangibles and can make or mar the organizational potential for prosperity.

What needs to be done...

- *Set agenda and take ownership*—Proactive HR needs to set agenda in perfect alignment with business development and partner the path of success rather than waiting for orders from the top for the next action. It is very important to understand what works when and where. The openness in HR leadership is often missing as many top HR leaders do not take courage in expressing their disagreements in the board room. It is often felt by many CEOs that HR leaders rather than being catalyst of change become road blocks to many promising projects.

- *Feel the pulse of people*—The pivotal research role through diagnostic studies has to be the leap forward for HR rather than sitting ideal on the data of information generated through various sources. Analytical mind in HR dwells deep into business realities vis-à-vis internal dynamics to strategize future forward action for the human capital development. HR needs to work as an intelligence arm of the business strategy, used with due diligence for making the business forge ahead.
- *Break the barrier of rituals*—There are many irrelevant policies, forms, formats, regulations that become redundant with time and situation. It is the responsibility of HR to identify those and restructure the policies, rules, regulations, etc., according to the need of the hour. HR cannot be paying a huge cost of losing its image by remaining silent in their comfort zone and by chanting, "it makes no difference." Small steps in pro action can bring lot of joy at the workplace and HR has to champion this cause.
- *Be the employer of choice*—Headhunting the right people on the right job and making them brand ambassador of the corporate image is undoubtedly the role that HR can craft a niche for. A few good recruitments lead to creating a wonderful workplace that may attract more talents from the environment. This is a double-edged weapon though, as one strong person leaving the organization can create a duckling syndrome whereby chain of talented people may leave along with the leader. Hence, all the more it becomes important to strike a right chord with the talent and to retain them, nurture them, and make them sprout new talents under their mentorship. Being the employer of choice does mean a strong work culture of diversity yet inclusiveness, empowerment yet performance, leading yet creating.
- *Sow the seeds of right people*—The recruitments at the lowest end are usually done in bulk and hiring young talents from campus recruitment is like a gala event for HR department, every year. One of the senior HR leaders from a multinational exclaimed once that we hire bucketful of talents every year but alas! The bucket is full of holes! The young turks have a huge potential, zeal, and energy to show what they can perform. However, if HR policies nip their enthusiasm, they lose their interest to share their wonderful ideas. Here comes the very important role of HR to constructively divert the synergy of young team and build a lot on their idea pool for the betterment of work and work-life. Getting the right people and making them do things right for the organization can yield a sustainable result for a longer period of time as these people develop a sense of belongingness with the organization and they perform beyond expectation.
- *Strengthening the learning and development (L&D) initiatives*—L&D is the most important part of people management techniques and the prudent L&D efforts never go waste. Most L&D initiatives look like a fashion statement in an industry and every small or big corporate tries to imitate understanding the relevance and need of it. It is very important to

understand the vision, mission, strategy, future of business, and available capabilities before HR jumps for opting an L&D initiative. It has to be tailor-made, similarly, L&D is not a ritualistic function where training budgets are literally exhausted only in the last quarter of the financial year, mostly which is an indication of a poor planning. There are many scapegoats ready to be sent for training programs as they are the easily separable employees, whereas the person who fits in the best for a specific training program may not be sent for obvious reasons. Training and learning man-days without the measurement of effectiveness utility and return on investment in performance and financial terms can be draining on the credibility of L&D functions; hence, it becomes imperative for HR to devise mechanisms and tools for such measurements.

- *Looking beyond the glass of performance management system (PMS)*—A large chunk of employees work time is attributed in meeting the requirements of PMS. Most of the performance management records of individual employees are used superficially for the promotion exercise and the 80 percent of the important information is seldom used for developmental purposes. If HR can bank upon these data sensibly, they can hit a huge potential for taking the organization into the next orbit.
- *Tapping the inclusive growth potential*—Cross functional teams and its synergy can be largely utilized for interventions that may create value proposition at all organizational levels. HR has to inculcate the bold HR leadership teams who can lead and make a huge difference.
- *HR as a communication hub*—HR leader needs to architect the people connect networks in such a fashion where people are heard and feelings are shared, achievements are applauded, difficulties are challenged, hurdles removed, and conflicts resolved. Yes, people should come first and then the strategy. In most of the cases, the communication connect is so weak that the harmony among the people and the organization is badly affected causing low morale and motivation. HR can make the town hall communications and can creatively use technology to support the connect in real time and witness wonders through them.

It is quite obvious that the responsibility, on HR, to prove its mettle, is in the hands of practitioners, professionals, academia, and trend setters. It is high time to respond to the wake-up call to reskill, retool, redesign, and restructure HR. As rightly exclaimed by Dr Anil Khandelwal in the workshop on Creative human resource management conducted by Academy of HRD that "Plastic Surgery is not the Answer to reskill HR the entire DNA needs to change to make HR move along and ahead of time."

References

Cappelli, Peter. (2015, July–August). Why we love to hate HR… and what HR can do about it? *Harvard Business Review*, 54–61. Retrieved from https://hbr.org/2015/07/why-we-love-to-hate-hr-and-what-hr-can-do-about-it (accessed on November 8, 2015).

Charan, Ram. (2014, July–August). It's time to split HR. *Harvard Business Review*, 34. Retrieved from https://hbr.org/2014/07/its-time-to-split-hr (accessed on November 8, 2015).

Charan, Ram, Barton, Dominic, & Carey, Dennis. (2015, July–August). People before strategy: A new role for the CHRO. *Harvard Business Review*, 62–71. Retrieved from https://hbr.org/2015/07/people-before-strategy-a-new-role-for-the-chro (accessed on November 8, 2015).

Global human capital trends: Engaging the 21st centuary workforce. (2014). A report by Deloitte Consulting LLP and Bersin by Deloitte. Retrieved from http://dupress.com/wp-content/uploads/2014/04/GlobalHumanCapitalTrends_2014.pdf (accessed on November 8, 2015).

Ram Charan (2014) It's time to split HR"; Harvard Business review, July-August, 2014, P34. ("https://hbr.org/2014/07/its-time-to-split-hr,%20%20downloaded%20on%20Novemebr%208" https://hbr.org/2014/07/its-time-to-split-hr, downloaded on November 8, 2015).

10 OD values in family-managed businesses in India

Keith C. D'Souza, Ravindra Dey, and Sheba Mathew

Gentle intellectual giant, Dr Udai Pareek, was a pioneer of the behavioral science profession in India. He exemplified by his behavior and operating style, the values that formed the backbone of the disciplines of organizational development (OD) and human resource development (HRD) (ODHRD), at least at the time that they made their appearance in theory and practice in India around the 1970s. Those of us who were privileged to be associated with him remember well the characteristics of openness, warmth, tolerance, and other "human" qualities combined with intellectual brilliance and prolific academic achievement, which distinguished him as a person and as a professional.

The framework of OCTAPACE—Openness, Collaboration, Trust, Autonomy, Proactvity, Authenticity, Confrontation, and Experimentation (Pareek, 1998, 2002)—which was one of Dr Pareek's numerous contributions to the field of ODHRD, neatly illustrates the key principles by which he lived and influenced so many others. It is, therefore, fitting that in a volume dedicated to his memory, we should discuss and reflect on the issue of ODHRD values in Indian organizations, many of which were beneficiaries of his professional contributions.

At a time when economic growth seems to be the predominant concern in society, discussions about values in the corporate business sector may appear somewhat quaint. As concerns about becoming or remaining economically viable and globally competitive grow, "softer" issues such as organizational values and culture tend to recede into the background. Organizational restructuring, downsizing, layoffs, retrenchment, and other "tough" measures have come to be accepted as inevitable corollaries of economic growth and development. In this context, it would be enlightening to examine how the basic values and principles that underlie ODHRD that, with rare exceptions among managements, appear "soft" and *prima facie* antithetical to business growth, are viewed by business.

This chapter examines the extent to which values that were traditionally seen as the backbone of ODHRD theory and practice, are considered relevant in businesses in India today, and the extent to which they are perceived as being actually practised. What the chapter tries to uncover is the extent of congruence between precept and practice in ODHRD.

We have chosen the family-managed business (FMB) sector as the site of our study because, apart from the ready access to the sample of respondents, the FMB sector is of particular importance in the Indian economic context.

Family-managed business sector in India

Management and OD literature, at least in India, do not appear to give much attention to the status of businesses as family-managed entities. This is a significant omission because family history, traditions, and values may have an important bearing on business practices, organizational culture, and HR strategies.

Family businesses ranging from small mom-and-pop stores to large conglomerates account for as much as 95 percent of all Indian businesses. About 73 percent of the top 500 companies listed on BSE are family-owned, while almost half the companies on NIFTY are second or third-generation family businesses (Godrej, 2015).

A few facts and figures about FMBs in India (PWC, 2013):

- Family companies account for two-thirds of India's GDP.
- They contribute to 90 percent of the gross industry output.
- 79 percent of organized private sector employment is generated by family businesses.
- 27 percent of overall employment is generated by family firms.

Given the number, size, and scale of FMBs in India, it is important to understand the ODHRD-related values of their managements, because such values would be instrumental in determining how FMB organizations are run and how they manage and develop their human resources.

ODHRD and the values issue

As a specialized discipline in the management of people and organizations, OD was born from the thinking of social scientists strongly influenced by humanistic psychology traditions in organizational behavior and management on the one hand, and democracy and liberalism in the sociopolitical context, on the other hand. OD emerged as an organized approach to managing and developing people and organizations, distinctly different from other conventional approaches founded largely on the scientific management and rational economic models (Cummings & Worley, 2005).

OD has played a seminal role in the growth and development of economic and social systems across the world, through the promotion of institutions with modern, progressive outlooks congruent with social and economic development. As a prominent writer on OD puts it, "It was OD that helped make the shift from command and control-oriented bureaucracies to high-involvement organizations. The approach incorporated values of human integrity, democracy, and performance" (Worley, 2014, p. 68).

Values play a critical role in organizations. They quietly but strongly influence the organization through the choices and decisions that managements make. Organizations such as the Tata Group, the AVB Group, and others are known to be guided by a strongly-held set of values that underlie their strategies and culture.

As an approach to developing and improving people and organizational effectiveness, OD distinguishes itself by its strong humanistic and normative-value emphasis. One of the best known and most used textbooks in India on OD—*Organization Development: Behavioral Science Interventions for Organization Improvement* by Wendell French and Cecil Bell (1999)—articulates the kind of values and assumptions that have been the basis of OD work: "OD values tend to be humanistic, optimistic, and democratic. Humanistic values proclaim the importance of the individual: respect the whole person, treat people with respect and dignity, assume that everyone has intrinsic worth, view all people as having the potential for growth and development" (French & Bell, 1999, p. 62).

HRD as an approach to organization and human resource management emerged in the 1970s in India as a kind of hybrid version of OD, stemming from the work of Udai Pareek, T. V. Rao, and a few others who were among the earliest pioneers of behavioral science-based interventions in India. S. K. Bhattacharya, Udai Pareek, and T. V. Rao, in the course of a mammoth organizational restructuring consulting assignment for Larsen & Toubro Ltd in India, were the first to advocate the setting up of a department of Human Resource Development, as distinct from the conventional personnel, welfare, and industrial relations departments (Pareek & Rao, 1981). The philosophy, values, approach, and methods of HRD were strongly influenced by the background of the consultants who were themselves profoundly influenced by the work of the NTL Institute of Applied Behavioral Sciences in the USA and the Tavistock Institute of Human Relations in the UK. Underlying the work of the NTL and the Tavistock Institutes were the new findings and principles of management and organization that emerged from the paradigm-changing contribution of Maslow (1954) and the human relations (Mayo, 1960) and the human resource (McGregor, 1960) schools of organizational behavior.

Much of the OD work in India too, especially between the 1970s and the 1990s, was shaped by these values. Many of the pioneering initiatives in OD were carried out in the 1970s and the 1980s in public sector organizations in India such as HMT, BHEL, BEML, and others, which themselves were under pressure to achieve not just financial and economic objectives, but also social and developmental ones (Pareek & Rao, 2006).

The OCTAPACE framework conceptualized by Pareek (1998) is a distillation of the key values and normative assumptions that underlay most of the pioneering work in ODHRD between its inception in India from the 1970s up to the early 1990s. The framework has been widely used in ODHRD research and practice in a large number of organizations in India and abroad.

A shift in the values implicit in ODHRD in India seemed to come about with liberalization and globalization of the Indian economy from the late 1980s. Traditional ODHRD values based on liberal-humanistic principles came under considerable strain when companies began to face the pressure of global competiveness. Managements were hard-pressed to reconcile the imperatives of economic growth and profitability in a competitive environment, with people-oriented humanistic values. In the conflict between the imperatives of economic growth and human development, the concerns of economic growth seemed to invariably prevail. Organizations began to undergo radical transformation toward being more growth and results-oriented (Khandwalla, 2002) and, consequently, perhaps materialistic and utilitarian. Managements and organizations began to view and treat people more as "human resources" than as human beings.

A more recent and also now widely used textbook on organization development by Cummings and Worley points to a distinct shift in the value orientation of OD in more recent times: "Traditionally, OD professionals have promoted a set of values under a humanistic framework, including a concern for inquiry and science, democracy and being helpful. They sought to build trust and collaboration; to create an open problem-solving climate; and to increase the self-control of organization members. More recently, OD practitioners have extended those humanistic values to include a concern for improving organizational effectiveness (for example, to increase productivity or to reduce turnover) and performance (for example, to increase profitability). They have shown an increasing desire to optimize both human benefits and production objectives" (Cummings & Worley, 2005, p. 83).

The *coup-de-grace* in academic literature which seems to have brought a drastic shift in what is perceived to be the effective ODHRD may perhaps be Dave Ulrich's (1997) trend-setter book, *Human Resource Champions*, which in more recent times seems to have become the bible for many HR professionals.

Ulrich (1997) suggested four key roles of HR professionals: employee champion, administrative expert, change agent, and strategic partner. While propounding that the management of HR in today's context requires that HR professionals play all the four key roles, the tenor of Ulrich's arguments and the highlighting of the extreme pressures of the business environment on organizations indicates a clear bias in favor of the strategic partner role. Indubitably, the implication is that for HROD professionals to be effective in today's context, they need to shed the "nice-guy" approach and join hands with the "tough-guy" business managers to ensure organizational survival and growth.

For managers as well as ODHRD practitioners, the shift of goals and priorities and their associated values and beliefs compelled fundamental changes in their approaches to ODHRD. During the time of recent economic downturns, many managers had to concede to, or even initiate, actions for reorganization, downsizing, manpower reduction, etc., in ways that went against the grain of their values and beliefs. They were compelled to do so with the belief that such actions were essential to their roles as business managers to improve organizational efficiency, profitability, and economic goal achievement.

In many business managers, therefore, one would expect a fair degree of scepticism about the relevance and practicality of values and assumptions that, though integral to ODHRD philosophy and approach, may seem not very relevant and practical in the current business context.

Objectives of this study

The potential tensions between the imperatives of growth and of people development in business organizations make it interesting to examine which values are seen as relevant in today's FMB organizations and which are seen to be actually practised, and to obtain an idea of the gaps between ODHRD precept and practice.

Data collection

Based on a survey of literature and the authors' own experiences with OD, a questionnaire, *OD Values Survey*, was constructed (D'Souza & Dey, 2013). The questionnaire listed 25 common values and beliefs that have been explicit or implicit in ODHRD. The name of each value or belief was accompanied in the questionnaire by a brief phrase or statement elaborating the meaning of the value or belief. The questionnaire also included spaces for respondents to list any other values and beliefs, in case there were any that they thought had not been listed.

Respondents were required to rate each value with respect to the extent to which they thought the values were relevant in today's context, as well as the extent to which they thought the values to be actually practised in OD today. This was done using a seven-point rating scale ranging from 1= not at all relevant/practised, at the lowest extreme of the scale, to 7 = very relevant/ practised, at the highest end of the scale.

The questionnaire was given by hand to about 125 owners and managers of FMBs, enrolled in the post-graduate program in Family Managed Business (PGP-FMB) of a reputed institute of management in Mumbai. The group represented a heterogeneous mix of people of different genders, age groups, educational qualifications, organizational hierarchy levels, type and size of family business, etc. Forty-five of the questionnaires returned were found complete and usable for analysis.

Analysis

Our analysis at this stage was confined to examining the mean ratings for the 25 value variables, with respect to their perceived relevance and their perceived practice, in order to see which values are considered most and least relevant, and which most or least practised. The tables at the end of this chapter present the relevant means and standard deviations.

The mean ratings of the 25 variables from the survey of this survey of 45 respondents were compared with the mean ratings obtained from another online survey of 75 respondents carried out about a year and a half ago (D'Souza & Dey, 2013), which comprised a heterogeneous mix of managers and HR professionals. The mean ratings obtained of those 75 are used to compare the mean ratings of this survey of 45, as a rough benchmark.

Findings and discussion

Table 10A.1 ranks orders of the mean scores of the 25 values based on their relevance scores of the FMB group. The top five and the bottom five in the rank ordering are highlighted in the table. Table 10A.2 similarly shows the rank order of mean scores of the values seen by the respondents to be most practised in reality.

Most relevant and most practised values

Of the 25 values, the five perceived to be *most relevant* by the FMB managers are:

1. Achievement of goals
2. Confidentiality
3. Honesty and integrity
4. Professionalism
5. Respect for all individuals

Of these values, *honesty and integrity*, and *respect for all Individuals*, are values that FMB managers share with managers from other sectors as shown in Table 10A.1.

The values perceived by FMB managers to be *most practised* in reality in order of mean ranks:

1. Profitability and economic growth
2. Achievement of goals
3. Professionalism
4. Empowerment
5. Respect for all individuals

Achievement of goals and *respect for all individuals* are found to be among the top five values to be perceived by FMB managers to be both most relevant and most practised in their businesses.

Least relevant and least practised values

The values perceived to be least relevant by FMB managers are

1. Diversity (rank 21)
2. Community (rank 22)
3. Power and influence (rank 23)
4. Confrontation (rank 24)
5. Patriotism and social identity (rank 25)

The values perceived to be least practised in reality are:

1. Diversity (rank 21)
2. Experimentation (rank 22)
3. Patriotism and social identity (rank 23)
4. Human development (rank 24)
5. Creativity (rank 25)

Diversity and *patriotism and social identity* are perceived by FMB managers to be both least relevant as well as least practised in their businesses.

Values common across FMB and other managers

Comparing the mean scores of the FMB respondents with those of managers from other sectors, the following are found to be commonly perceived by both sections of mangers:

Most Relevant (common among the top five in both groups):

1. Respect for all individuals.
2. Honesty and integrity

Least Relevant (common among the bottom five in both groups):

1. Community
2. Confrontation

3. Patriotism and social identity

Most Practised (common among the top five in both groups):

1. Respect for all individuals
2. Profitability and economic growth
3. Achievement of goals

Least Practised (common among the bottom five in both groups):

1. Patriotism and social identity

Gaps Between Relevance and Practice of Values (Table 10A.3):

The largest gaps (difference in mean scores) between relevance and practice of the ODHR values, in the case of FMB managers, are seen in the case of the following values:

1. Professionalism
2. Human development
3. Authenticity
4. Creativity
5. Empowerment

Of the above, only *authenticity* is common to both groups of managers in terms of the largest gap between relevance and practice.

Discussion

A reflection on the findings seems to suggest that there is some gap between the values considered relevant and those seen as being practised. It is not surprising that the expectedly business-oriented FMB managers score *Achievement of Goals* as the highest in terms of relevance, and that *Profitability and Economic Growth* are seen as being the most practised. What may be interesting is the high degree of relevance ascribed to values such *as Honesty and Integrity*, and *Respect for All Individuals*. Indian FMB managers evidently ascribe a high degree of importance these values in precept.

When it comes to practice, however, *Honesty and Integrity* together with *Confidentiality*, are replaced by other values namely *Profitability and Economic Growth*, and *Empowerment*. This may suggest that when it comes to practice, pragmatism may be more important than idealism—a finding that may reinforce the stereotype of the typical Indian family business management style.

What is noticeable and particularly interesting is that *Professionalism* and *Respect for All Individuals* are seen as not only among the top five relevant values, but also among the top five practised values.

Another noteworthy finding is that with respect to *Diversity*. We live in times when diversity has come to become a dominant value in the society as well as in organizations. The finding that FMB managers do not seem to share this value may suggest that they may be behind the times when it comes to this value.

On the other hand, with respect to *Patriotism and Social Identity*, the value emerges clearly as the least relevant as well as least practiced, suggesting that the same managers may be well ahead of the current times with respect to the being relatively free of identity prejudices and being universal in terms of their values. At a time when Indian industry is reaching out across national borders and becoming more global in reach, this value augurs well for Indian FMBs.

A major red-flag for ODHR professionals is the finding that *Human Development* obtains only a moderate rank in terms of relevance in the perception of FMB respondents, even while it is among the top five in the perception of managers in the larger sample of 75 managers. In practice, *Human Development* is seen to be among the lowest five priorities for FMB managers.

Implications for ODHRD

Given the number of FMBs and their role in the economy, the findings of this study suggest that there is both great need and great opportunity, for ODHRD professionals to make a mark in Indian industry by orienting FMB owners and managers toward the values, goals, and practices of ODHRD. The challenge of "Make in India" may well have to begin by making a difference in the minds and hearts of FMB managers.

References

Cummings, T. G., & Worley, C. G. (2005). *Theory of organization development and change.* New Delhi: South-Western.

D'Souza, K. C., & Dey, R. (2013, March 21). *Relevance and practice of traditional OD values in a globalizing business environment.* Paper presented at the International Conference on Globalization: The Way Ahead, Xavier Institute of Management and Research, Mumbai.

French, W. L., & Bell, C. H. (1999). *Organization development: Behavioral science interventions for organization improvement.* New Delhi: Prentice-Hall of India.

Godrej, A. (2015, February 9). Family managed businesses are the 'unsung heroes' of the Indian economy. *The Hindu.* Mumbai Edition. Retrieved from http://www.thehindubusinessline.com/news/family-managed-businesses-are-the-unsung-heroes-of-the-indian-economy-adi-godrej/article6875268.ece (accessed on July 25, 2015).

Khandwalla, P. N. (2002). Effective organizational response by corporates to India's liberalization and globalization. *Asia Pacific Journal of Management, 19*(2/3), 423–448.

Maslow, A. (1954). *Motivation and personality.* New York: Harper.

Mayo, E. (1960). *The human problems of an industrial civilization.* New York: Viking Press.

McGregor, D. (1960). *The human side of enterprise*. New York: McGraw-Hill.

PWC. (2013). PWC's family business survey 2012–13: Family firm—The India perspective. Retrieved from https://www.pwc.in/assets/pdfs/family-business-survey/family-business-survey-2013.pdf (accessed on July 25, 2015).

Pareek, U. (1998). Studying organizational ethos: The OCTAPACE profile. In John E. Jones (Ed.), *The Pfeiffer Library* (Vol. 15, 2nd ed., pp. 167–169). Misenheimer: Jossey–Bass/Pfeiffer.

———. (2002). *Training instruments in HRD and OD*. New Delhi: Tata McGraw-Hill.

Pareek, U., & Rao, T. V. (1981). *Designing and managing human resource development systems in Indian organizations*. New Delhi: Oxford and IBH.

———. (2006). *Designing and managing human resource systems*. New Delhi: Oxford and IBH Publishing.

Ulrich, D. (1997). *Human resource champions: The next agenda for adding value and delivering results*. Boston, MA: Harvard Business School Press.

Worley, C. G. (2014). OD values and pitches in the dirt. *OD Practitioner*, *46*(4), 68–71.

Appendix

List of values

Sr. No.	Description of Values
1.	***Human Development***: Developing and helping individuals and groups to realize their potential
2.	***Organizational Growth***: Developing and helping the whole organization to realize its potential
3.	***Professionalism***: Committed to being knowledgeable, skilled, and up-to-date in the profession
4.	***Achievement of Goals***: Fostering and enabling achievement of goals and objectives
5.	***Respect for All Individuals***: Treating all people from all backgrounds and levels, with value, respect, and dignity
6.	***Inclusion, Collaboration, and Participation***: Working together and involving all key stake-holders
7.	***Openness***: Being free and candid with sharing relevant information
8.	***Honesty and Integrity***: Standing up for the truth, even when it is uncomfortable
9.	***Authenticity***: Genuine behavior with no hidden agenda or politics
10.	***Inquiry***: Having a questioning attitude of enquiry to find out what is going on
11.	***Community***: Bringing organizations and people with common interests together for mutual benefit
12.	***Diversity***: Respecting differences and encouraging a wide variety of perspectives, even if they are conflicting
13.	***Experimentation***: Being willing to try new methods and approaches and encouraging individuals and organizations to do the same
14.	***Optimism***: Helping people think positively about themselves and their future
15.	***Confidentiality***: Honoring confidences, even when it is difficult to do so
16.	***Socio-economic Justice***: Equality of opportunity and fairness for all sections of society
17.	***Empowerment***: Developing the capacities of people to act and manage themselves
18.	***Creativity***: Discovering and developing new and innovative ways of doing things
19.	***Profitability and Economic Growth***: Achieving financial and material gain
20.	***Concern for the Environment***: Protecting and preserving the environment
21.	***Change***: Promoting and facilitating change and continuous improvement
22.	***Power and Influence***: Exercising power and influence in a positive and responsible manner
23.	***Peace and Harmony***: Ensuring harmonious relations and absence of conflict
24.	***Confrontation***: Questioning and challenging things that seem contrary to one's own convictions, values, and beliefs
25.	***Patriotism and Social Identity***: Strengthening loyalty and commitment to one's country, state, or ethnic/social group

Table 10A.1 Comparison of mean scores for relevance between FMB and other managers

	FMB: Mean scores for Relevance ($n=45$)				Other Managers: Mean scores for Relevance ($n=75$)		
Order	Parameters	Mean	Std Deviation	Order	Parameters	Mean	Std Deviation
1	Achievement of goals	6.38	0.89	1	Respect for all individuals	6.56	0.70
2	Confidentiality	6.33	0.67	2	Human development	6.36	0.65
3	Honesty and integrity	6.27	0.89	3	Organizational growth	6.32	0.77
4	Professionalism	6.24	0.88	4	Change	6.31	0.94
5	Respect for all individuals	5.96	1.04	5	Honesty and integrity	6.28	0.76
6	Authenticity	5.93	1.25	6	Achievement of goals	6.24	0.79
7	Empowerment	5.93	1.05	7	Authenticity	6.20	0.99
8	Optimism	5.89	1.07	8	Professionalism	6.19	1.23
9	Profitability and economic growth	5.87	1.04	9	Inclusion, collaboration, and participation	6.19	0.95
10	Organizational Growth	5.82	1.39	10	Socio-economic justice	6.15	0.97
11	Inclusion, collaboration, and participation	5.82	1.25	11	Empowerment	6.15	0.90
12	Socio-economic justice	5.78	0.95	12	Creativity	6.12	1.08
13	Change	5.76	1.42	13	Diversity	6.08	0.82
14	Human development	5.71	0.89	14	Profitability and economic growth	6.03	1.40
15	Peace and harmony	5.71	1.62	15	Openness	5.99	0.85
16	Concern for the environment	5.67	1.11	16	Experimentation	5.96	0.85
17	Experimentation	5.62	0.91	17	Optimism	5.96	1.08
18	Creativity	5.56	1.42	18	Concern for the environment	5.95	1.16
19	Openness	5.51	1.39	19	Power and influence	5.95	1.14
20	Inquiry	5.51	1.47	20	Confidentiality	5.87	1.12
21	Diversity	5.42	1.45	21	Community	5.81	1.09
22	Community	5.36	1.21	22	Inquiry	5.63	1.17
23	Power and influence	5.33	1.46	23	Confrontation	5.61	1.28
24	Confrontation	5.20	1.62	24	Peace and harmony	5.25	1.50
25	Patriotism and social identity	5.16	1.69	25	Patriotism and social identity	5.20	1.62

Table 10A.2 Comparison of mean scores for practised between FMB and other managers

	Mean Scores for Practised ($n = 45$)				Mean Scores for Practised ($n = 75$)		
Order	**Parameters**	**Mean**	**Std Deviation**	**Order**	**Parameters**	**Mean**	**Std Deviation**
1	Respect for all individuals	5.27	1.10	1	Profitability and economic growth	5.32	1.31
2	Profitability and economic growth	5.04	1.54	2	Achievement of goals	4.99	1.08
3	Inclusion, collaboration, and participation	4.87	1.12	3	Professionalism	4.67	0.76
4	Socio-economic justice	4.71	0.82	4	Empowerment	4.41	1.01
5	Achievement of goals	4.69	1.22	5	Respect for all individuals	4.36	1.51
6	Honesty and integrity	4.62	1.07	6	Confidentiality	4.36	1.19
7	Confidentiality	4.51	1.38	7	Change	4.33	.28
8	Optimism	4.42	1.27	8	Organizational growth	4.29	1.09
9	Organizational growth	4.36	1.33	9	Socio-economic justice	4.13	1.44
10	Power and influence	4.31	1.53	10	Human development	4.07	0.92
11	Professionalism	4.22	1.15	11	Community	4.05	1.11
12	Openness	4.22	0.90	12	Peace and harmony	4.05	1.01
13	Inquiry	4.16	1.82	13	Inquiry	4.00	1.29
14	Change	4.13	1.34	14	Inclusion, collaboration, and participation	3.99	1.21
15	Empowerment	4.11	1.13	15	Creativity	3.99	1.20
16	Authenticity	4.04	1.19	16	Optimism	3.85	0.98
17	Concern for the environment	3.96	1.48	17	Honesty and integrity	3.84	1.25
18	Confrontation	3.96	1.81	18	Experimentation	3.84	1.23
19	Peace and harmony	3.93	1.50	19	Diversity	3.76	1.17
20	Community	3.89	1.50	20	Confrontation	3.76	1.23
21	Diversity	3.87	1.31	21	Power and influence	3.73	1.26
22	Experimentation	3.80	1.38	22	Patriotism and social identity	3.63	1.51
23	Patriotism and social identity	3.80	1.47	23	Authenticity	3.61	1.51
24	Human development	3.78	1.28	24	Openness	3.47	1.22
25	Creativity	3.67	1.54	25	Concern for the environment	3.36	1.45

Table 10A.3 Comparison of gaps between relevance and practice scores between FMB and other managers

	Gaps between Relevance and Practice ($n = 45$)					Gaps between Relevance and Practice ($n = 75$)		
		Relevance	Practised	Gap			Relevance	Practised
Order	Parameters	Mean	Mean	Diff.	Order	Parameters	Mean	Mean
1	Professionalism	6.24	4.22	2.02	1	Authenticity	6.20	3.61
2	Human development	5.71	3.78	1.93	2	Concern for the environment	5.95	3.36
3	Authenticity	5.93	4.04	1.89	3	Openness	5.99	3.47
4	Creativity	5.56	3.67	1.89	4	Honesty and integrity	6.28	3.84
5	Empowerment	5.93	4.11	1.82	5	Diversity	6.08	3.76
6	Experimentation	5.62	3.80	1.82	6	Human development	6.36	4.07
7	Confidentiality	6.33	4.51	1.82	7	Power and influence	5.95	3.73
8	Peace and harmony	5.71	3.93	1.78	8	Inclusion, collaboration, and participation	6.19	3.99
9	Concern for the environment	5.67	3.96	1.71	9	Respect for all individuals	6.56	4.36
10	Achievement of goals	6.38	4.69	1.69	10	Creativity	6.12	3.99
11	Honesty and integrity	6.27	4.62	1.64	11	Experimentation	5.96	3.84
12	Change	5.76	4.13	1.62	12	Optimism	5.96	3.85
13	Diversity	5.42	3.87	1.56	13	Organizational growth	6.32	4.29
14	Organizational growth	5.82	4.36	1.47	14	Socio-economic justice	6.15	4.13
15	Optimism	5.89	4.42	1.47	15	Change	6.31	4.33
16	Community	5.36	3.89	1.47	16	Confrontation	5.61	3.76
17	Patriotism and social identity	5.16	3.80	1.36	17	Community	5.81	4.05
18	Inquiry	5.51	4.16	1.36	18	Empowerment	6.15	4.41
19	Openness	5.51	4.22	1.29	19	Inquiry	5.63	4.00
20	Confrontation	5.20	3.96	1.24	20	Patriotism and social identity	5.20	3.63
21	Socio-economic justice	5.78	4.71	1.07	21	Professionalism	6.19	4.67
22	Power and influence	5.33	4.31	1.02	22	Confidentiality	5.87	4.36
23	Inclusion, collaboration, and participation	5.82	4.87	0.96	23	Achievement of goals	6.24	4.99
24	Profitability and economic growth	5.87	5.04	0.82	24	Peace and harmony	5.25	4.05
25	Respect for all individuals	5.96	5.27	0.69	25	Profitability and economic growth	6.03	5.32

11 Transformation in a large organization: How intangibles drive tangibles*

Anil K. Khandelwal

Udai Pareek had a major influence, both as leading academic and as a mentor in my journey in human resource development (HRD) organizational development (OD). I met Udai and T. V. Rao in a workshop on HRD at Indian Institute of Management Ahmedabad around 1979, when they presented draft of their book *Designing and Managing Human Resource Systems* based on their classic work at Larsen and Toubro. I was quite fascinated about the new flavor in the people management area. I had sessions of long debates with Udai and T. V. Rao on this subject. These sessions helped me to begin my professional journey in HRD.

I remember Udai for the keen interest he took in introducing me to Indian Society of Applied Behavioural Science (ISABS) and, in some sense, in hand-holding me to complete my professional membership. I must say that it gave me tremendous insights into process work and development of process bias, in undertaking transformation of the organizations which I led.

Udai exemplified a true HRD missionary who helped many like me to discover their own potential and created an urge for continuous experimentation of new innovative methodologies in organizations. As a chief executive officer (CEO) of Bank of Baroda (BOB), I also had the privilege of having him for some time as a process consultant, during our transformation program and we immensely benefited from his insights. Despite his academic stature, he was always curious to learn. His quiet humility and commitment to build people and institutions was unique in many ways.

My career journey in BOB started in January 1971 as a probationary officer and ended as its chairman and managing director (CMD) in 2008. After completing my chemical engineering degree, a bank job was the last thing on my mind but a chance response to an advertisement landed me a job right next door to my residence at Jaipur, thus began my odyssey in BOB.

---

* This chapter heavily draws from my book *Dare to Lead* (SAGE, 2011).

The bank job, in fact, disillusioned me in many ways, more particularly about the prevailing work culture, militant trade unionism, hapless management, and disgruntled customers. In fact, the entire social system within the bank posed a number of question marks about the wisdom of my decision. A sympathetic personnel specialist, one Mr L. B. Bhide, Head of Personnel at Corporate Office, helped me to induct into Personnel job, apparently to deal with some of these issues by being part of the system. I learnt a lesson early in my career that human problems can be resolved by immersing oneself into the system and helping find solutions within the system itself.

I continued my journey into Personnel within the bank with two major interruptions: one that involved corporate-level job in Personnel in a private bank (1976–1980) and another as a Senior Core Faculty in human resources (HR) area at BOB Staff College (1980–1993). In both these jobs, I was engaged in Personnel/Industrial Relations (IR)/Training jobs. It was in 1995 that I was transferred to Banking Operations where I had opportunity to manage two difficult zones in the bank, Western Uttar Pradesh and West Bengal operations, between 1995 and 2000.

It was in the year 2000 that I was appointed to a board level position as the executive director (ED) in BOB by the Government of India. It was in this role that I initiated major reforms in the bank's industrial relations systems which was a major Achilles heel in managing the workplace and offering a hassle-free customer service. Despite many trials and tribulations, I could bring about paradigm change in the industrial relations policy of the bank and restore the authority of the line managers in managing the workplace. In March 2005, I was appointed as the CMD of the bank for three years, after a brief stint as CMD of Dena Bank in 2004. Returning to BOB as CMD was homecoming for me. Of my total career of 34 years, I had spent as many as 30 years with BOB in different capacities, witnessing its ups and downs. At a personal level, I felt quite excited to lead the bank that had given me my first nursery lessons in HR and banking. When I joined the bank in 1971, I could never imagine that someday I would reach up to such a level. This role was culmination of my lifelong learning and leadership journey in various roles. Here I had an exciting opportunity to use it in transforming the bank.

Context of transformation

With 97-year-old history and an enviable track record, known not only for its prudence and conservatism but also equally known for its innovations, BOB had a record of posting profits, every single year of its existence. After riding high in the late 1990s and reaching number one position in the nationalized banks in 2000, it started slipping so much so that in 2004, its ranking slid to number four in just four years. Its market share, in both deposits and credit, was on the decline. While peer banks showed good growth in credit, between 25 and 30 percent, BOB's credit growth was abysmally low, at less than 2 percent. The bank had lost nearly 1.1 percent share on the assets side. This was the largest share loss among nationalized banks. The slippage had been dramatic across segments. In fact, the slippage had been the maximum in the most attractive business segments—in retail assets and current accounts.

At a time when banks were going whole hog on retail and capturing new customers, BOB's share was lowest among peer banks at 2.2 percent. The bank's overall growth was below average, be it in total income growth, fee income growth, or asset growth. A further disturbing trend was performance in Gujarat, historically a stronghold for the BOB. Although the economy of Gujarat had been performing robustly and Gujarat operations were valuable part of the bank's franchise, the bank had been losing its position to other banks and had lost its market share substantially from 22.3 percent to 20.1 percent. More concretely, the state of the bank in 2005, when I took over, was pretty dismal.

Another worrying feature was that specialized corporate credit branches as also other large branches headed by assistant general managers/chief managers performed below their potential. The overall contribution of these branches to the bank's balance sheet had been on the constant decline.

Besides, the bank was perceived as poor in terms of technology. Its technology upgradation plans had not taken off. Over this period, some peer banks such as Punjab National Bank (PNB) had undertaken two transformation projects and had made major investments in information technology (IT). In 1999–2000, PNB, with the help of outside consultants, had initiated many steps for improving their market positioning and credit growth. By 2005, they were able to roll out 1,000 branches under core banking solution (CBS). Around this time, Union Bank of India and Bank of India had also kick-started CBS implementation. Although BOB had in all seriousness thought of an IT-driven business transformation strategy and hired one of the best in world—Gartner Group (US-based firm) right in the year 2001, to guide this effort, the project could not make headway on account of a variety of issues, but more particularly on account of the failure of management at that time to aggressively push the transformation agenda. This cost the bank dearly in terms of its competitive positioning in the industry and was left behind its peers.

BOB was yet to decide on the system integrator for implementing CBS. In such an environment, the bank also failed to initiate many other changes, such as expanding ATM network, to capture the retail boom. Staff morale was low and managers complained about high attrition of customers because of the absence of alternate delivery channels such as ATMs and also nonavailability of the CBS.

The bank's market image was continuously sliding and its market share continued to dip including in the western belt of the country that traditionally had been its stronghold. Analysts commented adversely on the bank. They tore the bank apart because of its sluggish pace of credit growth, low retail growth, low technology, and heavy reliance on treasury profits.

In an article dated April 7, 2005, Tamal Bandyopadhyay, Banking Editor, Business Standard, while analyzing the performance of public sector banks, observed:

> Bank of Baroda has virtually stopped growing.... It has clocked the lowest growth (35.4%) in assets in 4 years and its market share has fallen from 5.45 to 4.31 percent.... In 2003–04, its advances were flat (grew by 0.71%).... Bank of Baroda is one of the 4 banks that has shown single digit deposit growth over the four year span. (Bandyopadhyay, 2005)

Other analysts, who had been keenly following the bank since its initial public offer of shares in 1996, recommended "SELL" for the bank's shares. They criticized the bank's claim that it was making good profits as most of the profits during that time were not from core banking operations but from treasury operations, which in any case were not sustainable. I was quite concerned about banking analysts' observations labeling the bank as an underperformer.

The overall climate among the employees of the bank was one of low motivation and apathy. Fear psychosis prevailed among managers which hindered decision-making in credit area. I had a feeling that the general rank and file had a picture of the bank as an eagle sitting on a high and comfortable perch from where it was impossible to be dislodged. They had the least idea that we were more an elephant than an eagle. "The elephant not only had to be moved but also made to dance."

I began to realize that I had a formidable task on hand and would have to fire on many cylinders right from day one.

Challenges

The challenge before me was not only to restore the credibility of the bank with various stakeholders in the short term by re-establishing business but, in fact, to take such foundational steps that would sustain the bank in the long term and create resilience against any setback. As an insider, I was aware not only of the problems but the causes also. As ED of the bank between 2000 and 2003, I witnessed the decline and felt strong urge and need to stem the rot. The main problem was: the existing ways of doing things which lacked alacrity in handling customer issues, tolerance with mediocrity at various levels, obsession with running day to day "operations" so much so that important problems were ignored and adhocism ruled. The other issue was related to Industrial Relations policy that pampered trade unions who ruled the roost and were generally apathetic to any new change in system and policies. They always insisted that nothing should be done without discussions with them and were able to stall many long-term changes.

In the new scenario, it became difficult to follow this. Unlike past when unions successfully stymied implementation of restructuring or recommendations of consultants' reports proposing change in working, we did not want technology to become a matter of collective bargaining although we were prepared to discuss and brief them about the nature of technology and how it will impact our operations. The trade unions, however, scandalized the entire issue of bank engaging an internationally eminent firm Gartner as consultant to drive technology-driven Business Transformation project. This led to major IR problems. As ED and member of the Board, I had to take personal initiatives to redefine relationship with trade unions. During this time, I was at the center of major criticism including hate campaign unleashed by the unions. I was seen as an architect of new approach in IR and it was not wrong. As an HR specialist, I was helping the bank carve out a new framework of HR management which was perceived as management consolidating its hold on the work place.

We stood firm and initiated many steps to restructure IR and rid the bank and the work places from the daily interference of unions, especially in policy matters. In this process, the technology implementation was delayed by three years (2001–2004). One of the biggest achievements, however, was the regain of control in managing the work place. This was seen by operating managers as a great driving force to manage the branches without daily hassles and regain business.

Although the bank had done some initial work like identification of Hewlett Packard (HP) as a system integrator, the contract could not be signed because the outgoing CEO apparently wanted to remain out of any controversy as the contracted amount was quite high. After a brief stint in Dena Bank (February 2004 to February 2005) when I returned my main priority was to kick start technology by signing contract with HP. I was aware of the frustration of everyone around in losing about three years in CBS implementation due to IR problems.

In the new competitive environment, we wanted to quickly restore the image of the bank as an innovative and technologically driven bank. We wanted BOB to be a front-running India's international bank that is progressive, technologically modern, and providing absolutely hassle-free services to customers. We also wanted the bank to be proactive and innovative in products offerings and service levels. I believed that this was my greater priority than to look at the next quarter business. I had to think beyond my tenure. I wanted to create strong pillars of what I call intangibles such as technology, tackling internal bureaucracy, building new capabilities in people, creating leadership at operational and strategic levels, rebranding of the bank, and creating speed and execution as our competitive advantage. In some sense, I put business as a second priority in relation to my priority to rejuvenate the drivers of business.

The key priorities for me were:

1. Quick decision to start core banking project
2. Quick acceleration of credit portfolio

3. Turbo charging retail portfolio
4. Building the long-term health of the bank, going beyond my tenure
5. Building a new culture of service and innovations
6. Rehashing HR function to make it facilitative to seek employee engagement and build new capabilities
7. Building leadership for competitive times ahead

As mentioned earlier, the immediate challenge was, however, to kick start the technology and restore the morale of general employees, which had touched the lowest ebb due to delayed implementation of technology and consequent jibes of customers directed at the front line employees.

After taking over as the chairman of the bank in March 2005, we signed the contract with HP for implementation of core banking in next four weeks, thus initiating a breakthrough step in the life of the bank. This cheered everyone and I believe trade unions also whom I had invited in the contract-signing function. They congratulated me for signing the contract. They appeared equally committed to implementation of technology now but had taken a different line earlier. This quick step was the turning point in giving hope to customers, as well as employees who appeared desperate to remove logjam on account of technology.

Reaching out to employees

Soon after my appointment, I addressed the corporate staff (about 500) and in a passionate pitch shared with them the problems and called upon them to join my endeavor to create a "Happening Bank"—a bank that is responsive, proactive, innovative, and determined to reclaim the lost customers sooner than later. I called upon them to shed bureaucratic ways of dealing with field problems. Thereafter, I wrote to all 40,000 staff about the problems of the bank and sought their engagement in resurrecting the bank to its past glory. I also undertook a tour of several large centers where I addressed many town hall meetings sharing with the employees about the real problems such as slide in business, reports of deteriorating service levels, and exit of customers. I also assured them that we can restore the lost customers and the lost opportunities in business, if we collectively and passionately worked together. I also shared our transformational plan and urged upon them to work an extra-mile to achieve this. I also assured them of our openness in listening to field problems that hindered business growth and promised to solve them without loss of time.

I was very encouraged by the positive response of the staff across the cadres. I was very clear in my mind that the contemplated changes could not be achieved only by the management staff and everyone needed to understand the context of changes and commit to perform and deliver at a higher pace than ever. Speed was the new competitive advantage: speed to innovate, speed to respond, and speed to deliver. This was my main theme in my interactions with the employees.

I also designated each corporate general manager as a guardian of a particular zone to enable them to mentor and guide the process of business growth in the assigned territory, coordinate with other functional heads in regard to problems of the territory, and be spokesperson of that territory at the corporate office. This was with a view to quickly responding to field issues and providing expert guidance.

Redefining roles

Large organizations such as banks have tendency to get the CEO sucked into routine. I had the daily challenge to deal with the mounds of papers that continuously flowed from various functional departments. Occasionally, I wondered whether I was the CEO of a "paper factory"!! The transformational agenda needed my fulltime engagement and it was necessary for me to reorganize my work and in some sense to redefine my role.

Although Board lays down the financial and sanctioning powers of CMD, EDs, and general managers (GMs) and I did not fiddle with that, but there are whole lot of issues on day-to-day basis which come for discussion and problem solving at the level of CEO. Sometimes, we all end up doing more of the same thing. I redefined the top roles in a manner that most operational decisions could be taken up by EDs and GMs. This would include most visits and meeting the customers at their level except top corporate customers, whom I met personally. Each functional head was empowered to take decisions within the broad policy framework. In any case, I did not have any personal need to micromanage day-to-day operations.

I defined my role mainly to focus on building "intangibles" such as employee processes, customer processes, technology, business transformation program, building leadership, and also creative destruction of processes that inhibited innovations and creativity. I also focused on creating internal coordination as well as also ensuring extraordinary speed in implementation of various projects. To be able to focus on these roles, I created additional time for myself by nonparticipation in routine meetings that could be competently handled by my EDs. My absence also gave them freedom to exercise their own discretion and experience to deal with issues. Similarly, I would spend minimum time with issues of accounts, auditing, and inspection, which my operational colleagues handled brilliantly.

In order to focus on customer and employee issues, I created two new cells in my secretariat namely customer service cell and employee care cell, which exclusively dealt with two hotlines for customers and employees, which I had set up. Feedback from these put me in touch with real issues and install quick system to resolve problems which in turn helped restore our credibility to a great extent.

Building a new culture of performance and responsiveness

Banks are large geographically dispersed organizations. They are managed through defined hierarchies, roles, delegated powers at different levels, which is part of the system's risk management and conservatism, but this also becomes a major reason for working in silos, emergence of excessive internal bureaucracy, hassles, delays that cumulatively dilute expected level of service and consequently dissatisfied customers. Our bank was no exception.

One of my key challenges for implementing our vision and pursuing the transformation agenda was to improve the prevailing internal environment characterized by constraints, red-tapism, internal bureaucracy, interdepartmental conflicts, and control-driven communication. As such, the environment was not conducive to growth in competitive times.

In order to implement an ambitious growth and vision-driven agenda, we needed an environment that facilitated and promoted new ways of working in response to competitive pressures and designing a new future for the bank. As Einstein said, "We cannot solve problems by using the same kind of thinking we used when we created them."

Everyone in the management, especially at the top, had to learn new habit of collaboration and grow from controllers to leaders. Collaboration was not simply desirable, it was inevitable. In the competitive reality where speed dominates the business outcome, our traditional bureaucratic ways of working and communication style could be the biggest handicaps in our transformation program. We needed to change faster than ever before.

My earlier experience of the functioning of the corporate office while being a zonal manager weighed heavily on me. There was lack of coordination among many corporate office functionaries, who for most of the time seemed to be working in silos and engaged in their own little "turf wars." Many critical matters did not move at the desired pace and there seemed to be no accountability for delay or failure in facilitating business performance of the zones and regions. It was like pushing a giant wheel.

My objective was to make the functioning of the corporate office more business friendly and collaborative in dealing with executives at the operating level. Corporate office had to align with the business priorities of the field. This was necessary to drive better business results. What was needed most in the given context, however, was the commitment of the entire top management to pursue the change agenda with grit and determination. They needed to see the larger picture. They needed to pull together as a team, reinvent themselves, and work in coordination with each other to deliver results.

Another challenge lay in the promotion of a responsive and collaborative culture where operating managers would feel sufficiently confident to discuss business matters more freely, with corporate office functionaries. There was certainly no time for interdepartmental conflicts, managerial politics, or procrastination on decisions. In short, the corporate team had to lead from the front and function as change masters.

We also wanted our customers to have new experience and feel the new management with new ethos. For this, we needed to act collectively as a team, challenge the old ways of working, change the status quo mindset, and set in motion processes to significantly improve our response to them. This had to happen across the bank—at zonal level, regional level, branch level, and frontline counters. How could we rid the corporate office of the culture of red tapism that had engulfed its working? How could we build accountability for performance and execution?

I knew we could not achieve these objectives by an internal circular or periodical prodding with management team. On daily basis, we needed to review, reflect, learn from live events, and build a culture of diagnosis, ownership, and innovative solutions to problems to replace a culture of fault finding and post-mortem. In a dynamic setting of transformation, we needed to hone our judgment about decisions and apply collectively to move ahead. How do we do this? Would the conventional way of resolving these issues through "meetings" serve its purpose?

In my career, I had attended scores of meetings convened by my superiors. Some meetings were informative, some others were entertaining, and many were, to use a soft term, utterly boring and served no purpose. In other words, I had become cynical about meetings.

I often wondered if there was a way to make meetings purposeful. I looked for ways and means to make meetings interactive, lively, and purposeful, the meeting where everyone would have a stake.

Morning meetings: Our school of learning and a crucible for change

With these thoughts in mind, I decided to introduce the system of "morning meetings" as a vehicle of constructively engaging the top management team and setting the tone for vigorously pursuing the mammoth transformation agenda we had on our plate.

These meetings were organized with a strict discipline of starting sharp at 9.30 am and closing exactly at 10.30 am. They were designed essentially as agenda-less meetings. The idea was to defreeze the group and encourage everyone to raise issues in a spontaneous manner. Membership was confined to GMs (after some time deputy general managers were also invited), advisors, and my executive secretary, besides CMD and EDs. These meetings were serious events, where attendance and participation was mandatory. Except on the days I was out of headquarters on work, I always attended the meeting at the appointed time. In my absence, the meetings were conducted by the EDs.

In these meetings, we devoted more time in discussing the processes, systems, our communication styles, and customer grievances. I normally discussed issues that were referred to me during my field visits. One of my key concerns was to ensure superior level of guidance from the corporate office to the field functionaries. We needed to usher in a new culture of timely response, build mutual trust between corporate and field functionaries, and a hands-on approach to problem solving. In many cases, our discussions led to reorganization of work in the department, building new competencies in the executives, relocations of executives and officers and number of layers required, staffing pattern in relation to business forecasting, talent gaps.

We developed an ambitious 100 days agenda through consensus to achieving quick wins, and demonstrating a collective endeavor to make positive changes with speed. Many initiatives during this period were debated and discussed in these meetings and the success achieved was owned by each one of the participants.

Visioning process

The next step after dealing with the immediate priorities in the first few months was to think collectively about the long-term plans. We had to prepare the bank for the future. I had twin challenges ahead of me. The first was to resurrect the traditional business lines such as improving market share in deposits and advances, accelerating credit growth, and improving customer services. The second was to architect the bank as a financial supermarket offering all kinds of banking and financial services under one umbrella. To meet these twin challenges, it was imperative to embark on process improvements and innovations.

We asked many questions to ourselves. How different will Indian banking be in the future? What was the nature of changes happening in the banking and customer universe? Which of these changes will have a significant impact on BOB? What measures would the bank need to take to prepare it for these changes?

Therefore, we initiated discussions in the morning meetings around the theme of future banking, competitors' ways of working, their marketing strategies, bottlenecks in our way of working, and a whole lot of things that needed change in our internal processes, as well as new competencies and capabilities in our people that were required to take us into the new orbit of banking. Above all, we discussed the issues of culture, internal governance, and branding.

Morning meetings became the forum for setting in motion, the spade work needed for the vision exercise (Vision 2005–2010). Initially, functional heads developed a vision for their function by involving their key managers and thereafter, each GM presented his vision paper in the morning meeting and after clinical analysis, the paper was given final shape, ready for presentation to the Board. In this discussion process, each member had the opportunity to contribute and participate in finalizing vision paper for each key function. This positive alignment helped in raising the bar of discussions at a qualitative level and enabled everyone to see the larger, holistic picture. After series of in-house discussions and drafts, crack teams were commissioned to fine-tune and align the individual function-wise vision with the overall larger vision of the bank. Through this process, the ownership of vision was ensured. The draft vision document became the background note for the Board to discuss, deliberate, and arrive at the Vision 2010. The Board set out:

- To retain the leadership spot in the PSB landscape by 2009–2010
- To double the global business size in next three years (2005–2008)
- To double retail assets
- To acquire at least 2 million customers every year
- To transform the top 500 branches into best of the breed "sales and service" through improved ambience, processes, people, and technology

Collective problem solving

Morning meetings helped us in finding creative solutions to problems during implementation in many areas but more particularly in the matter of implementation of core banking solution (anytime, anywhere banking). None in the top management realized or anticipated some basic customer service problems when we switched to core banking network. In the initial period of migration to new systems, a large number of problems started pouring from both branches and customers. Collectively putting our heads together, practically on daily basis, we solved the issue of migration to new system. Besides this, we also discussed a number of key issues referred to by field managers by using everyone's wisdom to find a solution to specific problems. I often brought emails of customers to these meetings and discussed the issues raised by them. The whole idea was that every person in the group exactly knew about the problems and issues in the context of our processes; and pulled together to find solutions. It was never seen problem of a functional GM but that of the Bank. This helped break the silos and created alignment across functions. It also created an opportunity for daily learning through real-life problems.

Promoting strategic thinking

Another important priority in the morning meetings was to promote strategic thinking. We laid great emphasis on diagnosis as the starting point for problem-solving, as lack of diagnosis or wrong diagnosis itself can become a major problem.

Therefore, in our discussions on any live problem, our emphasis was always on some of the following diagnostic questions:

- Did we analyze the root cause of the problem?
- Can we look at the problem from the lens of the customer?
- Did we provide a bandaged solution or did we take permanent corrective steps to avoid the recurrence of problems again?
- Did we show a knee-jerk reaction to a situation or were there alternate ways of handling the problem?

- Did we hand-hold with the operating managers in solving the problem?
- How are we giving confidence and support to line managers?
- Is our communication to customers defensive or authentic?
- Is our communication with operating mangers sermonizing or supportive?
- Do we have enough capabilities to deal with an issue? Do we have no. 2 and no. 3 ready to chip in when needed?
- Why does a particular type of lapse continue to take place again and again? What could be permanent solutions?
- What are we doing to entice young customers and what steps are being taken to integrate young employees in the bank?

Apart from business issues, we often discussed emerging issues in leadership, developments in global economy, etc. I would often circulate good articles that I read. The subsequent sparks of creativity during discussions were evident in the creative ideas that always helped in implementation of a new initiative. The collective creativity of our colleagues in the morning meetings helped us to come out with many new terminologies for our product offerings and innovations such as retail loan factory, small and medium-sized enterprise (SME) loan factory, gen-next branch, and happy hour banking.

A forum of business review

Morning meetings also became the platforms for us to periodically review our progress. As we started moving in improving our processes and systems and reaching a level of collaboration among various functional heads, we regularly reviewed various parameters in business such as deposits, advances, low-cost deposit growth, recovery, retail growth, SME growth, and growth in agriculture credit. Reviews were also done on the pattern of growth in rural, semi-urban, urban, and metro areas. Merits and demerits of our products were critically examined in comparison to competitors' products and suitable changes and modifications needed were discussed. Alongside, process issues related to business always occupied our priority.

These sessions were very intense and I benefited a lot from the arguments and counter arguments in the group. There was an increasing openness in the group members. An occasional tinge of humor livened up the atmosphere. Overall, these meetings became an important forum where anyone could table a business issue, a process problem, a customer complaint, or an HR issue.

A forum for execution

During 2005–2008, we had launched a number of projects that included the implementation of Vision 2010. This called for doubling the business within three years, the implementation of CBS, expansion of the ATM network, the launch of new product lines, the rebranding of the Bank, special events for centenary year celebrations, the expansion of international operations, and the restructuring of our Indian subsidiaries.

These projects required impeccable treatment and time-bound implementation. It also required a high degree of coordination among various functionaries across the Bank. We had a mammoth agenda to be pursued and the time was moving fast. While the technological initiatives were proceeding at breakneck speed, we had to kick start new business strategies, design new business models, and put them on fast-track implementation. We had to come up with strategies to double the business size and stand out in the crowd. In the morning meetings, we collectively discussed the implementation details about each project identifying the owners of the project, the support required, and implementation schedule. We also needed to pursue new revenue streams. We therefore decided to seek expert advice of external consultants, who would have the market knowledge and would help us in designing new business models and also in internal restructuring wherever necessary. We appointed McKinsey and Co. as our consultants. The transformation project was code-named Project "Parivartan," which soon became popular among employees. In morning meetings, we regularly reviewed the consultants work and hand held them in the implementation process.

Although it took sometime but, over a period of time, our morning meetings became a forum where our functional heads ventured to come out from their functional silos to participate in a larger dream to create a bank for the next century. As they say, "None of us is as smart as all of us" and this could not be more relevant than at the time of transformation because problems during transformation are too complex to be solved by one person.

Our morning meetings also created an environment of esprit de corps in the ranks of management and helped stitch many loose ends in our policies. It invigorated and energized the top management and, to a considerable degree, de-bureaucratized the working of the bank. It enforced the discipline of execution and created accountability. The morning meetings played a key role in implementing all these projects within stipulated deadlines and in a flawless manner. No project got delayed from its specified target date.

But what is most important is what one executive confided in me:

> "Morning meetings were a school of management where I learnt my lessons in management and leadership."

Our morning meetings functioned as a learning organization where one could clearly experience collective passion for a new future for the bank, where new and expansive pattern of thinking was nurtured, where collective aspiration was set free, and where we continuously learnt how to learn together.

Personally for me, morning meetings were sessions of exploration, discovery, coaching, guiding, and persuading everyone to raise his aspirations. I truly feel that much of what we could achieve in the three years of my tenure can be attributed to our engaged sessions in the morning meetings. I personally benefited a lot as these meetings provided me deep insights into many operational and organizational issues. Truly speaking, morning meetings were my "sapiental circles" (the term was first used by leading American anthropologist Margaret Mead to mean knowledge-generating groups). Morning meeting became a major instrumentality of culture change and driver of transformation.

Action and reviews of key decisions taken in morning meetings

1. First 100 days agenda

    a. Expanding ATM Network—500 ATMs including launch of 200 ATMs on a single day
    b. Launch of 550 8 am to 8 pm branches and 24-hour human banking
    c. Rebranding of the bank—logo change, appointment of Rahul Dravid as brand ambassador

2. Finalization of Vision Document (2005–2010)
3. Core banking (anywhere anytime) Project
4. Launching of Centenary Year celebrations
5. Launch of Retail Loan factories and SME Loan factories
6. Joint ventures with foreign collaborations in Life Insurance and Mutual Funds
7. Revamping International operations strategy
8. Revamping Gujarat operations
9. Project Parivartan—Business Transformation Program
10. Various HRD initiatives

Building human capital

One of our key beliefs was that we need to build a new ecosystem of human resources, to meet the challenge of competition and create a unique brand positioning for the bank. We wanted our frontline staff to be smart, helping, engaged, and service oriented. They being face of the bank across the thousands of our branches were our brand ambassadors. They are also the ones who would present to the customers and the larger world, our new intent to serve them and be proactive in solving their problems. To achieve this, we wanted our staff to be sales and service oriented. We also wanted our staff to have the confidence and trust in the management that we respect them as individuals and their insights into customers and their problems. We also wanted to be a "caring management."

We also wanted our managerial staff to be re-skilled in managing people and managing relationships with customers. At senior level, we endeavored to create leadership that can deliver in competitive times and a team of rapid fire, business enthusiastic, and go-getters. We also wanted our systems and processes in HR to be hassle free and service oriented. In short, we wanted to completely rehash our human resources to deliver our new vision and to reposition the bank as a vibrant entity.

Some of the key decisions taken in human resources

1. Setting up the HR steering committee of the board and ensuring its regular meetings
2. Developing a written policy document on HRD
3. Restructuring of HR function to respond to developmental functions in a significant way
4. Creating an employee engagement cell in the CEO secretariat under direct supervision of CEO
5. Demystifying HR administration through technology
6. Initiatives for developing in-house excellence in training systems

    a. Setting up a special college for training staff in the new technologies.
    b. Grooming 500 credit officers, 100 international bankers, and 100 treasury officers
    c. A massive training effort for front-liners in "sales and service training"

7. Revamping HR for overseas operations
8. Introduction of a new appraisal system for 13,000 officers
9. Introduction of a new appraisal system for workmen staff
10. Implementation of personnel audit across the branches and offices
11. Performance improvement plan for slow pacers
12. Setting stage for leadership development—hiring consultant for grooming 300 leaders for the future
13. Talent identification and development—launch of "Khoj"
14. Employee engagement program

    a. Sampark: Helpline for employees (to reach CEO directly in cases of grave emergencies)
    b. Paramarsh: Employee counseling centers
    c. Baroda Manthan: Employee conclaves (large group events)

15. Developing a pipeline of young managers
16. Re-skilling HR officers in process-work and orientation to developmental agenda
17. Executive retreats in leading management institute for learning and growth

With many initiatives in business and a major strategy to change front line into sales and service staff, the employee response was very encouraging so much so that they took out rallies on their scooters and motor cycles to kick start many new initiatives. They accepted many changes voluntarily and their passion to work with new technology was phenomenal.

During my three years' tenure, I virtually worked in twin roles as HRD chief and CEO and enjoyed connecting HRD with business priorities. HRD became a key driver of our business. Creative communication and innovative initiatives in engaging people made the HRD–business connect come alive. I had the satisfaction of initiating some major shifts in our HR policies, creating a new culture of performance and service, across the bank and recognizing employees as the main instruments of change. The institutionalization of all these initiatives always remains a big challenge by the successor management. I believe that our endeavors in HRD turned out to be true value drivers to put the bank on a fast-track journey of growth and excellence.

Business outcomes

Despite the fact that we had opened many fronts such as technology, rebranding, restructuring of both domestic and international operations, and new outfits for retail, the business growth did not falter and in fact showed tremendous growth even in the short term which became a trend in later years. The highlight of the business between April 2005 and March 2008 included:

- Total business of the bank doubled from ₹1247.34 billion to ₹2587.35 billion (107%)
- Credit growth, a key concern of the analysts, grew from ₹434 billion to 1067 billion (145%)
- Gross non-performing assets reduced from 7.30 percent to 1.84 percent
- Net profit grew from 6.76 billion to14.35 billion (112%)
- Business per employee increased from ₹31.6 million to ₹70.4 million
- Net profit per employee more than doubled from ₹17.1 million to ₹39 million
- International operations of the bank also recorded more than 100 percent growth and a record number of 15 new offices including offices in new territories such as Australia, Trinidad and Tobago, Ghana, and Singapore were opened
- Over 1,700 branches were put on line in a record time
- All targets under the Statement of Intent signed with the Government of India were achieved
- Number of new customers added is 8 millions
- Ratings from major equity analysts changed to "outperformers" with "buy" recommendations

Our focus on building intangibles for a stable and long-term growth of the bank delivered excellent business results even beyond my own expectations. It was a conscious decision that we were willing to sacrifice near term growth for our investment in building the basic pillars of growth such as technology, rebranding, customer-centric innovations as we believed that our firing on all cylinders may not yield business results in near term. We were proved pessimistic and, in fact, the business started growing from the very first year itself. We believe, for good reasons, that when an organization is set on transformational fire, everyone rises above their normal performance levels and this collectively shows up in extra-ordinary performance.

Transforming a large-sized organization: Some lessons

1. CEO as a central figure

For the successful conceptualization and execution of any transformation, the CEO is the central figure who pilots and gives color and shape to the entire process. His motivation, energy, personal belief system, and capacity to take risks determine the success of the entire exercise. His own adaptability to change and the willingness to bring about change in his style of work plus allocation of time and energy are all very critical to the transformation process. Transformation in large, legacy-driven organizations is like moving a big flywheel. It requires intense stamina and a collective endeavor. It calls for a humongous effort on the part of the CEO to mobilize, motivate, and galvanize the management team as well as the foot soldiers. The CEO has to sensitize everyone from his senior management team to his front line about the "cost" of not undertaking the transformation. He has to raise the vision of every one, provide the necessary confidence to entire staff to stretch and work the extra miles in their own interest.

In the BOB case the CEO redefined his priorities and consciously spent about 70 percent of time on issues directly concerning transformation, such as building employee and customer centricity, technology implementation, rebranding, HR repositioning, leadership development, reorganization of domestic and international operations, and working with consultants on business transformation project. He traveled extensively, covering the length and breadth of the country so that he could directly talk to different groups of people, listen to them, and appeal to their conscience. He shared through power point presentations the decline that had set across all parameters of the business. He firmly believed that telling the truth to employees is important to seek their involvement.

2. Development of a compelling vision: A key starting point

The transformation agenda for an organization is always triggered by its contextual setting. This needs a clear articulation of vision. The vision envisaged has to provide a clear sense of direction leading to the long-term sustainability of the organization. Development of a compelling vision has to bear in mind the emerging environment of the business, changing customer profile, the new competencies to be developed and capabilities in employees, and the evolving technology concerns. Setting of vision has to be a rigorous exercise at the management and the Board level, undertaken after considering all the strengths and weaknesses of the organization. The vision should reflect an ambitious and doable agenda for change.

The CEO initiated a process to involve the entire top team to discuss the future vision of the bank in an open and engaged manner through several morning meetings. Each functional head was encouraged to prepare vision for his function in consultation with his team. All presentations were critically analyzed and a final document was prepared for the board. CEO also persuaded board to spend two full days to discuss the proposals from each function who after intense discussions finalized the vision document. One of the key directions provided by the board was to double the business in the next three years. Udai Pareek was invited to be present in the board meeting as a process consultant to give his observations during the discussion and at the time of summing up.

3. Building the top team, aligning them to the vision, and developing accountability

Any transformation exercise cannot be a solo act by the CEO. It requires collective genius of the organization to own the vision, develop a new mindset to move out of its comfort zone, and challenge the existing ways of doing things. It also needs to recast the structure and processes that can free the organization from excessive impact of its internal bureaucracy, dysfunctional impact of hierarchies, and related silo working. Collaborative problem solving, building new perspectives, and daily learning from issues and problems has to be the hallmark of the new way of working of the top management.

Our daily morning meetings were our daily learning nursery in leadership.

4. Culture change

Transformation is not a one-shot affair or some program that can be led through a set of instructions, sermons, or exhortations. It essentially aims at changing the DNA of the organization. Large organizations carry the burden of legacy and are hamstrung into the obfuscating culture. The transformation must aim at creative destruction of such a culture and sow the seeds of a new thought process, challenging the status quoist order. To my mind, the objective of transformation is far beyond the immediate business goals. Its main objective is to take all such steps that can help the organization to revitalize itself and to perform to its potential. If these foundational steps that go to the very root of creating revitalization are ignored in the name of operational exigencies, the organization can become fundamentally weak to fight the competition. Thus, taking such steps, to my mind, is the most strategic role of the man at the top.

Our focus, at the bank, during transformation was to challenge the existing "operational culture" at every level that was not conducive to innovations. In this culture, the problems were always postponed, lower service levels accepted, and critical issues suppressed, all in the name of operational exigencies. Our focus on changing the prevailing culture was to take on some issues head-on more so the delay—causing internal culture, non-responsiveness, and staff attitudes. We believed that operations are the key to business, but we must clearly separate the operational roles from the strategic roles. We cannot have a system where everyone seems to be doing more of the same thing.

Considering our focus on building a facilitative culture and a high performing environment across the bank, we underplayed the near-term financial performance and focused on the essential issues, the drivers of business. We solved many outstanding problems that acted as a barrier to business growth and customer service, such as delays in decision-making, critical staffing issues, and availability of trained manpower in critical branches. We owned up the problems and developed accountability within the management to solve those problems. We micromanaged excellence and eventually developed a culture that promoted goodwill with customers as well as with the employees. We also promoted a culture of response and learning across the organization, through various initiatives. Many of our initiatives helped us to create a culture of innovations and recognition of merit. It was due to investment in creating new culture that bank achieved extraordinary results. A key learning is that culture can be changed fast, provided bottlenecks are tackled, accountability is fixed, people capabilities are renewed, and processes and systems get constantly reviewed and challenged internally.

5. Relentless focus on execution

In large organizations, there are many bottlenecks that can frustrate the process of execution. Owning up responsibility and giving speed and momentum to the process of implementation is a key challenge. We believed that speed is a new competitive advantage and we wanted to demonstrate that we can deliver fast and surprise our stakeholders. Execution requires identifying champions for specific projects, competent team members, and making available all possible support to achieve the project goals.

The CEO developed an ambitious, 100-days agenda for the bank in consultation with the top team and the same was successfully implemented. This gave the team a sense of confidence and set the stage for many changes at a much higher speed. Our setting up of 500 ATMs, 550 8 am to 8 pm branches, and rebranding of the bank, all in the first 100 days, could be achieved by removing bottlenecks, raising the motivation level of the team, periodically reviewing the progress at the highest level, and engaging the entire top management and concerned employee in owning the project. We also rewarded "Execution Czars' through incentives like overseas training.

6. Gaining and working on field wisdom

In large organizations, the corporate and the field level are often separated by the web of hierarchy, rules, procedures, and lack of accountability for timely response. It is the field managers and other front line employees who directly interact with customers on a day-to-day basis and possess a big repository of customer insights and competitors' strategies. These insights need to be captured in time and worked upon to bring about necessary policy changes. This requires creating an open and unrestricted system of

communication between the field and the top management. Policies, based on contributions of the field staff, need to be highlighted and show-cased.

During transformation in the bank, we developed strong field orientation. The CEO and the top team extensively traveled and collected many insights from the employees and customers. Many problems were identified that obstructed effective delivery of services. We worked upon many insights gathered by us. In many cases, we changed product features in our retail loan products on receiving insightful suggestions and feedback from our front line staff. We also created a special email ID "Ideas@bankofbaroda.com" to encourage ideas and insights from employees. Many such ideas helped the bank to bring about changes in our products, system, and processes.

7. Engaging employees

For transformation and implementation of vision, engagement of employees becomes the cornerstone of success. It is they who deliver the services, who make customers experience their moments of truth, who bear the brunt of irate customers, and also who suffer the pressures of work. They need to understand the logic and rationale of the change, the cost of not undertaking such a change, and the benefits of change on their careers and quality of life. They also need to feel confident about management intent and compassionate attitude. Finally, they would want to be sure about a fair system in place, in dealing with their career and personal issues. Thus, engagement of employees involves engaging them through their minds and hearts.

In the transformation journey of the bank, employee engagement was our core value. We posted a bright HR manager in CEO secretariat to focus on employee engagement programs such as Sampark (direct hotline to chairman), Paramarsh (professional counseling services), and employee conclaves (large group events). It was because of engaged employees that the bank achieved dramatic business results and the perception of analysts about the bank changed from "A bad apple to a valuable brand."

8. Building strong communication architecture

Communication with the stakeholders, and more particularly with employees and customers, is the lifeblood of a pulsating organization. Most problems in organizations are experienced on account of deficit communication, which can lead to indifference among employees. Building a communication culture with employees, customers, as well as other stakeholders accelerates the pace of transformation. In the transformation of a large organization, it is only through open communication channels that an organization can create alignment of its vision across all its hierarchies.

During transformation in the bank, the CEO organized and attended dozens of town hall meetings to excite the staff about the new vision and to listen to their concerns. The CEO benefited a lot from the insights received

during such interactions. Besides this, monthly letters, individually addressed to 40,000 staff explaining the context of initiatives, were sent. Employee conclaves (large group events) were organized, the house journal was made more employee-focused, and a new communication culture was created, which aimed at responding to every single communication from employees and customers.

9. Building human capital

No transformation can be sustained without a constant focus on developing human capital. In the changing environment, continuous innovations in designing new products, service delivery, and also using technology smartly are necessary. Thus, emphasis on building human capital cannot remain merely a matter of good intent. It must find its expression in developing a concrete agenda in creating human processes that facilitate development of new capabilities, new perspectives, and a new mode of thinking. This requires relentless efforts by the CEO and the HR team, an investment of time and money in designing careers, having reliable systems of performance management, and creating opportunities for personal growth.

In the bank, we kept a rather obsessive focus on developing new capabilities in the emerging areas such as technology, corporate credit, treasury operations, and international banking. Utmost attention was given to developing excellence in our in-house training establishments and exposing thousands of our staff to training in new technology and the software which the bank had adopted for providing anytime anywhere banking. The bank also trained its front line people in building a "sales and service" culture. The bank also commenced on a project for grooming 300 leaders for the future.

10. Tough love

Finally, the transformation involves a certain degree of toughness and determination in dealing with the evolving problems. Past legacies sometimes create tight comfort zones, which tolerates mediocrity and average and incremental growth. Factors inhibiting performance and customer services need to be identified and tackled. Urgent attention has to be paid to create a hassle-free working environment. Transformation should challenge factors inhibiting performance, through a combination of persuasion and tough actions in appropriate cases. A new environment of performance has to be quickly put into motion. Moving toward a performance and productivity culture can never be without problems and a tender-minded approach can be counterproductive, more so when the tenure of the CEO is short and agenda for change is large. However, an environment of performance and a rapid pace of transformation can only be sustained when there exists a culture of compassion in dealing with genuine employee problems.

During our transformation program, we initiated a number of programs to deal with the emergent problems faced by the employees. As mentioned above,

two such programs, Sampark and Paramarsh, helped create confidence among employees about the bank's priority in dealing with the life and death issues faced by the employees with speed and compassion. Our tough love policy gained support from majority of the employees and the recognized workmen union.

Conclusion

This chapter succinctly details how organizational transformation has been brought about in a century-old public sector bank in India through use of HRD values, process-oriented interventions blended with business direction, and by creating and promoting a facilitative environment for turbo-charging the change process. Mobilizing people's passion by dynamic engagement process, creating an architecture of intangibles and a culture of responsiveness, and re-energizing everyone for a new future of the bank have been the hall mark of the leadership effort. This has truly been an organizational transformation covering not only business growth but also a complete business overhaul in terms of business direction, change initiatives, and the intensity of change. A downswing has been converted into an upswing. It is a classic case of an organization where human resources have been successfully harnessed and HRD has contributed significantly to the entire transformation exercise.

The success of transformation, however, in public sector companies depends more on the passion, courage, and integrity of the CEO. Long tenure of the CEO becomes a positive factor only when these factors exist. The risk factors are many in public sector enterprises/banks such as lack of support from government and bureaucracy, role of investigative agencies in commercial decisions, trade union resistance, political interference, and a complaint culture. What looms large in the mind is, there are no rewards for all the efforts put in but there exists a risk that problems can chase someone even after superannuation. So, the general tendency is to stay clear of opening too many fronts (changes) and to focus on day-to-day performance issues. The general adopted style is to keep unions happy (by not taking any steps that can change status quo), keep staff and others happy to ensure that no complaints are sent, to avoid any measures that can disturb comfort zone of management staff and trade unions, and to remain on the right side of the government and bureaucracy. In short, risk-averse attitude is normally prevalent.

Transformational initiatives can make a CEO vulnerable—emotionally, intellectually, and physically. This perhaps explains that barring some extra-ordinary individuals who dare to transform, most settle for a path that avoids any major trouble from any such sources. They work very hard on performance issues but often fail to undertake decisive steps to focus on the long-term health issues of the organization and transform the organization.

In the present case of BOB, the CEO, being an insider, understood the problems that stymied the growth of the bank. The real problem was the prevailing IR situation that often handicapped operating managers to pursue business initiatives.

The CEO was also an IR/HR specialist, who had done his PhD. on the bank's IR. His doctoral study provided him deep insights into the dynamics of IR in the bank. Unlike most of his predecessors, who avoided dirtying their hands in this messy function, the CEO was professionally competent to deal with the problem and dared to do so. The IR problems were tackled head-on and many initiatives led with courage and commitment, extricated the bank from daily IR problems. This gave freedom to the operating managers to manage. Alongside, a huge effort in building employee centricity through engagement and capability building was unleashed. This led to a breakthrough in setting the stage for a new performance culture, which eventually led to successful implementation of technology, many customer-centric innovations, and rapid business growth for the bank.

Transformation can be successful if the main problems blocking performance are identified and resolved. Transformation is not transplantation of the new on to the old. Old ways of doing things, old structures that stifle decision-making , old bureaucracy without accountability, old attitudes that dictate and control from the power of position, all these need to be changed and people need to be shifted from their comfort zones. Transformation requires replacing the old lenses and raising everyone's sight to a higher level to see the world. This requires extreme engagement of the CEO, which calls for a vision beyond business. Quick results in measurable matrix of business alone cannot be goal of transformation. The goals of transformation are: to make an organization healthy to face the tsunami of change and to deliver during both good and not so good times. It should provide stability and competence to grow year after year, better than the best. It also readies the organization to undertake new ventures with determination and grit.

A final personal note

After transformation of BOB, I have had the privilege of addressing many audiences of top managers, company boards, as well as students. A question often asked is: How is it that in spite of many risks in the public sector and no particular mandate of the government to undertake such a multidimensional transformation, why did I venture to undertake such a transformation and invite the ire of many vested interests? What was the trigger for my actions? What if I had messed up and failed?

My spontaneous response has been that my angst was the trigger for transformational initiatives: Angst with status quo; with the way bank surrendered to trade unions and put most operating managers at their mercy; against the internal bureaucracy, against lack of response to customers and employees, and a collective conspiracy for complacency (where everyone failed to acknowledge the problem even when the bank slid to no. 4 position from no. 1 in just four years). This angst was a driving force to push change.

Then, a final question that I asked to myself was: Why am I here as no. 1 among 40,000 employees? Am I here to pass my time and guard myself against all possible odds? It was more a matter of conscience that I was willing to try and fail instead of not trying and later bemoaning the opportunity lost. I wanted to deploy myself completely and mobilize others. Soon, you begin to realize that your own enthusiasm ignites infectious energy in others, it then expands, and you find more and more people join you in the journey. It is a different kind of momentum that charges the foot soldiers as much as the generals.

The transformation appears imminent and its consummation not far away. I am reminded of what Robert Kennedy said:

> Each time a man stands up for an idea, or acts to improve the lot of others, or strikes out against injustice, he sends forth a tiny ripple of hope and crossing each other from a million different centers of energy and daring, those ripples build a current that can sweep down the mightiest wall of oppression and resistance.

Despite all the challenges, occasional personal humiliations, and constant hounding by vested interests during this journey, we kept going. Personally, here was an opportunity to make a difference in the life of others through the organization and therefore, everything else became secondary compared to my own aspiration to seize the moment to step forward without fear of consequences. We eventually survived to see BOB become a valuable brand in just 900 working days. Thankfully, it continues to double business every three years, a trend set by the current transformation. The final learning is: Investment in "intangibles" works!

Appendix

No	Intangibles	Focus	Mechanism
1	**Compelling Vision**	A. Conceptualizing and Building a Vision for the Future	• Draft vision document arrived at in the morning meeting sessions • Board after elaborate discussions, finalized the Vision document • Sharing the Vision across the bank through structured communication • Developing accountability at all levels for execution • Mechanism for periodic review and Board and management level
2	**Employee Centricity**	A. Engagement	• Employee Conclaves, • Monthly letters individually addressed to employees
		B. Caring	• Hotline to CMD (Sampark); Professional Counseling services (Paramarsh)
		C. Listening	• Town hall meetings • idea@bankofbaroda.com
3	**Customer Centricity**	A. Listening to Customer Issues	• Hotline for customer grievances; customer meet at all levels
		B. Speed Retail/SME	• Retail Loan Factory, SME Loan Factory • Time deadlines for sanction of loans
		C. Innovation in Service level	• 8 am to 8 pm Banking, 24 hr Banking, Gen Next branches
4	**Communication Culture**	A. Building Communication Culture across bank	• Town hall Meetings with Employees –“ Manager to Messenger” program across the country
		B. Creating Culture of Responsiveness	• Guaranteed reply to any communication from Employees
		C. Reaching out to Customers and Employees	• Regular communication with customers & all other stakeholders
5	**Leadership Pipeline**	A. To create operational Business and Strategic Leadership	• A project to develop 300 Leaders • Consultants Hired to undertake and implement the project.
6	**Speed**	A. Challenging Timelines for Technology Implementation	• 500 ATMs in first 100 days • First CBS branch in just 8 months after signing the contract

		B. Quick response to all stakeholders	• 100 days challenging agenda accomplished § 550 branches § 500 ATM § Logo changed and new logo revealed § 200 ATMs on a single day
		C. Creative destruction of delay producing process and New Mindset	Continuous review of processes. Reduced layers in Credit sanctions
7	**Leveraging Technology**	A. To set in motion all processes for implementing the Technology project including CBS and commissioning of Data Center	• Interconnectivity of the Branches • Commenced CBS implementation • Operationalized 1700 branches on CBS • Training of more than 15,000 staff • Setting up of a specialized IT Training Center • Upgrading of existing IT superstructure • Promotion of digital literacy across the organization • Global Data Center
		B. Quick expansion of the ATM Network	• Commenced ATM Expansion
8	**Building Human Resources**	A. Capability Building of new environment (Sales/ Marketing/ Digital)	• Creation of a new training outfit to create digital literacy in the front line. • Marketing and Selling Skills for frontline
		B. PMS and talent search internally	• New PMS • Introduction of Performance Appraisal System for workmen staff • Building Performance Culture • Khoj: Talent identification Program
		C. De-bureaucratize HRM	• Introduction of single web enabled HRMS • HRness and Employee Payroll System for global operations

(Continued)

(Continued)

No	Intangibles	Focus	Mechanism
		D. Structure & Systems of HR	• A new organizational structure of HR • Reinvigoration of HR Steering Committee of the Board
9	**Leadership Excellence**	A. Promoting Strategic Thinking, Emotional intelligence and reflective ability	• Daily Morning meetings with strong focus on "Processes" to build depth and breadth of thinking, develop sensitivity to others point of views. • Forum of collective reflection.
		B. Behavioral Competencies & Marketing Skills	• Identification of Leadership Competencies • Comprehensive Leadership Development
		C. Developing Young Leadership	• Identification of different stages of Leadership Development • Implementation
10	**Ethics and Governance**	A. Creating an architecture of ethical functioning	• Quick action in cases of reports of incidents of unethical working, training in preventive vigilance
		B. Zero tolerance for ethical violations	• Introduction of whistle-blower mechanism • Setting up anti Sexual Harassment committee
		C. Getting Board engaged and using their expertise	• Setting up of Board Committee on Human Resources, which also included issues in ethical governance
		D. Authentic reporting of issues and problems and ensuring transparent discussions at Board Level	• Board always informed about achievements as well as challenges and failures.
11	**Branding**	A. Building New Image and repositioning as India's International Bank	• Rebranding of bank involving Logo change, appointment of Youth icon and cricketer Rahul Dravid as Brand ambassador • Ads and TV campaigns especially created to reach out to youth
		B. Projecting bank as a Universal Service provider and multispecialist bank	• Refurbishment of Branches • Creation of Gen Next Branches

12 High-performance work systems: An emerging dimension of strategic human resource management

*D. M. Pestonjee and Naresh N. Mehta**

Introduction

In the 20th century, one of the major challenges faced by practicing human resource development (HRD) managers is to understand business strategy of the organization. As a result, it limits their ability to design and execute an appropriate people strategy and bring alignment with business strategy to create synergies. An appropriately designed strategic human resource management (SHRM) system creates an environment for improving productivity, creativity, and innovation by the people of an organization for developing business models, strategies, processes, technologies, services, and products. The long-term success and sustainability of the organization can be attributed to improved productivity, creativity, and innovation carried out by the people within an organization on a continuous basis. Synergized people and business strategies enhance resilience power of organization in global economic conditions of "VUCA" (volatile, uncertainty, complexity, and ambiguity). While business managers design and execute business strategy to create sustainable competitive advantage for the organization to win the battle of fierce competition, the role of a practicing HRD manager is to design and execute people strategies in alignment with a business strategy to achieve sustainable competitive advantage through people. Usually, business strategy contains the elements of product and service differentiation, unique positioning of organization in market place, and

* The authors would like to gratefully acknowledge conceptual and analytical inputs from Dr S. M. Khan, Psychologist, Indian Railways and Daanish D. Pestonjee, doctoral candidate at University of Arkansas, USA.

identifying target groups for their products and services, etc. Since people have taken central role in the organization, the practicing HRD managers have to design and execute people strategy covering following elements:

1. *The human resources must be of value:* people are a source of competitive advantage when they improve efficiency and effectiveness of organization. Customer value proposition is enhanced when people find ways to reduce cost and provide something unique to customers or combination of both.
2. *The human resources must be rare:* people are a source of competitive advantage when their skills, knowledge, and abilities are not equally available to competitors.
3. *The human resources must be difficult to copy:* people are a source of competitive advantage when people's capabilities, contribution, and teamwork cannot be copied by competitors.
4. *The human resources must be organized:* people are a source of competitive advantage when the talents of people can be combined and deployed to work on new assignments at short notice.

But, the first question raised by practicing HRD managers is: How do we comprehend the SHRM systems, policies, and practices as a part of people strategy?

SHRM systems, policies, and practices as a part of people strategy

Becker and Gerhart (1996) and Schuler (1992) observed that strategic human resource (HR) activities may be conceptualized along several levels of analysis. At the lowest level, HR practices reflect specific organizational actions designed to achieve some specific outcomes. At a higher level of abstraction, HR policies reflect an employee-focused program that influences the choice of HR practices. An HR system operates at an even higher level of analysis and reflects a program of multiple HR policies that are espoused to be internally consistent and reinforced to achieve some overarching results. For example, any business conglomerate comprising various businesses that are in to different stages of their life cycle, like some strategic business units (SBUs) in their incubation phase, stable and matured phase, growth and expansion phase, and/or in their decay phase. In such a wide-ranging situation, the practicing HRD managers have to design and implement comprehensive strategic HR systems at corporate level, which is the highest level in the organization. But at the SBU level, the practicing HRD managers have to design and implement an appropriate HR policy to drive various employee-focused programs that influence the choice of various HR practices. Further, the practicing HRD managers have to design and implement appropriate HR practices at departmental and employee levels to achieve specific outcomes in each SBU (e.g., cost reduction skills of people working in SBU which is in decay phase, merger and acquisition competencies of people working in SBU which is in growth and expansion phase, etc.). The practicing HRD managers have to measure effectiveness of HR systems, policies, and practices periodically to realign people strategy with business strategy on a continuous basis. The authors have standardized and published one such research instrument named "Strategic Talent Management Practices Scale" (Mehta, Pestonjee, & Khan, 2015). Let us take another example. An organization is engaged in designing, manufacturing, and selling stand-alone engineering equipment for a long time, and now the organization changes its business strategy to reposition itself in market place as a total solution provider. This change in business strategy calls for reviewing its people strategy in terms of vision, mission, culture, mindset of people, skills and competencies of people, business systems, and processes. The practicing HRD managers have to realign people strategy (e.g., new skills for solution designing, selling skills for solution providing, project management capability, etc.) of the organization in view of new business strategy. Let us take one more example. An engineering business conglomerate, knowing the market potential and core competency of the organization, now decides to launch a new venture in the space of renewable energies as a part of their inorganic growth strategy by acquisition. The question to the practicing HRD managers is that the cultures of two different organizations need amalgamation or stay as two different cultures. The practicing HRD managers will have to realign people strategy from the viewpoint of multiple cultures of existing and newly acquired organizations that the people will now live their lives under one umbrella (e.g., induction program, sensitize the people of newly acquired company about values systems, culture of existing organization, etc.). One other example worth taking note of is of a business conglomerate operating in the space of B to B (business to business), B to C (business to consumer), and C to C (consumer to consumer) through various SBUs altogether. This business conglomerate would have offerings of engineering products, consumer products, banking, financial and insurance services, health care, hospitality, pharmaceuticals, information technology (IT) and information technology enabled services (ITES), buying and selling on Internet, infrastructure and power, steel and mining, automobile, etc., under just one umbrella. For such business conglomerate, vision, mission, and culture would also be different for each SBU. In addition to this, their corporate strategy, business models, and business strategy would be different for each SBU. In such a wide-ranging business environment, the practicing HRD manager working at corporate level will have to take a bird's eye view to design and execute people strategy in alignment with corporate business strategy at corporate level. Not only would this, but the chief of the HRM will have to design broad-level SHRM systems at corporate level. Followed by this, the practicing HRD managers working at various SBU levels will have to customize HR policies most appropriate to their SBU along with various HR practices at departmental and individual employee levels.

The second question raised by a practicing HRD manager is: Do we have a continuum for SHRM systems, policies, and practices and how their conceptualizations vary across the continuum to achieve desired objectives?

Continuum for SHRM systems, policies, practices, and its conceptualization

SHRM systems are conceptualized as follows. First, HR systems are often implied to span a continuum of two extremes ranging from high performance or commitment-oriented to more control-oriented HR systems (Arthur, 1992, 1994; Delery & Doty, 1996; Guthrie, 2001; Huselid, 1995). Essentially, HR systems are oriented toward high performance through investment in employees or toward a more administrative or controlling approach to manage employees. "The goal of control human resource systems is to reduce direct labor costs, or improve efficiency, by enforcing employee compliance with specified rules and procedures and by basing employee rewards on some measurable output criteria" (p. 672). Guthrie (2001), for instance, created a continuum HR system index with high scores reflecting high involvement, and with low scores reflecting a more control-oriented HR system. However, some researchers have implied that many types of HR systems may not be limited to performance versus a control dichotomy (Lepak & Snell, 1999, 2002; Youndt et al., 1996). The second implication is that HR systems may be designed to achieve various objectives. For instance, some authors conceptualize these systems as comprising HR practices that focus on enhancing employee commitment (Arthur, 1992), and others conceptualize these systems as a focus to use certain HR practices to maximize employee potential and other practices to maximize administrative efficiency (Youndt et al., 1996). Others conceptualize these systems as a function of the degree to which different HR practices are oriented toward maximizing organizational performance (Huselid, 1995). Bowen and Ostroff (2004) noted that the content of HR systems "should be largely driven by the strategic goals and values of the organization," and "to be effective in terms of content, the foci of the HRM practices must be designed around a particular strategic focus, such as service or innovation" (p. 206). Variations and conflicting conceptualizations of the same SHRM practices exist for different systems. For example, Dyer and Reeves (1995) noted that incentive bonuses were a component of the "control" HR system proposed by Arthur (1994), and a part of the "flexible" production scheme proposed by MacDuffie (1995). Becker and Gerhart (1996) cited differences in the use of variable pay by Arthur (1994), Huselid (1995), and MacDuffie (1995). Low emphasis was placed on variable pay as part of a "commitment" HR system in Arthur's (1994) study, and greater emphasis was placed on variable pay as part of HPWS in studies by Huselid (1995) and MacDuffie (1995). Huselid (1995) and Pfeffer (1995) described the use of internal promotions and access to formal grievance procedures as part of high-performance HR systems.

However, Arthur (1994) and Ichniowski, Shaw, and Prennushi (1997) included these as elements of more rigid HR systems. Becker and Huselid (1998) termed these two practices as components of "bureaucratic HR" systems when viewed individually. From a theoretical perspective, a challenge is to identify which SHRM practices should be included or excluded to achieve desired strategic objectives. However, we do not have a well-accepted conceptualization of SHRM systems, policies, and practices.

The third question raised by practicing HRD manager is: Which SHRM systems, policies, and practices are best suited to organization from time to time? Which SHRM practices must be included or excluded and how these practices are interrelated. Is there a multiplicative or an additive effect that must be examined when such practices are implemented simultaneously? Are some SHRM practices redundant or complementary to others? These questions have no single answer!

Fitment of SHRM systems, policies, and practices to the need of organization

Improving on the arguments of Batt (2002), Delery and Shaw (2001), Huselid (1995), and MacDuffie (1995), it was posited that Human Resource (HR) systems consist of three distinct HR policy domains that are oriented toward influencing employee knowledge, skills and abilities, employee motivation and effort, and provide opportunities for employee contribution. The set and structure of SHRM practices for individuals and organizations, used separately in practice, fail to reveal potential significance of unmeasured SHRM practices when used together. Studies have documented organizational benefits derived from specific SHRM practices used in isolation but not synergistically. Therefore, a systems view is more relevant. Delery (1998, p. 291) argued that, "the basic assumption is that the effectiveness of any practice depends on the other practices in place. If all of the practices fit into a coherent system, the effect of that system on performance should be greater than the sum of the individual effects from each practice alone." There are many variations of SHRM systems available in the literature, such as HPWS (Huselid, 1995), human capital enhancing HR systems (Youndt et al., 1996), high involvement HR (Lawler, 1992), sophisticated HR practices (Koch & McGrath, 1996), and commitment-oriented HR systems (Arthur, 1992; Lepak & Snell, 2002), HR systems for occupational safety, HR systems for customer services, etc., to name a few. Inconsistencies in the SHRM structure and several conceptualizations have been observed in the literature. A lack of consistency and consensus regarding SHRM systems, policies, and practices limits the ability of practicing HRD manager to truly understand their features to develop a cumulative body of knowledge for organizational results. SHRM has a long list of SHRM practices. Misunderstanding about SHRM systems, policies, and practices by practicing HRD manager results in poor performance of the organization.

The fourth question raised by a practicing HRD manager is: What is the importance of HPWS in twentieth century to achieve organizational strategic objectives?

HPWS: An emerging dimension of SHRM

The high-performance work systems (HPWS) is an organizational architecture that brings together work, people, technology, and information in a manner that optimizes the congruence of fit among them to produce high performance in terms of the effective response to customer requirements and other environmental demands and opportunities (Nadler, Gerstein, & Shaw, 1992, p. 118). The consensus about HPWS (Huselid, 1995) or high-involvement HR systems (Guthrie, 2001) exists with unclear relationship among them for their advantages. However, the question is: Is there a single overarching SHRM that is most effective or are there various SHRM that are effective within and across organizations for achieving different objectives? Organizational climate literature draws attention to different types of climates achieving different objectives. A climate has been defined as organizational members' perception of formal and informal organizational policies, practices, and procedures (Reichers & Schneider, 1990). In particular, organizational practices, policies, and procedures are argued to influence the organizational climate, which influences employees' collective attitudes and behaviors, which in turn influence organizational effectiveness (Ostroff & Bowen, 2000; Ostroff, Kinicki, & Tamkins, 2003). The policies and practices, included in HR systems, have been argued to be particularly influential in shaping employees' climate perceptions (Klein & Sorra, 1996; Schneider, 1990). Conceivably, the HR system that has received most attention in the literature is HPWS. As noted by Huselid (1995, p. 635), "high performance work practice can improve the knowledge, skills, and abilities of a firm's current and potential employees, increase their motivation, reduce shirking, and enhance retention of quality employees while encouraging non-performers to leave the firm". As noted by Zacharatos, Barling, and Iverson (2005), HPWS encompass elements of both the high-commitment and high-involvement HR system approach, but are broader in scope. These systems emphasize the potential competitive advantages that might be realized by employees through HR practices that treat workers with respect, invest in their development, and foster trust in management and commitment toward achieving organizational goals. Specifically, it consists of nearly all types of best practices including selective staffing, individual and group incentives, benefits, intensive training and development, performance appraisal, teams, employee involvement, work–life balance programs, and information sharing. Researchers have shown the use of HPWS to be associated with employee turnover, as well as financial and market-based measures of organizational effectiveness (Huselid, 1995). The open system perspective states that successful organizations are good at transforming inputs into outputs. However, it does not identify the subsystem characteristics that distinguish effective organizations from others. Consequently, an entire field of research has blossomed around the objective of discovering the best "bundle" of organizational practices that offers competitive advantage. This research has had various labels over the years, but it is now most commonly known as high-performance work practices (HPWP) (Appelbaum, 2000; Benson, Young, & Lawer III, 2006; Sels et al., 2006; Zacharatos, Barling, & Iverson, 2005). The HPWP perspective is still developing, but it already reveals important information about specific organizational practices that improve the input–output transformation process. Still, this perspective has been criticized for focusing on shareholder and customer needs at the expense of employee well-being (Godard 2001; Murray, 2002; Harley, 2005). Organizations are more effective when they consider the needs and expectations of any individual group or other entity that get affected by the organization's objectives and actions. This approach requires organizational leaders and employees to understand, manage, and satisfy the interest of their stakeholders (Friedman & Miles, 2006; Barnett, 2007; Freeman, Harrison, & Wicks, 2007). The stakeholder perspective personalizes the open system perspective; it identifies specific people and social entities in the external environment as well as within the organization (the internal environment). It also recognizes that stakeholder relations are dynamic; they can be negotiated and managed, not just taken as a fixed condition (Eden & Ackerman, 1998).

In the twentieth century, SHRM has almost become synonymous to strategic talent management system (STMS) and the organizations have started to adopt some of the latest strategic talent management practices (STMP) like talent acquisition of strategic job families, development of high performing and high potential strategic talent, strategic performance management based on balanced scorecard (BSC) and/or management by objectives (MBO), and strategic compensation and benefits, which are aligned to business strategies. It is observed that HPWS is yet to be adopted by practicing HRD managers working in small and medium-sized industries, some of which aspire to transform them into large corporation. Therefore, it is suggested to the practicing HRD managers to review literature of past and current research to design appropriate people strategy for their organization with a reference to HPWS.

Several researchers have documented advantages and pitfalls of adopting HPWS as a part of their people strategy in various organizations. However, researchers are yet to examine several hidden aspects of HPWS to achieve strategic objectives from the viewpoint of synergized people and business strategies.

In the year of 2014, in a research on the issue, it was decided to examine impact of HPWS on innovation and firm's performance in small and medium-scale manufacturing industries of private sector in India. The objectives of the research study were to establish significance of relationships between top management commitment, HPWS (strategic talent management practices, job/work design), innovation, and firm's financial performance and thus to understand the impact of variables on each other individually and synergistically.

The objective was also to examine the moderating role of job/work design on innovation. The research was carried out in 14 small and medium-scale manufacturing companies situated in various parts of India. Qualitative and quantitative research methodologies were adopted for the purpose of triangulation. The sample comprised 725 individuals working in 14 companies. A customized survey questionnaire consisting of 86 items was used. Construct validity and reliability, mean, standard deviation, and correlation matrix were worked out. Cronbach's Alpha of 0.94 was found for the whole survey questionnaire. In addition to this, a customized survey questionnaire was administered to collect data on top management commitment and financial performance. Interviews of top management and HR heads were conducted for the purpose of further analysis and triangulation. The companies were classified from collected data as high and low top management commitment toward HPWS and their impact level on HPWS was analyzed. Inter-correlation between HPWS (strategic talent management practices and job/work design) and innovation was analyzed with direct and moderating role of job/work design. Impact of innovation on company's financial performance was analyzed. Synergistic impact of all the variables on company's financial performance was also analyzed. The authors developed and tested the model given in Figure 12.1:

Figure 12.1. Theoretical model for high-performance work systems

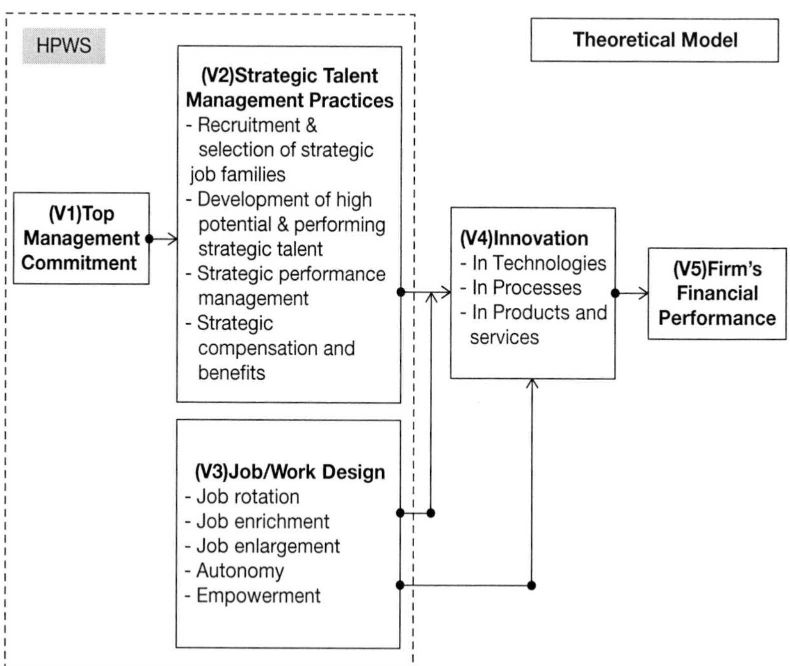

Source: Mehta, N. N (2014).

In context to the above-mentioned theoretical model, the different dimensions identified for each variable were: HPWS—talent acquisition, performance management, compensation, training and development, strategic benefits, skill variety for the variable of strategic talent management practices: job characteristics, task identity, autonomy, job empowerment, job enlargement, skill multiplicity for the variable of job/work design: thinking space, innovation entrepreneurship, idea management, innovation culture, innovation technology, inbuilt innovation for the variable of innovation from the viewpoint of multicolinearity. Descriptive statistic was deployed for the analysis of gender, age group, experience, and qualification. The Pearson Product Moment Correlation (zero order) was deployed to examine impact and correlations between variables. Multiple liner regression analysis was applied to summarize the data, as well as to study relationship between single-criterion variable and many predictor variables and goodness of fit of model. Hierarchical (moderator) multiple regression analysis was applied to examine moderating effect of job/work design on the relationships of STMP and innovation. Structural equation modeling (SEM) was deployed to examine confirmatory and exploratory model fit to proposed theoretical model.

The outcomes of research in brief were: (a) STMP were positively and highly correlated with job design and all its dimensions (job characteristics, task identity, autonomy, job empowerment, job enlargement, skill multiplicity). STMP were also positively and highly correlated with innovation and all its dimensions (thinking space, innovation entrepreneurship, idea management, innovation culture, innovation technology, and inbuilt innovation). The magnitude of relationship of STMP was studied with job design and innovation in reference to high and low committed top management. It was observed that the magnitude was smaller on autonomy, job enlargement, skill multiplicity, and overall job design for low committed top management in comparison to high committed top management. The magnitude of relationship of STMP with innovation was smaller on idea management, innovation culture, innovation technology, and overall innovation for low committed top management in comparison with high committed top management. (b) The multiple linear regression analysis revealed that proposed variables are significant predictors of innovation in case of both high and low committed top management. (c) The result confirms that there was no overall moderating effect of interaction of STMP and job design on innovation. There was no moderating effect of interaction of STMP and job design on innovation in case of high and low committed top management. However, the probability to reject the hypothesis was lesser in case of high committed top management in comparisons to low committed top management. This leads us to infer that there is a low moderating effect of interaction of STMP and job design on innovation. (d) Job characteristics, task identity, and job design (in combination with skill variety and STMP) influenced the innovation ability of employees. At the same time, many other factors/variables of even greater influence were clearly operating to determine the innovation and resulting financial performance and were beyond the scope of the study.

Conclusions

Alignment of people and business strategies is extremely important for the practicing HRD managers to create sustainable competitive advantage through people, develop synergies, and build resilience power of organization in competitive business environment. While lots of ambiguity and non-consensus about SHRM systems, policies, and practices exist in the literature, HPWS is now drawing the attention of the practicing HRD managers. We hope that this article will bring insights for them. It is suggested to practicing HRD managers to undertake thorough literature review on SHRM and HPWS in particular and to include or exclude relevant HPWS practices while developing people strategy for their respective organizations. Further, it is also suggested to practicing HRD managers working particularly in small and medium-scale industries to adopt an appropriate HPWS as part of their people strategy and carryout research on HPWS implemented in various formats in their organization for adding value.

References

Appelbaum, E. (2000). *Manufacturing advantage: Why high-performance work systems pay off.* Ithaca: Cornell University Press.

Arthur, J. B. (1992). The link between business strategy and industrial relations systems in American steel minimills. *Industrial and Labor Relations Review, 45*(3), 488–506.

———. (1994). Effects of human resource systems on manufacturing performance and turnover. *Academy of Management Journal, 37*(3), 670–687.

Barnett, M. I. (2007). Stakeholder influence capacity and the variability of financial returns to corporate social responsibility. *Academy of Management Review, 32*(3), 794–816.

Batt, R. (2002). Managing customer services: Human resource practices, quit rates, and sales growth. *Academy of Management Journal, 45*(3), 587–597.

Becker, B. E., & Gerhart, B. (1996). The impact of human resource management on organizational performance: Progress and prospects. *Academy of Management Journal, 39*(4), 779–801.

Becker, B. E., & Huselid, M. A. (1998). High performance work systems and firm performance: A synthesis of research and managerial implications. In G. R. Ferris (Ed.), *Research in personnel and human resources management* (Vol. 16, pp. 53–101). Greenwich, CT: JAI Press.

Benson, G. S., Young, S. M., & Lawer III, E. E. (2006). High-involvement work practices and analyst's forecast of corporate earnings. *Human Resource Management, 45*(4), 519–537.

Bowen, D. E., & Ostroff, C. (2004). Understanding HRM-firm performance linkages: The role of the "strength" of the HRM system. *Academy of Management Review, 29*(2), 204–221.

Delery, J. E. (1998). Issues of fit in strategic human resource management: Implications for research. *Human Resource Management Review, 8*(3), 289–310.

Delery, J. E., & Doty, D. H. (1996). Modes of theorizing in strategic human resource management: Tests of universalistic, contingency, and configurational performance predictions. *Academy of Management Journal, 39*(4), 802–835.

Delery, J. E., & Shaw, J. D. (2001). The strategic management of people in work organizations: Review, synthesis, and extension. In G. Ferris (Ed.), *Research in personnel and human resource management* (Vol. 20, pp. 165–197). Greenwich, CT: JAI Press.

Dyer, L., & Reeves, T. (1995). HR strategies and firm performance: What do we know and where do we need to go. *International Journal of Human Resource Management, 6*(3), 656–670.

Eden, C., & Ackerman, F. (1998). *Making strategy: The journey of strategic management.* London: SAGE.

Freeman, R. E., Harrison, J. S., & Wicks, A. C. (2007). *Managing for stakeholders: Survival, reputation, and success.* New Haven, CT: Yale University Press.

Friedman, A. I., & Miles, S. (2006). *Stakeholders: Theory and practice.* New York: Oxford University Press.

Guthrie, J. (2001). High involvement work practices, turnover, and productivity: Evidence from New Zealand. *Academy of Management Journal, 44*(1), 180–192.

Harley, B. (2005). Hope or hype? High performance work systems. In B. Harley, J. Hyman, & P. Thompson (Eds.), *Participation and democracy at work: Essays in honour of Harvie Ramsay* (pp. 38–54). Houndsmills: Palgrave Macmillan.

Huselid, M. A. (1995). The impact of human resource management practices on turnover, productivity, and corporate financial performance. *Academy of Management Journal, 38*(3), 635–672.

Ichniowski, C., Shaw, K., & Prennushi, G. (1997). The effects of human resource management practices on productivity: A study of steel finishing lines. *The American Economic Review, 87*(3), 291–313.

Klein, K. J., & Sorra, J. S. (1996). The challenge of innovation and implementation. *Academy of Management Review, 21*(4), 1055–1088.

Koch, M. J., & McGrath, R. G. (1996). Improving labor productivity: Human resource management policies do matter. *Strategic Management Journal, 17*(5), 335–354.

Lawler, E. E. (1992). *The ultimate advantage: Creating the high-involvement organization.* San Francisco: Jossey-Bass.

Lepak, D. P., & Snell, S. A. (1999). The human resource architecture: Toward a theory of human capital allocation and development. *Academy of Management Review, 24*(1), 31–48.

———. (2002). Examining the human resource architecture: The relationships among human capital, employment, and human resource configurations. *Journal of Management, 28*(4), 517–543.

MacDuffie, J. P. (1995). Human resource bundles and manufacturing performance: Organizational logic and flexible production systems in the world auto industry. *Industrial and Labor Relations Review, 48*(2), 197–221.

Mehta, N. N. (2014). *Impact of high performance work systems on innovation and firm's performance* (Unpublished Fellowship Dissertation). Ahmedabad: AHRD.

Mehta, N. N., Pestonjee, D. M., & Khan, S. M. (2015). *Strategic talent management practices scale (STMPS–MNPDKS) Hindi/English.* India: National Psychological Corporation.

Murray, G. (2002). *Work and employment relations in the high-performance workplace.* London: Continuum.

Nadler, D. A., Gerstein, M. S., & Shaw, R. B. C. (1992). Organizational architecture: Design for changing organizations (1st ed.) San Francisco: Jossey-Bass.

Ostroff, C., & Bowen, D. E. (2000). Moving HR to a higher level: HR practices and organizational effectiveness. In K. J. Klein & S. W. J. Kozlowski (Eds), *Multilevel theory, research, and methods in organizations: Foundations, extensions, and new directions* (pp. 211–266). San Francisco, CA: Jossey-Bass. *A Conceptual Review of Human Resource Management Systems 255*

Ostroff, C., Kinicki, A. J., & Tamkins, M. M. (2003). Organizational culture and climate. In W. C. Borman, D. R. Ilgen, & R. J. Klimoski (Eds), *Handbook of psychology: Industrial and organizational psychology* (Vol. 12, pp. 565–593). Hoboken: Wiley.

Pfeffer, J. (1995). Producing sustainable competitive advantage through the effective management of people. *Academy of Management Executive, 9*(1), 55–69.

Reichers, A. E., & Schneider, B. (1990). Climate and culture: An evolution of constructs. In B. Schneider (Ed.), *Organizational climate and culture* (pp. 5–39). San Francisco: Jossey-Bass.

Schneider, B. (1990). The climate for service: An application of the climate construct. In B. Schneider (Ed.), *Organizational climate and culture* (pp. 383–412). San Francisco: Jossey-Bass.

Schuler, R. (1992). Strategic human resources management: Linking the people with the strategic needs of the business. *Organisational Dynamics*, *21*(Summer), 18–32.

Sels, L., Winne, S. De, Delmotte, J., Faems, J., & Forrier, A. (2006). Unravelling the HRM-performance link: Value-creating and cost-increasing effects of small business HRM. *Journal of Management Studies, 43*(2), 319–342.

Youndt, M. A., Snell, S. A., Dean, J. W., & Lepak, D. P. (1996). Human resource management, manufacturing strategy, and firm performance. *Academy of Management Journal, 39*(4), 836–866.

Zacharatos, A., Barling, J., & Iverson, R. D. (2005). High-performance work systems and occupational safety. *Journal of Applied Psychology, 90*(1), 77–93.

13 Habitual mindsets to mindfulness: Wise approaches to change management

S. Ramnarayan

A construction agency made an offer to the members of a cooperative housing society. Two additional floors will be built in the apartment complex, and the new apartments on those floors would be sold to new members. The income would be utilized to enlarge and renew the apartments of existing members at no additional cost. The offer was unanimously and enthusiastically accepted. But the mood turned ugly as soon as the work started. There was noise, muck, and disruptions from the building construction work—the residents were not ready for this. At the "head" level, they liked the idea of larger, renovated apartment at no cost; but at the "heart" level, they were not willing to put up with the hassles of transition. This is one of the paradoxes of change—we like the idea of change when it signifies progress; but we do not feel like giving up the security and comfort that comes from status quo. That is why change leaders need to go beyond the world of logic, reason, and analysis, and deal with the challenges of mobilizing people's emotional energies and generating trust and credibility in relationships.

How do effective leaders formulate the right change agenda and bring about successful change? What approaches are required to make meaningful change happen? Drawing not only on my experiences with a wide variety of organizations, but also on the extensive literature on the subject, I have outlined the approaches that I consider significant.

We often notice that ineffective managers are ineffective because they do not sufficiently vary their approaches based on the specific context or given set of circumstances. Their thoughts and emotions become habitual, and they get entrenched in certain behavioral patterns. With habitual mindsets, behaviors become almost automatic. Researchers point out that when thoughts and emotions become habitual, they form neural networks, and then people start seeing things through colored lenses or filters. The more people repeat their behaviors and act in predictable fashions, the stronger the neural network gets.

So people feel convinced that they are absolutely right about whatever they believe. They have fixed ideas that have not been examined to check their relevance to the given context. Such an approach is highly dysfunctional when a manager is navigating through the world of change characterized by complexity and uncertainty.

To be effective in the world of change, you cannot be a prisoner of habitual mindsets. You need to be mindful. Mindfulness is a state of active, open attention to the present. Mindfulness means living in the moment and awakening to experience. When you are mindful of the context, people, and organizational purpose, you are able to identify appropriate vision, strategy, structure, and culture to make change effort successful.

Clarity, focus, and emotional connect with a larger purpose

When Anna Hazare returned to his native village of Ralegaon Siddhi after his early retirement from the Indian Army, he felt distressed with the economic and social backwardness of the villagers. His deep dissatisfaction with the status quo drove him to bring fundamental changes to the social fabric of the village. In a similar vein, though many capable individuals had preceded Rajiv Jaruhar in the position of Member (Engineering) in Indian Railways, it was Jaruhar's intense feeling that his organization was constantly letting down freight customers that fueled his efforts to change axle load against formidable odds.

Many managers at middle or senior levels see themselves as doing a job that starts at 9 am and finishes at 5 pm. Their habitual mindset puts them perpetually in a reactive mode. As a result, they do not have much energy for the pulls and pressures of change. Certain individuals are in the habit of preferring an accommodating approach in a bid to be perceived as a "nice" human being. When you are mindful, you notice whatever is happening around you. You feel excited about a new approach or a possibility of making life around you better. For example, when benchmarking data is brought in, mindful managers pay attention to why their own organization seems to be lagging behind on certain yardsticks. With habitual mindsets, others do not feel inclined to give up the security that existing standards or targets of quality or pace offer. Their social consciousness may be low, creating impediments to change.

Thus, a strong emotional connect with a larger purpose is a key starting point. It translates into a deep desire for change that pushes these persons to go the extra mile when tackling heavy workloads and meeting tight deadlines. The renowned economist Lord Keynes has pointed out, "A large proportion of our positive activities depend on spontaneous optimism.... If the animal spirits are dimmed and the spontaneous optimism falters, leaving us to depend on nothing but mathematical expectation, enterprise will fade and die".

In business organizations, emotional connection may be demonstrated through passion for products, services, and customers. In non-business organizations, it is evident by the missionary zeal that you bring to the role. Emotional connection is the source of intensity, energy, motivation, and commitment, without which you cannot achieve high levels of performance. Sensitive heart and passionate curiosity for knowledge lead to empathy, better communication, and understanding of the employee needs. Emotionally connected leaders are able to attract followers. As a leader remarked, "Everything must have a compelling purpose. Larger the purpose, greater the buy-in."

Choosing your battle and developing a thoughtful change plan

Change involves a journey through complexity and uncertainty. In dealing with such challenges, it is necessary for change leaders to explore goals, strategies, and contextual factors so that they have a framework for effective decision-making. In the absence of broader understanding, energetic and well-meaning actions alone are unlikely to pay off. Additionally, while there may be dissatisfaction with status quo and even agreement on the broad direction, there could be multiplicity of views with regard to problems to be tackled and pathways to be taken to bring about change.

With habitual mindsets, managers do not invest sufficient time and pay adequate attention to key stakeholders to create a shared understanding of the different dimensions of the problem and also the possible solutions. Mindfulness involves paying attention to both the formal organization of policies, structures, and systems and also the informal organization of values, attitudes, and beliefs.

Thus, it is necessary to focus on the right key priorities to bring about worthwhile and valuable change. Apart from energy, unless there is focus, it would not be possible to really make a meaningful difference. To develop a sense of perspective, you need to understand in detail the environment in which the organization operates. You must be aware of the mindsets within the organization concerning the environment. In other words, you need to understand how the organization is making sense of and interpreting the environment. You should be aware of the changes in the environment. You develop a sense of perspective by listening carefully to people with diverse views, reading widely, and reflecting on your experiences. Sensing of perspective is largely about mindfulness.

Another key quality is the ability to spot the right solutions to an organization's problems. This comes from an ability to see the big picture and an insight into the workings of an organization. Individuals with this quality look for pragmatic and practical, not ideal, solutions. They do not concern themselves with why something cannot be done. Instead, they focus on what can be done.

As we see in the story of Sudhir Kumar in his role as Sales Tax Commissioner of Bihar (Box 13.1), an analytical mind helps you question assumptions and beliefs. Sudhir Kumar's example shows that it is necessary to select right

Box 13.1 Choosing the right strategic thought

When Sudhir Kumar joined as the Sales Tax Commissioner of Bihar, the revenues from sales tax were pitifully low. He was given the mandate of implementing necessary changes for significantly enhancing the tax revenues. He realized that creating maximum impact would require him to select right strategy or choosing the right ideas. He began by making a careful and systematic analysis of the organization and the environment. He interacted with a wide variety of stakeholders with diverse views, read documents, and reflected on his own experiences. By questioning assumptions and beliefs, he developed insights into the functioning of his department and taxation issues. Then he considered the question of what could be the right solutions to the organization's problems.

His analysis revealed that enhancement of tax collection would require him first to plug tax avoidance. His thorough homework had given him insights into the loopholes through which tax was being avoided. For instance, hair oil was getting taxed as edible oil, mosquito repellent was getting taxed as pesticide, hybrid seed was getting taxed as food grains, and packing material was getting taxed as jute. The obvious next step became one of plugging all the loopholes and controlling tax avoidance. Steps were also initiated to ensure that these loopholes did not remain in law so that future tax avoidance was prevented.

Another important strategic idea was that it was extremely difficult to prevent tax evasion in unorganized sectors like spices or cashew nuts. It was only in organized sector that the first sale at the point of origin is in white currency. Kumar felt that by selectively going after specific product categories like scooters, motor cycles, or cell phones, he would be able to control tax evasion. This required him to build a reliable database.

He decided to target the 30 largest companies whose stocks made up the Sensex—the index of the Bombay Stock Exchange. When he found out that he could get funds for computerization, he computerized his own office and seven other key offices in Patna, the capital of Bihar. He used these offices to target the 30 companies. By utilizing the power of computers and information technology, he was able to plan and execute decisions with surgical precision. Thus, he focused on key areas, set concrete, challenging goals for himself and his people, and made the requisite day-to-day improvements. His steps increased the tax revenues four-fold in a very short time.

strategy or idea for creating maximum impact. If the basic decisions were wrong, there was no way that the organization could have achieved performance irrespective of the style or approach.

Leading change practitioner Prasad Kumar of the GMR group says, "It is important whether managers throw up problems, or do they make proposals. When proposals outweigh problems, potential is activated. Managers should be able to generate and navigate proposals. The quality of the proposal is important. It should be as precise as possible in terms of costs or risks. This requires that you know your job thoroughly. Your strong "will" should translate into a sound change plan that is focused on the fundamental and not peripheral challenges and appeals to key stakeholders."

Attention to the "will" and "energy" of relevant others

No single leader, no matter how brilliant, can bring about successful transformation single-handedly. There is a need to create a change coalition consisting of individuals with necessary expertise, relationship and communication skills, wide credibility, strong action orientation, and status in the system. For each major change, the coalition should work with a mission orientation.

As the FSL Foundation story (Box 13.2) shows, when you announce changes in technology process or delivery system, you are likely to run into active or passive resistance. The FSL subunit manager exhibited mindful behavior and so succeeded in managing this resistance effectively. With habitual mindsets, managers typically respond by pushing the change even harder. They apply the force of reason. They may choose to ignore resistance. Unfortunately these reactions only increase opposition. Resistance may assume passive forms, but it persists.

It is important to remember that people expressing resistance are not born resistors whose only mission is to ruin the organization's change initiative. Change management expert Rick Maurer (2010) states that there are three forms of resistance. At Level 1, people implicitly say, "I don't get it!" At Level 2, the unsaid statement is, "I don't like it!" At Level 3, the unexpressed sentiment is, "I don't like/trust you!" The first level pertains to the world of information—facts, ideas, analysis, or logical arguments. At the second level, we are dealing with emotional resistance. This arises from fear—loss of face, status, control, or even their jobs and future well-being. We confront personally oriented resistance at the third level. This reflects absence of trust and confidence in the change leader. That makes this resistance a hard pill to swallow.

In organizations, employees couch their Level 2 and Level 3 resistance using words of reason and logic for their self-protection. But when you listen with your head and heart, then you realize that there is more than what meets the eye. Thus the three types of resistance represent three different worlds of reason, emotion, and trust. As a leader, when you encounter resistance, you should diagnose the underlying concerns correctly. Ask yourself: "What are the causes of this opposition?" Resistance should be addressed in the right way. To deal with emotional or trust issue, you cannot provide information or make logical arguments. In emotional or trust issue, the more you push with reason, the more people will get embedded in resistance. On the other hand, when you listen to people, appreciate their stakes, involve them, and discuss what is in it for them, you would help people deal with their emotional resistance. It is important to remember that perception is reality. People may read risks wrong, but in their minds, the danger is real.

Box 13.2 How a manager built the "will" of internal stakeholders

FSL Foundation was in the process of making a change. The unit had planned to introduce a new computerized Project Management System (PMS) to replace an older and simpler system that had been locally developed and easy to use. The subunit manager in "location A" was entrusted the task of implementing the system in his subunit.

A day-long meeting of 30 to 35 FSL staff from location A was called to plan the change implementation process. Two senior persons from the headquarters also visited location A to be at the opening session. They made a presentation on the advantages of the new system and how PMS would help generate reports that are useful not only for effective monitoring of different projects being implemented by FSL Foundation, but also for generating periodic reports for the funding agencies. There were a few polite questions on the operational aspects of PMS, but there seemed to be little enthusiasm among the 30 to 35 members of the subunit who had gathered in the conference room.

The subunit manager felt concerned that there was very little energy or excitement about moving to new PMS. It seemed to him that the members had reservations, but they were not expressing them. He initiated a discussion on why the change was needed. There were a number of views expressed, and people recognized that the older system was inadequate to meet the emerging needs. But there still seemed to be misgivings about the new PMS system.

At this stage, the subunit manager tried what he termed a "reverse brain storming" session. He split the larger group into subgroups of about eight members each and asked them to brainstorm the reasons why the new PMS would fail. Though the group initially thought that it was a strange assignment, they quickly warmed up to the task. In an hour's time, different groups came up with their lists, and a total of 45 reasons were listed as to why the new change would NOT work. At that stage, the subunit manager informed the group that there was another important part to the exercise—the group had to remove overlapping items, and prioritize the concerns. This exercise led to five significant factors that posed major threats to the successful change implementation. Then, the manager asked the groups to work out a detailed plan to ensure that those factors were taken care of. This exercise helped the subunit identify and avoid a number of potential problems, and contributed greatly to effective change implementation.

Building will and motivation requires attention to certain interdependent factors: helping people see the need for change; being sensitive to their losses; learning to influence non-reporting relationships; and building partnerships and alliances with key stakeholders. While we discuss these aspects separately, they are somewhat related to one another.

Helping people "see" the need for change: You propose a change based on certain underlying assumptions and perspective. That perspective marks a radical departure from the narrow world view that people may hold based on their limited experience. It is, therefore, imperative that you are able to successfully show them another world. You have to address certain important questions. What is wrong? Why do you want to change? Why is it necessary to create resilience and prepare for a future that has not yet come? What do the present trends and environmental signals suggest? How do we know for sure? What is the burning platform? What is the business case? If you are proposing certain interventions, how are they a solution to our existing difficulties?

In a family business manufacturing plastic tubes, a young family member wanted the unit to change its approach to product promotion, but knew that the founder patriarch was not convinced about the need for change. So she quietly made a short and innovative video film by herself and showed it to a group of potential customers in the presence of the senior person. When the older man personally witnessed the excitement that the new approach had generated among the customers, he immediately approved her change idea. The youngster had succeeded despite lack of authority and resources. She had successfully created positive emotions. She did not say, "Our earlier approach is bad. It is not working. Without change, we'll be stuck in a rut." She found a way to engage fresh thinking and enthusiasm of her superior. She found a way to instill hope, optimism, and excitement in him. She found the feeling.

The change guru Kotter says that managers focus excessively on logic and rationality. But core of the matter is always about changing people's behavior (Kotter & Cohen, 2002). A logical argument may create understanding, but not change behavior. Most people assume that people analyze; they think; and then they change. But Kotter's research shows that analytical arguments do not overcome resistance. In successful changes, the sequence of change is "See–Feel–Change." When you are able to present evidence as disturbing look at the problem, hopeful glimpse of solution, or sobering reflection of your current habits, it speaks to people's feelings. By carefully thinking about what your stakeholders would see, you succeed better in making them feel energized, hopeful, creative, and competitive.

Sensitivity to "losses" of stakeholders: When a traditional family business hired bright youngsters from top institutions to champion its efforts to enter a new market, the older employees experienced decreased importance resulting in negativity (status threat). Sudden inflow of new people led to uncertainty about future promotions, work distribution, and organizational stability (loss of certainty). Creation of new teams led to feelings among older employees that they had less freedom, and this led to increased stress (loss of autonomy). As known faces were replaced by newer people on certain teams, newcomers and old-timers started forming separate cliques; there was perceived loss of support and reduced feeling of belonging (relatedness threat). Finally, newcomers were perceived to receive higher salaries and better designation (loss of fairness). Though the leadership wanted to make the change only for the well-being of the organization and had every intention of continuing to regard existing people as valuable members, people experienced "losses" all around.

Drawing extensively on social neuroscience studies, David Rock (2009) proposes a simple framework on the kinds of losses people experience when faced with change. His SCARF model is an acronym for five domains of human experience: Status is about relative importance to others; Certainty concerns being able to predict the future; Autonomy provides a sense of control over events; Relatedness is a sense of safety with others; and Fairness is a perception of fair exchange between people.

Researchers in neuroscience point out that perceived losses in these areas activate the same regions of the brain that signal physical pain, thirst, or hunger, and cannot just be wished away. As human beings, we are wired to be deeply affected by these "losses." It is, therefore, very important for change agents to orchestrate delivery of information and manage employee engagement in such a manner that emotional reactions of affected stakeholders do not subvert the change effort. In Box 13.3, we present the story of acquisition of Magadi Soda by Tata Chemicals. Though Tata Chemicals had intended it as a progressive move for internationalization and growth, employees in all units experienced a sense of loss.

Box 13.3 Change perceived all around as "loss"

Tata Chemicals Ltd (TCL) was in the middle of a major change program of internationalization. It had set its heart on a company that was in Kenya—the Magadi Soda Company. But Magadi Soda was a part of BMGL, a UK company, and BMGL had no intention of selling one of its "crown jewels." After a couple of years, BMGL was available as the private equity owner was looking for an exit, and TCL acquired the entire group.

Ramakrishnan Mukundan, CEO, described how the acquisition was misunderstood all around by employees. He said: "We had a site in India at Mithapur; we had a site in Magadi, Kenya; and we had a site in the U.K. When I went to Mithapur, they were very worried that we had made the acquisition. They said, "Now that you have acquired Kenya, your logical choice would be to shut India down because Kenya is more cost-competitive. Why would you want to put investment now in India?" When we went to Kenya they were highly concerned. They said, "You have acquired us because we were beating you in the marketplace. Now, you will shut us down because we were your competition. You have acquired us to stall us; you did not acquire us to grow us." When we went to U.K. they said, "You actually did not want us. You only wanted the unit in Kenya. You will shut us down very soon." Thus, we had all the three entities telling the business heads: "We do not belong to the family and you do not want us." We had all three units perceiving the change as "loss" and feeling demotivated.

To address this concern, change agents should be good at recognizing improvements made, and provide positive feedback and public acknowledgements to enhance status. Attempts should be made to ensure that goals, norms, and expectations are clear and there is transparency right through the change process so that sense of certainty is maintained. Greater involvement and flexibility can help with autonomy. Leaders employ the system of buddies or mentors and build team events to strengthen relatedness. Consistency between words and deeds and frequent and congruent communication help maintain a sense of fairness. Most important, leaders should anticipate possible losses that different stakeholders are likely to experience so that they are better prepared to deal with those.

Influencing non-reporting relationships and building alliances: Prabir Jha has championed changes in a wide variety of organizations. He says, "In leading change, perhaps your most important challenge is to be able to influence people who don't report to you. Before making, say a Board presentation, I prefer to schedule an informal session with key decision-makers. If the chief executive officer (CEO) or director has an objection or a concern, I would like to know that in advance. You cannot be blind to the politics of change. Like good and bad cholesterol, there is good and bad politics. Good politics is about stakeholder sensitivity, and knowing whom to influence, how to influence, and when to influence."

Quite often, you are expected to bring about change without having the authority to issue orders or give directives. You are not likely to have abundance of resources. You depend on other people to get the work done even if there is a certain amount of skepticism around you in the hierarchy. Your proposal may not be readily visible to organizational head. You may need a sponsor for the upward journey of your change proposal. Your own superiors, peers, and other relevant stakeholders may have a very different view of reality, and it is important to be able to influence their thinking and action. Unless participatory processes are handled effectively, the decision would often come down to the lowest common denominator.

A wide variety of methods, approaches, and styles are available to influence non-reporting relationships. Box 13.4 outlines a few influence methods employed by officials of Indian Railways to reach closure on key decisions pertaining to change. Without change agents exhibiting patience, persistence, and proactive approach to influence relevant people across functions and disciplines, certain decisions have remained pending in Railways for several years.

For a manager involved with change, two things are very critical: ideas and friends. In Box 13.5, we describe how a change agent crossed organizational reporting lines and worked collaboratively with several individuals and groups to make change happen. In many cultures, you have to invoke relationships to get things done. It would often be insufficient or less effective to remind people about their roles or responsibilities. It is not very practical to merely take a stand that people should follow the system, perform their role, and take timely actions. Personal equations are very important. A manager has to get out of role boundaries and have productive person-to-person relationships.

Box 13.4 Some influencing methods used by Indian Railways officials

As the Indian Railways was a huge bureaucracy with strong departmental/functional silos going right up to the top, it was extremely difficult to reach a closure on any decision that cut across more than one discipline. Change agents in coordinating role employed certain methods to successfully influence non-reporting relationships and gain support. These approaches helped close long-pending and contentious issues.

Humility and patience: Typically, change agents were careful not to compel anyone toward a course of action. They listened to people, remained open to their ideas, and waited unwearyingly for the right moment when they could suggest a course of action. By consciously cultivating qualities of humility and patience, they created a psychological climate where people felt safe in sharing ideas and proposals for change.

Persuasion through persistent follow-up: To gain consensus on crucial change initiatives, the strategy was to contact individual members of the decision-making group on a regular basis till they reached a consensus. The general approach was not to ask for a specific decision but to request the group members to work together and reach a decision that was acceptable to all departments as per agreed timelines.

Ownership through involvement: For instance, when change agents came across an interesting idea or innovative proposal, they would actively involve important stakeholders in drafting the proposal or revising the drafts. Active involvement in creating or modifying change proposals made people to support those ideas.

Persuasion through framing: The term framing refers to the use of arguments, examples, and stories to create a favorable impression of the change issues facing the organization. Framing helps people to identify their interests that may be latent and diffuse. For example, when a change agent sought support for increasing the axle load, he articulated the issue not in technical or operational terms. He framed the issue as meeting the unmet and under-met needs of the Indian Railways customers who were involved in keeping the wheels of the nation's economy moving.

Attention to timing: In Indian Railways, high-priority events like annual railway budget were seen to set a rigid deadline that cannot be violated. Change agents utilized such events and deadlines to move the organization toward settling differences, reaching a closure on key decisions, and launching the change initiatives.

Foot-in-the-door approach: According to the psychology of influence, even if someone is not fully convinced about change, if they publicly make commitments to small changes, they are more likely to change their attitude toward change in order to remain consistent with their behavior. This approach of eliciting small commitments to build support for a larger

(Continued)

(Continued)

program is called the foot-in-the-door approach. It framed initiatives as "experiments" or "pilot projects" that diminished the risk of failure. Thus, it reduced opposition and resistance as consent was sought only for small changes. It also helped change agents gather data and make corrections, if necessary. For instance, the parcel business was outsourced first in smaller stations and then gradually extended to bigger stations over the four-year period. The carrying capacity of wagons was increased by six to eight tons over a four-year period in smaller increments.

Faith and self-belief: People leading change mobilized support by fostering positive emotions and relentlessly focusing on outcomes. They constantly communicated faith and self-belief, not only in themselves but also in the ability of people in the organization. The success of the transformation reinforced people's belief in their own ability to bring about positive improvements in the organization.

Box 13.5 Alliance with external agencies for successful change

In its efforts to renegotiate employment terms with the Contract Workers' Union, Dr Reddy's management was also concerned about the organizational change agenda of strengthening global standards of safety and quality. But this process required building alliances and partnerships with multiple stakeholders. Apart from key decision-makers within the organization at different locations and levels, there were external parties that had significant influence on the outcome. Unless the dependence on different stakeholders was proactively managed, the decision-making process would become too complex or too slow where huge hurdles can unexpectedly arise at the time of execution. The key managers took a number of initiates to involve these stakeholders as briefly outlined later.

District administration/Joint collector: The management actively sought the mediation of the Joint Collector as he enjoyed a high level of credibility with the union and contract workers, and at the same time understood the business requirements of operating in a competitive environment.

Police authorities: Key change agents were in touch with the police to ensure that the operations were buffered from external disturbances. The Deputy Superintendent of Police (DSP) played a key role in bringing a closure to the negotiation process by getting a disruptive element out of the negotiation process.

Media: Earlier the media interactions were centralized. But when this arrangement was found to pose constraints, the communication policy was changed, and an appropriate person was chosen to manage communication and relationships with the local media on a regular basis.

(Continued)

(Continued)

Surrounding village communities: Senior people from the organization visited the villages periodically and had meetings with village elders. As Dr Reddy's was involved with community development initiatives in these villages, the company had a long-term association that it could build on.

Political executives: The management sought out meetings with key political executives to communicate to them that the organization had progressive practices, and was committed to being a good Corporate Citizen. It sought the help of political executives in mobilizing the district administration to play a constructive mediating role to resolve the dispute amicably.

Other companies in the industrial estate: The management was also in touch with other organizations in the neighborhood to communicate to them that it understood the concerns and compulsions of these companies.

The company's success in making requisite changes for building a global organization was based on its successful alliance with these different stakeholders.

M. M. Murugappan of the Murugappa Group states, "Our strength in working with partnerships with a wide variety of organizations and agencies over several years arises from our deep commitment to the following principle outlined in the *Arthashastra* around 300 BC: 'The fundamental principle of economic activity is that no man you transact with will lose; then you shall not'. When we work with partners, we feel deeply committed to their interest and well-being. Unless we make emotional investments in relationships, we would have mere acquaintances around us, no real friends."

Planning for unintended consequences

Our actions have both intended and unintended consequences. It is necessary to consider risks inherent in the plan, and be alert to surprises or unintended consequences. With habitual mindsets, managers have a "fairy tale" model—they do not subject their change plan to a "stress test" or consider the hurdles that they might run into in the implementation phase. For such managers, when the plan does not work, there is frustration. There is a tendency to avoid looking at the possibility of bad news, say of plans going awry. There is little thought on what may be one or two things that can influence multiple factors and have a disproportionately large impact.

After developing a process for change, there is a need to anticipate what can possibly go wrong, and what preventive actions we must initiate or what back-up plans should be prepared. As we see in the story described in Box 13.6, Carborundum Universal (CUMI) managers had done this kind of thinking and this helped them achieve outstanding success. If you realize that things can go wrong, you can develop a back-up plan, allocate appropriate resources, and build resilience to strengthen the plan.

Box 13.6 Mixed results of ambitious change plans

Carborundum Universal (CUMI), an engineering company belonging to the Murugappa group, planned for the acquisition of a loss-making unit in Russia by clearly articulating what CUMI would do differently to achieve greater value. Carefully chosen technical teams were entrusted with the task of writing not one, but five growth stories. While every effort was made to develop thoughtful, innovative, and implementable strategic ideas to bring about a quick turnaround, the management recognized that success in all would probably not happen.

One of CUMI's growth stories was that they would take production of the Russian plant from 52,000 tonnes of Silicon Carbide to 65,000 tonnes. Even during the best of Soviet times when there was no disturbance, the plant never produced beyond 55,000 tonnes. CUMI's plan was based on several technical improvements, such as changing the design of the furnace and changing the furnace head. The technical teams had worked out what they would change, how they would do it, and how they will monitor.

Another growth story involved changing the mix of two categories of products—crystalline material and metallurgical material. Fifty-five percent of the 52,000 tonnes that they were initially producing was of the crystalline variety. As the price of crystalline material was 50 percent higher than that of the metallurgical material, CUMI prepared plans to take the crystalline material from 55 percent to 80 percent, and achieve a significant increase in the profitability.

When the acquisition was made, CUMI did a quick 100-day integration plan. Within the 100 days, these growth stories were rolled out by strong execution teams consisting of members from both CUMI and the acquired company. The teams were given necessary resources and support. Of the five growth stories, two worked brilliantly—the goals pertaining to production of Silicon Carbide and product mix were achieved. But despite committed efforts, the other plans did not work out. For instance, the Russian plant was making about 12,000 tonnes of bonded abrasives. They had a name plate capacity of 50,000, but even during Soviet times, they had made only 24,000 tonnes. CUMI's target was that they should move from 12,000 to 24,000 tonnes within two years. Teams of experts invested substantial efforts. But after two years of hard work, they had not even achieved 12,000 that was being made and sold before the downturn. As the company had anticipated and planned for such setbacks, CUMI still achieved its larger goal of effective turnaround of the Russian unit in record time.

It is necessary for managers to be alert to what is happening around, scan to pick up possible unintended consequences, and fine-tune the implementation plan on the basis of the ground realities. As the story of Delhi Metro in Box 13.7 illustrates, change managers should remain in touch with key stakeholders and address their concerns on a proactive and timely basis.

Box 13.7 Assuming responsibility for unintended consequences

When the Delhi Metro Rail Corporation (DMRC) undertook their massive project in the Nation's capital, the leadership realized that among other challenges, it had to generate external support from Delhi's residents, politicians, bureaucrats, contractors, and several other groups. The project owes its success to effectively managing the dependences on a wide array of these external stakeholders. To illustrate the leadership's approach, let us examine how the public support was mobilized.

The leadership team headed by E. Sreedharan saw the challenge not merely in terms of communication, but as winning people over to the corporation's philosophy and approaches. For example, a number of procedures were instituted at work sites to minimize disturbances. The procedures even specified that the vehicles should not be allowed to leave the work sites without their tyres being cleaned. As the other public utilities were notorious for their inefficiency, the interface with those agencies was handled by taking additional responsibilities to ensure that there was no public discontent. For instance, when traffic diversions had to be made, DMRC appointed additional personnel at important signals to help traffic police in maintaining smooth flow of traffic. Though it was clearly not its responsibility, DMRC also undertook road widening and road repairs where necessary to ensure that no road was closed at any time and people were not inconvenienced in any way. In the same way, power, water supply, sewerage, and other issues were also proactively addressed. Right through the process, there were regular community interaction programs. People were provided advance intimation and regular updates by using several media. Help lines were available to report difficulties. As a result, the project consistently enjoyed a great image in the eyes of the Delhi residents, and received their support.

Carl von Clausewitz, the author of a seminal work on the conduct of war, states: "In war, everything is simple, but it is simple things that are difficult." Changes often fail because leaders have not taken into account all the irksome conditions or frictions that have to be dealt with for the plan to succeed. Leaders should be able to anticipate frictions and deal with them quickly and effectively.

Tracking change and being results oriented

A progressive engineering organization made acquisitions in five countries across continents. For each acquisition, a very senior leader from the top management committee was formally appointed as a mentor to support the local management overseas. They sat on the boards of the respective subsidiaries, thus assuming responsibility to the larger company board for the overall success of the SBU or that international part of the business.

Though they were called mentors, they actually reviewed with the CEO of that unit, the P&L, the business plan, the capital expenditure, the HR plan, the growth plan, and the R&D projects. They helped integrate the unit with the rest of the system. For instance, if an overseas operation was found to be struggling with a particular problem of manufacturing, the mentor would find out which other plant in the organization has solved that problem. The mentor would arrange for the right employee from that unit to make a visit and run the operation for a short time to help the struggling unit. Thus, the overseas operation would not be left without support, and the mentor would actually become a helpful link to the larger organization. Apart from the business role of doing monthly review in terms of P&L, balance sheet, performance, the various indices, and identifying the gaps, the mentor in this organization actually got involved in helping the leadership team achieve those numbers. The mentor reinforced certain values, for example working together, celebrating together, or sharing good practices.

Change practitioner Prasad Kumar says, "When you manage change, you often report targets versus actual. Instead if you focus on the delta or the improvement, you get better motivation and results. What was the performance or achievement last week or last fortnight? Where are we now? Thus, we compare the actual performance or quality during the last planning period and the present performance or quality. That measure tells me more about how the change is progressing."

After gaining some familiarity with the change project, most managers develop a routine for data tracking, which remains quite stable. We may use the metaphor of "control panel" to capture this behavioral tendency. Each dial on the control panel provides signals from the system about developments in a given area. While the control panel serves to routinize data tracking, it influences what we are likely to consistently ignore. For example, we may not be tracking the energy and enthusiasm of the employees or certain developments occurring in hidden stakeholders. We get caught up in a reactive mode, and remain unaware of certain developments until there is a crisis. It is, therefore, important to keep alive the values of questioning, listening, and inquiry in our reviews. While formal systems have their place, we should be alert to what is happening through observations and informal conversations with a wide variety of stakeholders.

It is also important to be able to create visibility, buzz, or excitement around the change. That requires some drama or story. It makes change fun, rather than drudgery or serious head-level work. An organization created an HR exhibition to familiarize employees to the changes in HR policies and procedures. People were invited to an exhibition, played games, participated in quizzes, and won small gifts. That created much better communication of the new procedures.

Finally, we cannot overemphasize the quality of results orientation. Change leaders need to be constantly pushing for results. This quality is manifested in taking initiative and risks. Leaders with an action orientation do not procrastinate and remain aware of the value of time. They establish deadlines and push themselves and those around them to a achieve results.

Mindfulness and self-reflection

E. Sreedharan, who has successfully handled highly complex challenges, says that it is important to be mindful and self-reflective. In his words, "Constantly worrying about change can be a huge emotional load. It is, therefore, important for the change initiator to have other interests in life. While commitment to change is important, it is as important to not remain preoccupied with change challenges all the time." An effective change leader generates feelings of hope, optimism, and positivity.

A self-reflective leader is self-aware, emotionally intelligent, focused, energetic, and resilient. To be able to build these mindsets for making change happen, it is very important for leader to possess more than mere knowledge of facts, concepts, and relationships. One needs a deep appreciation of what one knows and does not know. When you handle different problems with different requirements and different time patterns and remain open to learning constantly from experience, you develop heightened awareness of the pitfalls and choice points in managing change challenges. You recognize choices and remain flexible with respect to these choice points. You do not get entrenched in unadaptable behavioral patterns.

To summarize, change leaders need to have faith in their ability to bring about positive change and they need to communicate this faith through actions and results. They need courage to challenge dysfunctional mindsets but must also be able to offer viable alternatives. They should understand their strengths and limitations and should be able to team up with other leaders having complementary strengths. They need to appreciate and respect the history and the strengths of the organization, even as they bring about changes where required. They need to be able to work hard, persevere, and be patient because there would be forces of inertia that make organizations sluggish toward change. Finally, they should be able to generate unity of purpose, respect, and appreciation of differences and honesty and integrity in dealings to achieve the potential inherent in the organization for successful change and adaptation.

References

Kotter, J., & Cohen, D. (2002). *The heart of change: Real-life stories of how people change their organizations.* Boston: Harvard Business Review Press.

Maurer, R. (2010). *Beyond the wall of resistance.* Austin: Bard Press.

Rock, D. (2009). *Your brain at work.* New York: Harper Collins.

14 Toward healthier HR

Visty Banaji

The human resources (HR) function in India is sick. Even globally some eminent academics have announced its demise or demanded its dismemberment. While such obituaries and amputations may be premature, we need to take active steps if they are not to be prescient.

Naturally, remedies cannot make sense before we share an understanding of the ailments. It must be stressed in advance that not all organizations in India have HR departments suffering from all or even most of these problems. Unfortunately, the profession gets tarred with the brush intended for its weakest members. What makes for juicy word-of-mouth are the failings of obnoxious HR jerks. And word of mouth is what is believed, not those flattering chief human resources officer (CHRO) profiles public relations (PR) departments get inserted into the media.

There are five maladies that most frequently afflict some members of our profession. Some of them appear to be mild indispositions. However, they all have the potential to become life-threatening, given the people-centric path to development India must choose and the heavy demands placed on HR as a result.

We have lost our wings

Before we turn starry-eyed about all the new territories HR needs to conquer, let us cast an eye over domains that we used to own but which many of us have abandoned.

It may be instructive at this point to recall the evolution and fate of the dodo. Its ancestor was a flying member of the pigeon family that moved to Mauritius in prehistoric times. Over thousands of years, the dodo lost the ability to fly owing to the lack of mammalian predators and the availability of abundant food at the ground level in Mauritius. Since it no longer had to cope with predators, the dodo was also fearless of humans when the first European sailors landed there around the beginning of the 17th century. This fearlessness and its inability to fly made the dodo easy prey for hungry seamen and led to its extinction.

When I started my career in HR (or "Personnel" as it was then called), the function was essentially industrial relations (IR) with a bit of administration thrown in. Training was mainly technical and usually kept separate from Personnel. Those of us who wanted to grow managerial talent or dabble in organizational development were relegated to small cells that existed on sufferance and frequently had a tense relationship with the mainstream IR stalwarts. HR, talent management, and all that we do today under its aegis have come of age since then. We have been gorging ourselves on the plentiful HR work that makes for easy pickings without having to go to remote factory sites where IR still makes a difference. The lack of predatory or militant trade unionists in many sectors has, over the years, pushed senior IR people, where they exist at all, into small, understaffed cells. Many of us have thus lost the ability to fly in the face of industrial action or employee unrest.

It took tragic and grisly acts of violence in different parts of the country to wake the general HR community up to the fact that labor militancy can recur. We suddenly remembered that employee unrest can take very ugly forms and realized that inadequate IR skills are a major part of the reason friction bursts into conflagration.

Of course, IR does no longer make the same demands as it did decades ago and is rightly renamed as employee relations (ER). The new nomenclature is not simply cosmetic. It indicates that we not only need to manage collective bargaining with unions but also build trust-based relationships with individual employees and between them and their line managers.

Today, every CHRO worth the name worries that IR/ER skills of a high order are scarce in the country. And let not our proud, service sector enterprises think they are exempt from the requirement. By the time they reach the painful realization that they have ignored building these skills, the gap will have become huge and the time to groom experienced ER hands in their type of environment will be totally inadequate for the challenges they face. Overdeveloped event-management muscles cannot substitute for the wing-muscles necessary to fly in the face of employee disaffection and unrest.

ER, of course, is only one example of the capabilities and attitudes that several HR departments have allowed to atrophy through neglect and disuse.

We are more loyal than the king

On December 29, 1170, four knights murdered Thomas Becket in Canterbury Cathedral because they thought King Henry II wanted him dead. There was a public outcry and Henry claimed he never ordered the killing.

Some of our HR leaders scurry to anticipate the wishes of their chief executive officers (CEOs) even when they are crossing the spirit of the law or the duty they owe to the welfare of employees. This means of currying favor is particularly prevalent when the CEOs concerned are expats, promoters, or fresh entrepreneurs. Perhaps, in these cases, the CEOs are themselves not sufficiently aware of or used to the constraints of Indian laws and culturally accepted norms. Moreover, as in the case of Becket's murder, the CEO may not have actually given such an order or will disown it when the matter explodes for one reason or another.

In recent years, we have seen several instances of patently unfair people decisions (exploitation of contract labor, dishonoring campus appointment letters, buying over of union leaders, forced VRS simply to unlock land value, and many others) taken by otherwise reputed and upright organizations. Many of these can be explained by the combination of eager-to-bend CHROs responding to the needs of ambitious business leaders with underdeveloped consciences and overly Nelsonic eyes. For every such decision that reaches public notice, there are thousands of instances of ad-hocism, favoritism, petty tyranny, dismantling of welfare measures, sacrificing long-term development schemes for ready-cooked lateral recruits (we can go on and on) that originate from the same blind following or anticipation of the boss's "*ishaara*" by HR. No amount of town halls and jazzified employee portals can counter such liquidation of the people practice goodwill acquired by an organization.

There are three absolute non-negotiables for HR in any organization. None of them can be traded off against the others or for any other reason. These are:

- Driving the organization's purpose to fruition
- Championing employees (both individually and collectively)
- Guarding the organization's core values (particularly those relating to fairness)

It is very doubtful if anyone would deny these. In reality, though, the type of HR professionals I have excoriated in this section pay these ideals only "lip service." Their true dedication is to a less impressive trinity for whose sake organizational health, employee interests, and values are blithely sacrificed:

- Blind service to the business leadership (there is a difference between being a business partner and being a business slave).
- Responding to power rather than organizational need (this is part of the reason why HR is seen as political and manipulative rather than even-handed and approachable by all).
- Advancing one's own interests (and wheedling exemption from the rules made for all).

But what if HR is not just amplifying hints and is actually instructed to do things that contravene long-term organizational health, employee interests, or core values? My answer is simple though not always easy:

- Find an innovative alternative way to meet the goal for which the contravention was suggested.
- Convince the senior of the long-term damage the proposed path will cause.
- If the anticipated damage is serious enough, take the matter up with the Board and, if that too fails, seek other pastures.

We have a limited repertoire of ritualistic mantras

A hypothetical CHRO is recruited to turn around the moribund people practices in a traditionally managed business. Without so much as waiting to catch breath after joining and without bothering to identify the organization's unique situation and needs, the new CHRO goes about implementing the following:

- A performance management system with the "bell from hell" to toll the demise of people who, till then, were perfectly adequate performers. This is frequently accompanied by money-median-magic in the compensation survey game that permits the organization to claim a percentile position as high as it chooses by the simple expedient of limiting significant spends to a small proportion of the population.
- A coaching program for senior executives, hard-sold with the unsaid but deluded promise that (just as coaching classes are believed to get children into IITs without parents having to worry about their studies) executive coaches will make the company into a CEO factory without the top management having to exert themselves about leadership development.
- A highly publicized gender-diversity initiative that corners resources and garners column-centimeters while ignoring, for instance, the lack of employment and growth opportunities for tribals in areas where company's plants are located or turning a blind eye to the promoters' bias against certain religious minorities.
- The list could go on… but you get the idea.

In days past not so long ago, when such a CHRO was pressed for the reasoning behind resorting to such a stock repertoire, the answer would, as often as not, be "Because GE does it." If the same question were asked today, the response (at least from those who have realized that GE does things differently now) is likely to be "Because Google does it." And a Google mantra is likely to substitute one of the above rituals.

This is not to say that differentiation or coaching or gender diversity is never needed. The criticism is about the rote manner in which they are applied and prioritized when the strategic demands of the organization might require something completely different. It is so much easier to tell CEOs that some US success story is being followed than it is to diagnose, research, innovate, and absorb the pains of pioneering, and that it is no wonder few of us adopt the latter course.

Many people have rightly bemoaned India's lack of world-class scientific discoveries or industry-shaking technological innovations. HR has joined in the general sorrowful head-shaking. But what if we were to turn the glare of these comments on HR? How well would our employer award winning corporate fare under such a criterion? HR frequently champions the cause of innovation in organizations. How about innovation in HR? Do we think people cannot see through the MNC hand-me-downs we wear while pretending to be emperors of innovation?

We have accumulated a lot of plaque in earlier years

The noble inscription on the Statue of Liberty reads: "Give me your tired, your poor,... The wretched refuse of your teeming shore...." The lines do not provide guidance to immigration officials manning any border post or visa counter in the US. HR in India, however, adopted this slogan long back to govern entries into its ranks. We, therefore, became the dumping ground for the "wretched refuse" that other functions were ejecting.

In much of the private sector at least, things have improved considerably. However, the plaque we collected, when "Personnel" had the least favored function status, continues to block the growth of the truly world class talent flowing into HR in recent years. Even today, in those enterprises where line managers are rotated to senior positions in various departments, HR can get people who do not merit more prized assignments.

I must immediately correct any impression I may have given that people coming from line management to HR at relatively late stages in their careers are all misfits. The majority of the most effective HR leaders it has been my privilege to know throughout my career have not qualified in HR or started their careers in the function. In fact, to deprive HR of this valuable source of talent would be to put it on a steep slope of terminal decline. But these successful career transitions were all of people who had a bent for HR and volunteered to come into the function after making a success of whatever they were doing.

The worry is when people who are mediocre or failures elsewhere are foisted on HR under the impression they cannot do too much damage here. Such complacency is dangerous even in the short term because such CHROs are bound to be insecure and, therefore, resort to "politicking" to shore up their fortunes. In the process, they can cause irreparable damage to the trust and credibility without which HR just cannot operate. Even more disastrous than the immediate impact are the poisonous fruit such incompetents leave behind, sometimes unintentionally. Other than the CEO, the CHRO has possibly the longest time span of discretion of any chief experience officer (CXO) role. The strategic errors made in such a role can, years later, prove toxic enough to threaten the organization's survival or, what is more difficult to spot, result in missed opportunities that can rarely be retrieved because most HR strategies have such long gestation periods.

In the controlled economy that India once had, where managing licenses and permits was the most valued core competence for many business houses, a below average "Personnel" department perhaps did not matter. This is a capability compromise organizations can no longer afford.

We seek salvation in separations

"A doctor can bury his mistakes, but an architect can only advise his clients to plant vines," said Frank Lloyd Wright. HR practitioners in some of our sunrise sectors go one better than even doctors. With soaring attrition rates flushing discontented elements out of the system, these human removal professionals seem to seek protection under the (now abolished) common law rule that states "no body, no murder."

HR departments in these new service sector industries publicize how greatly their employee retention has improved and some of them (misguidedly) make actual efforts in this direction. In truth, however, attrition is their best friend. In several cases, their business models seek to arbitrage India's surplus of overqualified people for mind-numbing, routine work with limited advancement. If they did not have significant employee wastage, the pressure of disgruntlement would lead the workforce to organize itself and take industrial action for claiming a much higher share of the arbitrage.

Not only does this approach do away with the need for ER wings (mentioned with the example of the dodo above), but most of the organs of a mature HR function are required only in their most primitive form. There are a couple of exceptions, though. Recruitment and onboarding, needed both for the rapid growth of such businesses and the insatiable replacement demand for people, are practiced at world-beating levels. The resulting HR organization is as ludicrous or frightening (depending on whether you are an observer or caught in its jaws) as a gulper eel with just its mouth hugely developed and most other faculties existing in a prehistoric version, if at all.

So what is wrong if unique sectoral demands have evolved an HR organization that is most efficient for its limited purpose, with none of the competencies and processes conventional HR departments spend time and energy to build? Just one thing. A day will come when the music of exponential growth will stop playing. Like all industries that have ever existed, these modern marvels too will have to face slowdowns and contraction. In such a scenario, steroidally grown staffing and induction processes will be of little use. Neither resources nor time will then be adequate for such HR departments to develop the new (for them) HR competencies that are essential for normal organizational existence.

It bears repetition that not all the above aliments afflict all of HR in India and there are some splendid HR teams that are free from all of them. On the other hand, the five prescriptions that are described below have near-universal applicability. While several of them do specifically target the ailments we have been considering, their impact on HR health is intended to parallel exercise and vitamins rather than surgery.

Treating people fairly and guarding values

What makes credible HR distinct from all other functions in the corporate world is that its goal-set is not limited to results that end up maximizing shareholder value or customer delight. HR also has the responsibility for effectively championing the needs of the people in the organization and guarding the organization's core values (particularly those relating to fairness).

Is not this too tall an order? Can the leadership of HR really face up to the might of the top management of the company on those (hopefully, extremely rare) occasions when they are being unfair to the organization's employees or turning a blind eye to its core values? The HR leadership is very unlikely to succeed if it plays David against the organizational Goliath at the moment of transgression. The cards could be very differently stacked, however, if the organization had in place a code of conduct, hopefully catalyzed in advance by HR.

Let us take a parallel from our body politic. It would be a rare judge who could take up the cause of wronged individuals and groups against an overbearing executive in the absence of a constitution that guaranteed fundamental rights as well as laws safeguarding these rights in specific situations. HR, in a very limited sense, can learn and benefit from this system of institutionalized checks that has evolved, over the centuries, in many countries. It is, therefore, very much in HR's interest to involve all levels of the organization as well as its board in shaping a code of conduct which the company is proud to declare publicly. Once such a code is in force, when HR champions individual cases or points out collective discrimination, it will not be seen as impertinence or insubordination but simply of a functionary discharging the duties enjoined by the code.

Does each organization need to sculpt such a code from bare rock? Not necessarily. At the prompting of the National HR Committee of the Confederation of Indian Industry (CII), I developed a "Fair Organization Code" that can provide a starting point for any organization wishing to create such a bulwark against unfair or unscrupulous practices.

The template of the "Fair Organization Code" has eight elements that essentially answer the key questions following each heading listed as follows:

1. *Ethical and sustainable business practices:* Does the organization deal honestly with external stakeholders, including customers, minority shareholders, Government authorities, the extended communities, and the ecological environment in which it operates, and give them their due? Does it desist

from any malpractice or subterfuge in these interactions? We start with business practices in general because it is not possible to create a culture that avoids unfairness to employees but is impervious to such concerns about other stakeholders.

2. *Living and leading the laws:* Does the organization implement laws (especially those affecting employees) relevant to the jurisdiction in which it finds itself, in both letter and spirit? Does it openly and honestly advocate changing laws that do not make business sense rather than seeking to subvert them or "manage" those administering the law?
3. *Honoring commitments:* Does the organization uphold the "word" given by its officials or documents without resorting to legal or "fine print" loopholes to wriggle out of commitments or established practices? When there is a difference in interpreting a commitment, does it give the benefit of doubt to the weaker party?
4. *Diversity and inclusiveness:* Does the organization have zero tolerance for discrimination against people (e.g., SCs, STs, physically challenged, religious minorities, and women) who have traditionally been denied opportunities at the workplace? Does it have affirmative action programs to restore those balances that have been most distorted in that organization's context? Are these efforts borne out by employee population and progression statistics across levels?
5. *People-valuing policies, practices, and treatment:* Does the organization (and its managers) treat with dignity all those working for it, regardless of the type of employment contract or level? Does it value each individual in her/his own right? Are there clear principles governing key areas such as differentiation, sexual harassment, data privacy, and data access?
6. *Equitable sharing:* How does the organization share the "pie" (and the pain) across levels? In particular, how do people right at the bottom of the pyramid (e.g., contract workers) fare in terms of remuneration, reward, facilities, and progression? What is the organization's impact on other stakeholders, particularly the communities around its zone of operation?
7. *Dealing with deviation, deficiency, and dissent:* How strictly and swiftly does the organization treat deviations from the code, regardless of the level of the transgressor? Are there sustained development efforts for dealing with work performance deficiencies and a transparent "due process" even for those employees not covered by the law or protected by unions? How well are whistle-blowers protected, given external recourse and how seriously is their feedback taken?
8. *Review and progress:* Does the organization have metrics for checking performance against the code? Are there mechanisms for reviewing and updating the code periodically?

Once some reputed corporate implement such codes, their employees will experience the benefits of working in an environment where they have an HR champion backed up by an openly declared code. This should make a substantial difference to those employers' value propositions and encourage other companies to follow suit. Just as certain birds choose mates based on the elaborateness of their plumage, employees will choose organizations based on the (increasingly progressive) codes of conduct they publicize and follow.

The plumage birds carry is costly even if it makes ultimate evolutionary sense. There is no reason to believe that the implementation of such a code will be effort-free. Much of that cost will have to be borne by HR as the code's initiator and guardian. Nor should organizations be enticed to adopt a code of conduct with the lure that it will always be profit-enhancing. No fairness code worth its name can be justified primarily on commercial grounds. It is possible, however, that in future some such voluntary codes may be the price organizations may have to pay for societies to permit them to continue functioning without legislatively imposing even more onerous restrictions in a typically ham-handed fashion.

Broad-based HR capabilities

While championing the cause of people and upholding the core values of the organization are the price of admission for a professional to the HR arena, only a sound and growing understanding as well as practical experience in applying a broad range of HR competencies provides the justification for the continued existence and progression of the HR professional in any organization.

A passenger began to experience a heart attack at 30,000 feet on a commercial flight from Canada to Hong Kong in 2014. An anaesthesiologist who was on board the flight saved the passenger with the help of a pharmacist and a policeman trained in first aid who were also traveling on the flight. All doctors get certain basic training and we would have been quite surprised if the doctor had said he only knows how to give anaesthesia but cannot help with a medical emergency. We would have been even more amazed if the doctor had offered, instead of assisting the patient, to help the pilot fly the plane since he liked playing flight simulation games. Yet we do not seem to mind when some of our HR colleagues refuse to take up basic HR tasks that do not fall within their subspecialization and claim their personal development efforts are mainly directed toward acquiring business skills. Well, the business hired them for their HR competencies. If they cannot display these in adequate and continually improving degree, the business has really no need for such HR professionals, thank you.

To remedy these deficiencies in the very raison d'être for HR's existence, the National HRD Network (NHRDN) began work on a set of critical competencies for HR professionals more than a decade back. HRSCAPE, as it is now branded, is probably the most comprehensive general HR competency model anywhere in the world. Admittedly, my proselytization is hardly unbiased since I have been associated with the development of the model since its early stages. Without going into details, I would like to describe the scope and flexibility of the model so that budding HR professionals can appreciate the extent and depth of our function. They can then decide how to remedy basic gaps that may exist and take a call on their specialization (hopefully with more detailed access, through NHRDN, to HRSCAPE itself).

Many of India's leading HR thinkers were involved in the crafting of this model and the final consensus was to identify eight technical competencies and four behavioral competencies. In both cases, the model limited itself only to competencies that were particularly demanded in HR roles and left out generic competencies that all corporate professionals might require.

The eight technical competencies demanded by HRSCAPE are:

1. *Strategic HRM:* Crafting an HR architecture and competencies that are aligned with the business strategy.
2. *Organization design:* Creating appropriate organization structures together with the supporting processes and culture to enhance business performance.
3. *Workforce planning and staffing:* Strategically responding to changes in workforce demographics, business models, and economic conditions while ensuring the availability of the right talent with the right skills and at the right time and cost, to support the organization's goals.
4. *Talent management:* Developing talent (particularly those with high potential), planning career trajectories, succession planning, and improving employee engagement.
5. *Learning and development:* Organizational learning strategies and models leading to integrated development plans, learning delivery though a variety of means and frameworks for its evaluation.
6. *Total rewards:* Design of jobs, compensation, benefits, and rewards (both monetary and nonmonetary) to attract, retain, and incentivize the appropriate talent to meet strategic goals.
7. *Performance management:* Design and deployment of effective performance planning and enhancement systems as well as performance-related dialogues that lead to and sustain a high-performance culture.
8. *Employee relations:* Industrial relations frameworks that yield high productivity through sound employment policies, trust between employees and their supervisors/management, healthy collective bargaining, and the welfare of employees and surrounding communities.

At least three out of the four behavioral competencies listed later are, in a sense, the observable individual behaviors that complement the prerequisite commitment and systemic support (i.e., the code) for fairness, described in the previous section. The behavioral competencies needed especially for HR are:

1. *Credible champion:* Gaining credibility through demonstrating high integrity in personal and professional dealings and standing up for what is right for employees.

2. *Managing change:* Diagnosing the need and then designing and implementing change programs for individual and organizational transformation.
3. *Service orientation:* Imbibing the needs and concerns of internal and external customers and responding to them readily, promptly, and effectively.
4. *Diversity and inclusion:* Policies, processes, and practices that result in the organization inducting diverse populations, facilitating their progress and respecting, valuing, and supporting their diverse perspectives for mutual benefit.

The HRSCAPE model projects these competencies at four proficiency levels that are self-explanatory:

- Basic
- Competent
- Advanced
- Expert

Depending on the individual HR job profile, a higher or lower level of proficiency in a particular competency may be expected. It would be difficult to imagine any HR job, however, that does not even demand the Basic level of proficiency for all the competencies.

The model and the evaluation of individuals against it are only the first steps. The more important and urgent tasks are to improve proficiency levels, individual by individual and competency by competency, to the point where they reach or exceed what is required for an HR professional's present and future anticipated roles.

Pioneering new pathways in HR

A good grasp of HR competencies makes for a good HR professional. A great one, however, must be able to abandon the beaten track and pioneer in the field. Clearly, not everyone has the capability or courage for thought and practice leadership. Despite the much higher quality of raw material opting for the HR profession in the last couple of decades, such creativity continues to remain rare. Perhaps the story of a pioneer *par excellence*, under whom I had the good fortune to work, can provide us with some clues about the kind of single-minded passion for achievement that fires pioneers.

Sumant Moolgaokar was one of the truly great corporate Titans this country has produced. Few can approach, leave aside match, his track record of creating (or transforming beyond recognition) a series of industries at a time when the climate for private enterprise was far from salubrious. Among his many unique characteristics was a fierce refusal to let lack of know-how (forced at that time due to foreign exchange and licensing regulations) prevent his industry-building ambitions from being realized. Every part of the organizations Sumant Moolgaokar commanded was not just encouraged but goaded into creating the missing intellectual capital indigenously. This does not mean he was blind to developments in the rest of the world. On the contrary, he encouraged his team to keep close track of them, entered into some of the most fecund collaborations the country has known, and insisted on importing only the best production equipment the world had to offer. But he used all these only as a launching pad for his enterprises to develop and manufacture, generation after generation of products and capital equipment uniquely suited to Indian needs and operating conditions. So much so that, after years of parallel development, the Indian product usually beat the collaborator's latest version in the Indian market. Did all this take longer than continually buying ready technology from abroad? Assuredly. Were more mistakes made along the way? Of course—and some of them were costly. But what an unstoppable innovation-belching dragon was created as a result: one whose momentum carried it far beyond Sumant Moolgakar's lifetime or the engineering projects in which he took direct interest. It fell to my lot (after some very worthy predecessors) to run with this philosophy for HR in the largest company Sumant Moolgaokar oversaw. The thrill my team and I got in pioneering people policies, processes, and sources that had never been tried before (at least in India) was heady. It created a lifelong addiction that has been the prime source of whatever accolades I may have got in the HR profession. Admittedly, there were occasions when it was extremely tempting to buy the ready-cooked-offerings that became available as more and more global consultants set up shop here. And our home-baked efforts were never packaged as glamorously. Time and again, though, the ingredients carefully chosen and uniquely prepared for an organization's specific requirements proved to be far healthier, longer-lasting, and competitively more advantageous.

Very obviously then, an organizational environment where pioneering is not simply permitted but prized, is a prerequisite. Cultural change is expected to be part of the HR tool-kit. In trying to mold a pioneering culture HR needs to examine whether some processes, like forced comparative rankings, discourage the risk-taking that pioneering innovation demands. HR leaders also need to do a lip-service-check to ensure they are not demanding original thinking from their teams while lumping them with consultants whose prime ability is to report or replicate what some other corporate has done.

Having set the organizational stage, what kind of approach and preparation can be suggested for individual HR professionals to capitalize on a pro-pioneering milieu?

Here are five prerequisites for those who would like to be known for pioneering in HR:

1. *Reasoning from first principles:* Countries that have made technological breakthroughs invariably have higher education systems that place great emphasis on basic disciplines like physics, chemistry, and mathematics. The eminently sensible logic behind this is that great leaps in applied

domains like engineering can only be the result of insights originating in the basic disciplines. For HR, of course, the bedrock is psychology, together with the related behavioral sciences and select humanities. Starting from the basics of motivation and organizational dynamics and progressing to the holy grail of designing excitement and purpose into work, the entire gamut of HR endeavor has research-supported starting points in these foundational bodies of knowledge.

2. *Seeding from divergent disciplines:* In *The Act of Creation,* Arthur Koestler argued that the essence of creativity lies in "the perceiving of a situation or idea… in two self-consistent but habitually incompatible frames of reference." He used the term "bisociation" to characterize this process. One way to "seed" bisociation is to acquire reasonable familiarity with the principles and salient subject matter of one or (preferably) more additional disciplines. It would appear that the specific choice of discipline is not as critical as the necessity that it possesses frames of reference that are at variance from the domain where one seeks to make the breakthrough. Thus, for HR, the seeding could come from learning about or recruiting from (see the section on "Diversified Sourcing" later) domains as diverse as the natural sciences, information technology, literature, or the fine arts.
3. *Forced frugality:* Referring back for a moment to the Sumant Moolgaokar illustration with which we started this section, let us remind ourselves that Government policies and economic constraints in the preliberalization years made indigenous innovation a necessity for any organization that wished to grow very rapidly over a sustained period. Those wishing to expand nonlinearly became pioneers—virtually out of compulsion. An even more recent and striking example has been the success of the Indian Space Research Organisation, which was also forced to strike out on its own. Contra-wise, if there is no scarcity of one kind or another, a key driver to path-breaking innovation is gone. For the last couple of decades, we in India have been spoilt for choice in terms of prepackaged technology, processes, and solutions from the most developed economies in the world. Recent business cycles have also been kind enough to most Indian corporate for them to be able to easily afford buying these off-the-shelf answers. Taking such a direction would, however, deprive us one of the most important spurs to pioneer. As Jeff Bezos said, "I think frugality drives innovation, just like other constraints do. One of the only ways to get out of a tight box is to invent your way out." So even if (or, rather, particularly when) an organization is doing well financially, HR must protect its pioneering pathways by hedging itself with constraints and self-imposed frugality.
4. *Scepticism about fashion while pioneering with a passion:* General George S. Patton once said, "If everyone is thinking alike, then somebody isn't thinking." What may have been right for organization A in situation X is very unlikely to be appropriate for organization B in situation Y.

Even if the basic remedy is sound, if it is already a fashion, it can hardly yield differential competitive advantage and, therefore, does not deserve prime focus. As C. K. Prahalad put it, "Best practices lead to agreement on mediocrity. I do not have much interest in best practices. Because all of us benchmark each other, we gravitate towards mediocrity in a hurry. What we really need is to ask [is] what is the next practice, so that we can become the benchmark companies, benchmark institutions around the world." Moreover, if the cure had been around for a while, several large consulting firms would have "productized" and packaged it. Consequently, any health-enhancing active agents present in the original solution would have been boiled out of the standardized and shrink-wrapped residue. Of course, when every alternate company in the industry is implementing that particular nostrum, the pressure to conform, not least from your own top management and professional peers, can be intense. In such situations, the path-breaking road can be lonely, open to ridicule, and sometimes even career-threatening. For a person who wishes to pioneer, therefore, the need to be distinct from the herd is paramount. Such an individual's credo can perhaps be best summed up by the concluding lines from one of Robert Frost's short poems:

> Two roads diverged in a wood, and I –
> I took the one less traveled by,
> And that has made all the difference.

While we can hope to have a growing number of highly competent HR professionals, the wizards of the field will always be in very limited supply. It is these stars that organizations lucky enough to have them should treasure (despite their concomitant rough edges) and for whom our professional bodies should reserve their highest recognition.

A major side benefit of pioneering programs is the potential they have for generating an India-specific organizational behavior knowledge base. Of course, preparing accessible documentation with the necessary rigor could require the involvement of academics and support associates. There could be few better ways, however, to pay a tribute-through-action to Udai Pareek who, for decades, provided one of the finest models of pioneering thought leadership in organizational behavior.

Sound foundations in fundamental disciplines

Difficult as it may be to imagine, at one time my education was even less adequate than it is today. Fortunately, I was nominated to the first program on the "Integrated Approach to Knowledge" designed by the National Institute of Advanced Studies—possibly the last brainchild of J. R. D. Tata to see the light of day. The five-week program was masterminded and moderated by

Raja Ramanna—an unequalled genius and polymath. It brought India's leading thinkers, in a variety of knowledge domains, to impart a small part of their vast learning and to engage with senior civil servants and private sector talent over a series of sessions. In a short span of time, I found individual, social, and organizational behaviors, which had seemed puzzling and irrational, make sense. Large-scale change lost much of its mystery—though not its anticipated complexity or difficulty. It has been my firm belief since then that there are some subjects without which it is virtually impossible to grasp the deeper trends and causes affecting organizations and people.

A cursory glance at the following headers might raise plenty of questions about subjects that have been excluded. Apart from the need to keep the list very short, I have purposely left out management subjects (such as strategy, finance, quantitative techniques, information technology, operations, sales, and marketing) that should be part of any manager's skill-set. A much more comprehensive elaboration of these can be picked up from the syllabus of any reputed management institute.

There are three disciplines that, in my opinion, have a particular connect with HR. For each I have provided some significant (but obviously not exhaustive) reasons why HR learners might find them useful.

1. *Psychology and the social sciences:* It might be asked whether organizational psychology does not already form a part of the MBA course curriculum, at least for those specializing in HR at our elite institutes. Whatever may be in the curriculum, few fresh MBAs seem to be able to bring it to bear on the job adequately. Moreover, there is exciting and HR-relevant new work taking place in sub-branches such as social psychology, evolutionary psychology, and neuro psychology which is a long way from finding even mention in the texts used in MBA campuses. The behavioral science knowledge gap in HR is so glaring and vital that I am going to make a plea for psychologists to be an essential part of the sourcing mix for HR (see the next section). In the meantime, psychology must head the list of remedial learning for HR professionals (who are always very eager to prescribe remedial instruction for improving the employability of others whom the Indian educational system has failed).

   Let us take another example. In polite circles in corporate India, the unmentioned elephants in the room include biases against depressed castes and certain religious minorities. HR cannot afford to assume a lotus-fed ostrich attitude to these prejudices if it wishes to take its responsibility for diversity and nondiscrimination seriously (see the section above on "Treating People Fairly"). Moreover, permitting bias to deprive an organization of the talent of substantial parts of the population is irrational apart from being unfair. Left unchallenged, "caste" mind-sets also creep into other areas (e.g., how we deal with different categories or sources of employees) and prevent the organization from taking optimal decisions about them. For all of these reasons, HR has to understand caste and the (frequently unconscious) biases most decision-makers possess about it. My own understanding of the subject was extremely superficial till I encountered André Béteille and read his work. Credit for my understanding of unconscious prejudice must go to Mahzarin Banaji and her seminal work on implicit biases. The point is, HR cannot cure fundamental problems by preaching, punishing, or pretending they do not exist. If the causes of dysfunctionalities at the workplace are rooted in the social ills of the country, as is the case with caste, sociology must form part of the understanding that precedes a solution.

2. *History:* The case studies that are generally used as a management education staple stretch back, at best, a few decades. Would not it be remarkably mind-enlarging to have case studies going back centuries, where the protagonists were faced with crises and challenges infinitely more consequential than any confronting a corporate emperor? And what if we had a "Case B" that spelt out what the person did and what then followed? For other management areas there may not be such a magical source of cases but for the study of leadership and organizations, history provides a far richer compilation than all the OB cases of the world put together.

   That, of course, is not the only use of history. There can be no better antidote to dogmatic thinking and a single interpretation of events than reading the history of the same period or happening penned by a variety of historians. And as we prepare our managers to operate in distant global outposts, there is no short-cut to understanding foreign cultures, with their attendant prejudices and sensitivities, other than through an appreciation of the history that gave them birth.

3. *Philosophy:* This inclusion is not intended as a sop to those woolly headed protagonists of spirituality in management, many of whom perfectly exemplify the role of the *Tapasvi* in film *PK.* Quite to the contrary, we need an unbiased study of philosophy in order to acquire the scepticism that alone can immunize us from the fashions and fads against which we need to swim. Stewart Butterfield, the co-founder and CEO of Slack Technologies and himself a philosophy major, said, "Studying philosophy taught me two things. I learned how to write really clearly. I learned how to follow an argument all the way down, which is invaluable in running meetings. And when I studied the history of science, I learned about the ways that everyone believes something is true—like the old notion of some kind of ether in the air propagating gravitational forces—until they realized that it wasn't true."

   Political philosophy can permit us to grasp the dynamics that govern mega-sized corporations, figure out which organizational principles are truly immutable, and wonder whether the kind of thought development that doomed authoritarian states will have any bearing

> on the progress of corporate democracy. Finally, it is ethics that provides us with choices of methodology and language to reason about our codes of conduct and values, as well as to weigh alternatives when fundamental values conflict, as they frequently must.

Budding HR professionals who are ready to leave earth orbit sometimes ask me what authors I would recommend. They are puzzled to hear names like John Rawls, Randall Collins, Margaret MacMillan, and Leda Cosmides. If some of these enquiring minds have read the preceding paragraphs, I hope my responses to them are no longer mysterious.

Diversified sourcing and structured development

Let me enter one final minefield before concluding. Even assuming that HR has stopped being the *poubelle* of the professions, it risks being devoured by the Hydra of homogeneity—with each of the nine heads having the same HR MBA stamp. HR does a great job of preaching diversity while operating the greatest clone factory since Star Wars Episode II.

The best HR departments have a rich mix of people who have spent their entire careers in HR and promising executives from other functions who have decided on making a mid-career switch or temporary way-point in HR. To my thinking, up to half the strength of HR should come from people with such varied experiences. A major proportion of these should ideally have been line managers. However, leavening with people who have worked in consultancies, NGOs, and even the armed forces makes for a dynamic team that is never short of new ways of tackling seemingly intractable problems.

Even the sourcing of people who spend their entire careers in HR should be diversified. At least a third of these entrants should not be HR MBAs. Obviously, the majority of these non-MBAs should have Behavioral Science degrees but I have seen extraordinary HR leaders emerge from fields as diverse as Statistics.

This is not the place for giving more than cursory attention to the development regimen that makes for healthy HR capabilities. Suffice it to say that fresh entrants to HR should have, where feasible, an exposure to Employee Relations, another Center of Excellence (COE) in HR and a Business Partnering (BP) role in the first phase of their careers (say, within the first 10–12 years). Thereafter, while progressing through more senior BP and COE roles, there should be at least one stint in a line or project management assignment. It may not be possible (though certainly it is desirable) for all corporate to weave in geographic, sectoral, or scale diversity. Of course, the availability of such opportunities provides a potential advantage for large business groups to exploit. What even medium-sized firms can certainly do is to expose HR talent to varied situations such as those demanding crisis management, recovery and rejuvenation, or vertiginous growth. HR professionals who are conscious of their development and who find themselves in companies that cannot or will not provide at least some of the opportunities mentioned above should look for making a change to those that do.

I am conscious that I have presented a hypochondriacal list of ailments and a demanding set of cures, some of which go far beyond anything we have tried so far. I am not inclined to water down either for a simple reason. HR was called center stage primarily because the nation's economic growth after liberalization depended on managing people much more effectively than before and we were the function that had responsibility for this resource. Already our inability to play the star role convincingly has seen us being nudged to the sidelines in several organizations. It is for us to decide whether we wish to exit the stage of action altogether or claim the pivotal role the nation demands of us.

I am privileged to have been invited to write a piece in this book dedicated to Udai Pareek, a teacher I greatly admired. I first met him at the Indian Institute of Management Ahmedabad when I went there for the 6-month Management Education Program in 1978. By then, I had been working in HR for several years and thought I pretty much knew all there was to know about the function. After attending Udai Pareek's classes for half a year and having many informal discussions with him, I came to realize that a lifetime would not suffice to fill the pit of ignorance I had on the subject. Udai, I am still trying.

PART II

Institution Building

15 Beyond management: Some conceptual contributions of Dr Udai Pareek to the modern world*

T. V. Rao

> A man literally is *what he thinks*, his character being the complete sum of all his thoughts. As the plant springs from, and could not be without, the seed, so every act of a man springs from the hidden seeds of thought and could not have appeared without them
>
> —James Allen

Great gurus like Dr Pareek have spent their entire life time professing certain things by self example, and living in certain ways that spread desirable values for the good of the present and future societies. Dr Pareek had done this for over 85 years. We are all fortunate to have had his presence and wisdom in this period from 1925 to 2010 and wish it was much longer. The thoughts shared in this chapter are extracted largely from his works and, particularly, *Beyond Management*. The thoughts expressed in this book are relevant not only for today's world, but also to the future of organizations (inclusive of public, private, governmental, nongovernmental), to our country, other countries, the global society, and the entire mankind at large. In my view what Dr Pareek gave us through his books and writings, as well as his thoughts, and actions is of relevance and significance to the future generations of all—not only various organizations in this country but entire humanity at large.

For most of you who do not know Dr Pareek, he did his BA from St. John's College, Agra (Agra University), 1944, B. T. Teachers' Training College, Ajmer (Agra University), 1945; MA (psychology), Calcutta University, 1950; MA (philosophy), Agra University, 1952, and PhD (psychology), University of Delhi, 1956. He also did his diploma certificate in research methods from Italy. As his biography would indicate Dr Pareek widely traveled and experienced

---

* Based on the first Udai Pareek Memorial Lecture for the Jaipur HRD Network Foundation, Jaipur, delivered on March 23, 2011, by Dr T. V. Rao.

living in most countries representing the modern world ranging from the USA to Europe to Asian and South Asian countries such as Thailand, Indonesia, Singapore, and Malaysia. His students are spread across continents of the world, Australia and Africa to name a few. He admired all cultures and people from different countries and for the same reason he had very close friends in all parts of the world.

I met Dr Pareek in 1968 for the first time at the Administrative Staff College of India after his return from the USA. Then he joined the National Institute of Health Administration and Education at New Delhi. I was in correspondence with him as a student of the Osmania University and tried unsuccessfully to get a fellowship to work with him in the USA. I was already in employment at the Andhra University, Waltair and started the Department of Psychology and Parapsychology with Professor K. Ramakrishan Rao. The year 1968 was about half way in Dr Pareek's professional life and was the beginning of my career. Since then, I am fortunate to have been associated with Dr Pareek for almost 42 years which constitutes my entire professional life. In these 42 years, we had co-authored or co-edited 15 books and also edited 2 journals (*Indian Behavioural Sciences Abstracts* and *Indian Psychological Abstracts*).

There are many contributions Dr Pareek has made. For this chapter, I would like to take up the themes that are dearer to Dr Pareek and his original contribution. The most important of them in my view are his concepts of extension motivation and role efficacy and his conceptualization of decision-making and institution building in educational institutions. Also are his related concepts of dependency motivation and other motives. His conceptualization of the principles underlying human resource development (HRD) is more relevant even today and is certainly ahead of our times. I will deal with them one by one and draw implications for future emulation, action, research, and dissemination.

Extension motivation and extension values

As early as in the mid-1960s, Dr Pareek proposed the concept of extension motivation in his seminal articles on the new paradigm for development published in the *Journal of Social Issues* and *International Social Science Journal* (Pareek, 1968).

To summarize, the concept extension motivation simply means a need or a desire to extend oneself or the ego to others and relate to a larger group and its goals. It means a motivation for helping others, working for larger goals that benefit larger groups or society. It also means an ability to sacrifice one's own comforts and desires for the sake of others. It is this powerful motivation that has led to many great people to make sacrifices for the good of the larger community. All great preachers and saints have led a simple life and taught people to lead simple life. Mahatma Gandhi said Earth provides enough to satisfy every man's needs, but not every man's greed—it elucidates that the nature earth has enough resources and means to meet the basic requirements of a man but it cannot serve the endless greed of man. Here, he also meant implied to help the needy people of the society. Dr Pareek proposed that it is extension motivation that causes any given society to develop. "A super-ordinate goal probably arouses this motive. Such goals may therefore be important not only in developing harmony but also in sustained motivation of people in development" (Pareek, 2002, p. 120). See Box 15.1.

Box 15.1 Superordinate goals

Superordinate goals are not ordinary goals. They are meant to serve a larger principle. In creating and working toward these goals, a person derives satisfaction from the feeling that she is existing for a cause. Perhaps she was born for that. As one starts doing good work, others begin to appreciate them. With every good deed the person gets more power, appreciation, and recognition and this has a tremendous force. Take the example of Sarath Babu, the Indian Institute of Management Ahmedabad (IIMA) graduate who decided to set up his own "idly factory" immediately after his studies at IIMA instead of taking up a high paid, secure job. He has become a success story and a role model for many management graduates to emulate. Recently, he narrated an incident about a troubled, young girl who was about to end her life. She came across Sarath Babu's story and was so inspired by his achievements that she decided to live and make something of herself. It is incidents of this kind that help build one's determination to work for superordinate goals even though they may be weak at first.

Superordinate goals give individuals a high sense of efficacy. Mahatma Gandhi worked for an independent India adhering strictly to the principle of nonviolence. This was the driving force that enabled him to make several sacrifice and it was this ability to make sacrifices, not seeking power for him and leading a simple, immaterial life that made him the greatest leader on earth. He derived his values from his family and what he read when he was young. He was most influenced by the stories of Harischandra and Shravana Pitrubhakti Natak. He experienced the power of ahimsa in the way his father reacted to his confession that he stole.

Dr V. Kurien is another example of a superordinated goal achiever. He worked single-mindedly to make India self-sufficient in milk production and was the architect of Operation Flood, the world's largest dairy development program. Mr Narayana Murthy of Infosys started with a long-term goal of liberating information technology and generating wealth to be distributed among a large number of those who work for Infosys. After creating Wipro, Azim Premji started devoting his time to nation building and has been promoting education in rural India. They are all examples of superordinate goal seekers.

Source: Rao, 2010.

All organizations in the corporate or voluntary sectors are concerned about motivating their personnel to work for organizational or social goals, which go beyond an individual's own-perceived interests, and to promote collaboration, and commitment to superordinate goals. Extension motivation addresses itself to this. Extension motivation is also reflected in cooperation with others for achievement of a common goal, faith and trust in members of a group, and involvement in goals, which concern not only oneself, but also large groups, community, or society. Various dimensions of extension motivation may be: helping people who need it, collaborating with others, empathizing, risking one's comfort or safety for others, sacrifice, patriotism, hospitality, etc. Experiencing, reflecting, hypothesizing, conceptualizing, and experimenting are the chief ingredients of the process of learning in a laboratory setting. The EM Lab will utilize experience-based learning as the main training method. Dr Pareek had designed and conducted a few laboratories using this concept.

I consider this as very appropriate in today's circumstances when both India, at macro level and at micro level, and the world at large are witnessing certain vents that threaten the mankind. These events include scams that have shaken the very basic fabric of a country, unethical and irresponsible dealings by some of the corporate sector leaders who sacrificed basic values and long-term interests of the company or the country for immediate financial gains. Incidents like these result in confusion in the minds of people and loss of faith on professionals and professionalism. In my view, the reason behind the scams and unethical activities like insider trading, fudging of accounts, etc., take place essentially because of greed and selfishness. This greed and selfishness is obviously on the increase. In my view, extension motivation and extension values are antidotes to such diseases. They work both as antidotes to cure and even prevent selfishness and unethical exploitation of others for short-term gains of a few.

In this context, I must mention that his long-term friend and colleague Dr Prayag Mehta has been doing a lot of work on similar lines. Dr Mehta's conceptualization of social achievement and development motivation is particularly relevant to note here (Mehta 1994, 1995). Prayag Mehta has observed in his book on social achievement that the pace of development has been slow. "Studies emphasize that importance of organization, participation, and motivation along with public action for achieving such development goals. People are motivated by the need for social achievement and for acting on the environment for obtaining better quality of life and work" (Mehta, 1994, p.1). I am not going into details but Prayag Mehta's writings are of great significance in promoting development motivation particularly among the government agencies and agents and social achievement among the poor. In both these concepts, Dr Pareek's extension motivation seem to be part, though Prayag Mehta's concepts go far beyond a single motive and have great implications for development interventions.

A few examples of persons who lived with extension values and motivation are Mahatma Gandhi and Mother Teresa, and in recent times, Abdul Kalam and Kiran Bedi also demonstrate extension values. See Box 15.2.

Box 15.2 Extension motivated leaders

We all know little bit of Mother Teresa. One incident that is narrated often is when Mother Teresa went to a bakery man for bread for her orphan children. The story goes that as she raised her arms, the bakery man just spat on her hand. Then Mother Teresa told him, I would keep this for me, but give me some bread for my children while holding her another hand. On that spot, the bakery man realized Mother Teresa's gentleness and become a main bread donor for her orphanage. She started Nirmala Bhavan to look after orphans and disabled, and now it is spread throughout India. All through her life she worked for the poor and destitutes.

Mahatma Gandhi lived a simple living and worked for the country following nonviolence principles. When Mahatma Gandhi was traveling on a train one of his shoes slipped off onto the track, and he could not have picked it up, he promptly threw the second one also onto the track. When asked why he did that apparently he answered that when someone finds the first shoe the second one will help him to have both and use them. That is reflective of extension attitude.

Dr Kalam took up academic pursuit as Professor, Technology and Societal Transformation at the Anna University, Chennai, from November 2001 and was involved in teaching and research tasks. Above all, he took up a mission to ignite the young minds for national development by meeting high-school students across the country. Dr Kalam is an extremely simple man. He is above seventy and a bachelor. He is a strict vegetarian and teetotaler. He is a "work alcoholic" who knows no holidays in the seven-day week. He works 18 hours a day. He is fond of music and spends his leisure hours practicing the lute (*veena*). He is a great lover of books. He is a voracious reader of both Bhagvad Gita and Koran. Dr Kalam quotes, "for great men, religion is a way of making friends; small people make religion a fighting tool."

Kiran Bedi has been devoting her life during her working life as well as now for social causes working with nongovernmental organizations. While in service wherever she was posted she showed high concern for larger causes and tried to serve the people courageously and assertively.

If I analyze the people behind the scams, insider trading, fudging of accounts, or cheating the public, that has largely come from those who were born in early independent India—in the 1940s and the 1950s and perhaps a few born in the 1960s. I believe those born around that time were born in an independent, yet insecure, India. They had high aspirations and a few of them perhaps did not have the patience to put in hard work. As an outcome of which, they started resorting to short-cuts. They were perhaps born in a "subsidy" country, a country of reservations and a country where the government is supposed to "give" than "take." Their ambitions resulted in greed out of which they indulge in unethical activities. Such greed is not the property of some but unfortunately many. If this greed characterized the youngest and upcoming generation (Gen Y or Gen Z), then the country has no great future. To prevent this epidemic from breaking out, we need a strong medicine. Extension motivation is not only a good medicine to cure but a great medicine to prevent. We need to inculcate the same right from childhood, through schools, colleges and education and corporate training programs. I also like to propose the term "extension value" to be added to our list of desirable values.

Dr Pareek had written his book *Beyond Management* in 1981. He revised it in 1984 and again in 2002 and re-titled as "Effective Organizations." He dedicated the first volume to Ravi J. Matthai, the second one to T. V. Rao and the third to Rolf and Ronnie Lynton a unique couple engaged in institution building across cultures. It is not unusual for Dr Pareek to dedicate his books to his students. He dedicated one of his books to his disciples—Dr Mahindra Singhvi and Dr Narendra Dixit. This is something to emulate and speak of the person.

In what forms does extension motivation figure out in Dr Pareek's writings?

Two articles of Dr Pareek appeared in the book on effective organizations (Pareek, 1967, 1968).

In his article on "motivational paradigm for development," Dr Pareek gives a simple formula that

$$\text{Development} = (\text{Achievement Motivation} \times \text{Extension Motivation}) - \text{Dependence Motivation}$$

If you want socioeconomic development to take place, increase achievement motivation and extension motivation and decrease dependence motivation.

Motivation occupies an important position in the dynamic process of social change. The paradigm Dr Pareek (1968) suggested is as follows:

> Motivation causes human Behaviour, and to that extent 'causes' changes in a community, like increased entrepreneurial activity or adoption of new methods. But I don't view it as a primary cause of change in human society. Motivation is a strong and important link in the dynamic causal cycle of human evolution. (p. 117)
>
> According to this paradigm, a specific societal system generates a specific pattern of motivation and system of values that strengthen it, sustain it, and ensure its continuity. Behavior of people in the society is caused by dominate motivation in that society which in turn has been generated by the social system. This is perhaps a mutually confirming or reinforcing cycle.

Extension motivation is defined as the need to extend the self or the ego and relate to larger groups and its goals. A superordinate goal probably arouses this motive. Such goals may therefore be important not only in developing harmony, but also in sustaining motivation of people in development.

Poverty: Using this paradigm, Dr Pareek explained poverty as a consequence of low-achievement motivation, low extensional, and high dependence motivations. He explained further to say that low achievement results in disproportionate risk taking, interest in chance and not control, lack of interest in feedback, high interest in friends and not experts, and low activity and initiative. Similarly, low extension results in lack of regard for others, lack of trust and faith in others resulting selfishness or self-centeredness, and lack of cooperation. High need for dependence results in avoidance, fear of failure, seeking favor from supporters, and aggressively rejecting authority.

Lessons from Dr Pareek's extension motivation for future of nation building and institution building

- We must create an extension culture in India where (a) people value sacrificing their own conveniences for the sake of others and the larger goals; (b) they are willing to sacrifice short-term gains for long-term good; and (c) they become considerate about future generations and leave the planet for their safety and healthy living of future generations.
- We make policies that are driven by extension motivation and future.
- Extension values and related family of values should be promoted and taught in schools, colleges, families, everywhere.
- Extension motivation and value-based films and stories that promote the work done by social reformers and others should be made available to the public at large. Some industry groups are already making efforts in this direction.
- Corporate sector is genuinely motivated by concern for the welfare of the larger society and undertakes corporate social responsibility (CSR) activities out of such genuine concern and values than as a business strategy.
- The government and various ministries make policies that truly reflect concern for the welfare of the society and long-term thinking rather than short-term goals. For example, the Finance Ministry should look at the extent to which tax laws are promoting development of the country than merely increasing tax collections in a particular year somehow to meet immediate needs. Other governance systems are made extension friendly.
- Award ceremonies to felicitate those who help others to be organized.
- Extension motivation and extension value to become a core value of the country and the education system.
- Recruitment to teaching, health, and other social service professions is based on extension value.
- Scholarships are instituted for that exhibiting extension motivation.

Institution building

In his book *Beyond Management*, Dr Pareek postulated, while tracing the history of management and differentiating it from Institution Building, the following: Administration was replaced by Management. Management should be replaced by Institution building. Administration has been concerned with successful maintenance of an organizing, and its running according to laid down rules and regulations.

Management brought in changed emphasis and is primarily concerned with efficiency, that is, accomplishing tasks with minimum resources. In recent years, in addition to concern for efficiency, organizations are striving for two other aspects: continued growth and development (self-renewal) and creating a larger impact on a segment of the society or the entire society. For public systems the importance of both these aspects is too obvious. The first edition of the book had 13 chapters, the second edition had 36 chapters, and the last one expanded to have 43 chapters. The intention of the book is to help transform organizations into institutions. While management is concerned with getting results, stability, quality, effectiveness, strategy, achievement motivation, and competence, institution building focuses on vision, future, trend setting, networking, culture building, mentoring, extension driven, and empowering.

Dr Pareek differentiated institutions from organizations. To him "Institutions are distinguished by their mission, values, and impact on society."

Institutions should be agents of change in the society and the community. Knowledge utilization is a focal point of institution impact on the community. Institutions have self-renewal process.

One main contribution of an institution is to generate new values in the society or into the field of its operation. The nine criteria suggested by Dr Pareek for Institution Building are:

1. Attention to process
2. Significance of goal or uniqueness of the filed-urgent social needs
3. Innovative nature
4. Autonomy
5. Generating new values
6. Impact
7. Multiplication of know-how
8. Linkages
9. Development of people

Dr Pareek proposed a new value framework as:

1. From elitism to populism
2. From percolation to growth
3. Centralism to decentralization
4. Isolated professionalism to dialogue

Dr Pareek identified the following frameworks for decision-making:

1. Feudal framework that is based on ownership by a few and dependence is the main motivation.
2. Bureaucratic framework that is based on rules, procedures, control, and the likes. Main motivations are control and affiliation, and relationships are political and clique formation.

3. Managerial framework is focused on efficiency. Relationships are task driven; motivations are achievement and power driven.
4. Institution-building framework where health and organization's growth are primary concerns. Collaboration, extensions, and creativity are the motivation patterns.

See Box 15.3 for illustrative stories of institution builders.

Box 15.3 Institution builders

In my view the physicist and founder of ISRO (Indian Space Research Organisation), Vikram Sarabhai, and Ravi Matthai, the founder of IIMA, are two great institution builders. Sarabhai built a number of institutions in different fields, most importantly in space and management. Ravi Matthai promoted his belief of professionalizing management and actively demonstrated in his own life how management should enter all sectors.

Take IIMA. The symbols that differentiate people and communicate that you work for a team or group are minimal here. Every faculty member gets the same-sized room and each one, irrespective of their designation, shares the same secretary and privileges. You charge the same consulting fee irrespective of the designation. You are addressed as professor, irrespective of whether you are a professor or an assistant professor. There are no departments and there are only areas. The term "area" signifies a broad categorization. You may be a member of more than one area, group, or center. You may also change areas. All these are organizational mechanisms to create a larger identity and bring down the overheads or transaction costs associated with management of the system. IIMA follows most of the principles Udai enumerated.

Source: Reproduced from Rao, 2010.

Lessons for future

Heads of educational institutions, vice chancellors, managing directors, chief executive officers (CEOs), ministers, and all those heading government and non-governmental agencies should understand these principles of institution building.

First, they should respect and imbibe this institution-building role as a part of their role. Their appointments, socialization, performance reviews, incentives, and continuance in office should be subjected to a review of the extent to which they understand and follow the above conceptualization and values of institution building. They should be superordinate goal driven, less dependency driven, and more achievement driven. Then only will we have great institutions built.

As recommended by Dr Pareek, institution builders and heads of institutions should be assessed and trained on the following roles and the extent to which they are discharging their roles:

1. Identity creation role or the extent to which they promote unique identity for their organizations and at the same time ensuring societal contributions of the institution;
2. Enabling role in which they develop a variety of resources in the institution including human resources;
3. Synergizing role in which they ensure collective contributions and collaborative culture make the organization integrate various resources systems and achieve more than the sum of its parts;
4. Balancing role where they balance conformity with creativity and short-term with long-term goals and activities;
5. Linkage-building role where the CEOs create linkages required with external agencies and subsystems;
6. Futuristic role where they develop the capability to anticipate the future and future changes and prepare the institution to meet these needs or create its own future;
7. Impact-making role where the institution makes both internal impact through its achievements and climate it creates as well as the external impact in terms of influencing the policies in the field in which it operates; and finally
8. Superordination creating role where the top executive gives a sense of fulfilment to its members by deeply connecting what they do to the larger good of the society.

Role efficacy

Extending the concept of sense of efficacy to the role, Dr Pareek formulated a term called "role efficacy" (Pareek, 1993). According to this formulation, individuals with high degree of role efficacy carry with them different perceptions and feeling of their role. Role efficacy is defined as the potential effectiveness of an individual occupying a particular role in an organization. It consists of making your role the way you like (role making), feeing important and central in the organization through your role (role centering), and linking various aspects of the role to make it stronger (role linking). The various dimensions of role efficacy include:

1. *Self Role Integration:* Where the role provides individual with greater opportunity to use his or her special strengths. Integration between self and the role leads to higher role efficacy, while distance between the self and the role leads to low role efficacy.

2. *Proactivity:* Proactive behavior (taking the initiative) contributes to higher efficacy. While reactive behavior (merely responding to the expectations of others) contributes less to efficacy. Lack of opportunity to take initiative leads to low efficacy.
3. *Creativity:* Opportunity for creativity and innovation increases role efficacy, while performing only routine tasks becomes harmful for high role efficacy.
4. *Confrontation:* Confronting problems and reaching a relevant solution contributes to higher role efficacy, while avoiding problems or shifting problems to others leads to low role efficacy.
5. *Centrality:* A person's perception of the role as central to the organization contributes to high role efficacy, while a person's perception of the role as peripheral is likely to lead to low role efficacy.
6. *Influence:* The more influence/power a person is able to exercise in the role, the higher the efficacy.
7. *Personal Growth:* Person's perception of the role as providing opportunity to grow and develop leads to higher role efficacy, while a perception that the role does not provide the opportunity to develop contributes to low role efficacy.
8. *Inter-Role Linkage:* Linking one's role with others' increases efficacy. Joint efforts in identification of problems, problem solving, etc., increase role efficacy.
9. *Helping Relationship:* Person's perception that help is available when needed leads to higher role efficacy, while a perception that respondents are hostile leads to low role efficacy.
10. *Superordination:* Opportunities to work for superordinate goals have the highest role efficacy, while a perception that performance in a role is of value to the organization leads to higher efficacy.

Factors contributing to role efficacy include a participative climate, higher job satisfaction, climate-promoting concern for excellence, use of expertise, and concern for larger issues and an innovation-fostering environment.

Consequences of high role efficacy are less role stress, less anxiety, and work-related tension. Persons with high role efficacy tend to rely on their own strengths to solve problems, use more purposeful behavior, and are active and interactive with people and environment. They persist in solving problems, are inclined to growth, and exhibit attitudinal commitment while adopting a positive approach. They have a high degree of satisfaction with their jobs and role in the organization.

Of all the things that make a manager successful is the self-image the person carries with him or her. The self-image is like a shadow. It is known by different terms and ways. The related concepts are self-respect, self-confidence, ego, sense of efficacy, self-concept, self-worth, sense of inner worth, etc. People with high sense of values and superordinate goals share a feeling that they exist for others, etc. See Box 15.4.

Box 15.4 Efficacy

Consider the following responses by two of IIM graduates, both Indian Institute of Technology (IIT) toppers and IIM rank holders. Both were employed as executive assistants to the marketing director of two different organizations in two different cities. Both of them were working in similar companies. When the author visited them six months after they were employed and asked them to describe their job and how they were enjoying it, the conversation was something like this:

Person A: I am not sure what to say. I must admit that I am not at all comfortable doing what I am doing. I am not doing anything significant. When the human resources (HR) head and one of the line managers came to the campus for recruitment they drew a great picture about this company and my job. I was promised a number of things and none of that has happened. I came here with the expectation that I will have an opportunity to use my talent. Being an IIT topper and IIM rank holder, I came with eagerness to participate in strategy formulation, making marketing policies, influencing pricing, deciding marketing budgets, and helping the company and the director of marketing implement them. I was given a nice office next to the senior vice president (SVP). It was a good cabin and had all facilities. I was happy in the beginning but soon I was quite disillusioned to discover that my job has nothing to do with policies and strategies. I sit next to the SVP to act more like his secretary than as a manager. I am supposed to maintain his appointments though a secretary assists him, and it almost amounts to my doing the job. I am supposed to coordinate the visits of various customers and vendors. You know what that means. It means booking their hotels, arranging their transport, making their appointments, arranging their meetings, settling their bills, etc.—all clerical jobs. Then, I am supposed to arrange the weekly and monthly marketing meetings for the department as well as for the other departments. It means booking the meeting rooms, setting the agenda, circulating the agenda, taking down the minutes, getting them approved by the boss, ensuring that coffee and tea are served during the meetings and incurring the wrath of the boss in case of delays, etc. There is no strategy and there is nothing here to learn. I get my salary on time which is of course a great thing. However, professor, I am wasting all my talent and keep feeling what a gap there is between what you all taught and what we are doing here. I feel that I should get out of this place after a year.

Person B: Thank you, professor. I am quite happy here and am learning a lot. I sit next to the SVP Marketing. He is a person with 25 years of experience and has worked in three companies before he joined here. He is not an MBA but more than an MBA in terms of his experience and thinking.

(Continued)

(Continued)

However, he is not exposed to systems and I help him a lot. For example, I am required to maintain his appointments. He does not have a secretary but I help him as his executive assistant. I have developed an electronic diary on Google and after two months, began to help him to analyze the way he was spending his time. It helped him a lot. He discovered that 30 percent of his time goes into unplanned activities. He now regularly consults me and asks for my feedback on how to a make his role more effective and strategic. I have also begun to mentor some junior executives which he requested me to do after my analysis of his schedule.

I manage the customers and other visitors to this company. It was a little difficult in the first two weeks as I was new to this company but it gave me an opportunity to learn about it. I went to each head of department and asked them to give me a brief so that I could brief the customers who visit us about the various functions of the company. Now I know the company perfectly. I book the hotel accommodation and transport for the visitors and customers. Each one has their preferences. I have explored all the hotels in the city and now have all their details, the concession they give, etc. I also know the customers' preferences of hotels and have knowledge about the taxi system and various vendors of taxi services. This task of arranging their transport and hotels gave me an opportunity to explore the city. I am supposed to book the meeting rooms and ensure that the discussions are minuted. I kept quiet and observed for the first three weeks. After the fourth meeting, I pointed out to the boss that the decision taken at the meeting is problematic as the pricing they decided on ignored some variables. My boss appreciated my input and started involving me in the discussions. Now he consults me on the agenda and also relies totally on me to maintain and manage the minutes and follow-up of the decisions taken. I have also gained a lot of his confidence. This company gives me a great opportunity to learn and use my capabilities. I am happy and am learning and almost feel that I am the SVP Marketing.

Which of them is likely to be more effective?

The answer is obvious. Person A views everything negatively and expects learning to come to his doorstep, while person B takes initiative and sees an opportunity to learn in everything including administrative tasks. He is confident, takes initiative, applies his knowledge anywhere including minor issues like maintaining the schedule of his boss, booking hotel and transport for customers, and managing minutes of departmental meetings and booking meeting rooms and support services. The first one has a fixed mind and treats everything as a burden. Obviously, the second person is likely to be more effective.

Source: Rao, 2010.

Lessons for action

Corporations and institutions should relook at the role efficacy of their employees to ensure that their role efficacy increases in order to create the right atmosphere in the organization. Particular attention should be paid to the dimensions of "superordinate goals" and "helping," "growth," "proactivity," and "creativity" parts of role efficacy.

As Robin Sharma in his recent book on leadership observed organizations should link pay cheque with purpose. Organizations should interpret their goals and objectives properly and highlight for the benefit of their employees the larger impact they are trying to make on the society.

Human resources development and organizational development

Dr Pareek identified the following principles underlying HR systems (conceptualized in 1975 and I had the good fortune of doing this with him jointly for Larsen & Toubro, see Pareek & Rao, 1998).

1. *Focus on enabling capability:* The main focus of human resource services (HRS) should be to help the company increase what may be called its "enabling" capabilities.
2. *Integrating the development of people with OD:* Any HRS should aim at the development and multiplication of human resources.
3. *Maximizing individual autonomy and growth through increased responsibility:* Learning experiences get maximized when individuals have opportunities to experiment and also hold the responsibility for outcomes.
4. *Decentralization through delegation and shared responsibility:* People who work together or close to each other know each other better than those who do not have such opportunities to be together.
5. *Participative decision-making:* HR should attempt maximum participation or representation of people who are likely to be involved in its implementation or are affected by it.
6. *Balancing adaptation to and changing organizational culture:* While designing HRS, it is necessary to take into consideration the nature and culture of organization for which it is being designed and at the same time it should be designed to suit the prevalent culture of the organization and the intended direction of change.
7. *Balancing differentiation and integration:* With increase in expertise in HRS, the three distinct functions—personnel administration, organization development and training, and industrial relations need to be differentiated and at the same time integrated.
8. *Balancing specialization and diffusion of the function:* While HRS is a specialized function, its special identity should not result in lack of involvement of line people in various aspects of HRS.

9. *Ensuring responsibility for the function:* Unless the HRS is introduced at a high level in the organization, with a very senior person is in-charge of HRS, the insightful and creative leadership, which is required to make the function effective, may not be provided.
10. *Balancing linkages within and with other functions:* HRS should be designed to strengthen various functions that are important in the company.
11. *Building feedback and reinforcing mechanisms:* The various subsystems in HRS should provide feedback to strengthen one another.
12. *Balancing quantification and qualitative decisions:* While attempts should be made to continuously improve and quantify several variables and even to design computer storage of various kinds of data, the qualitative and insightful decisions will always be necessary and desirable.
13. *Balancing internal and external help:* A company that is completely closed to external resources and is doing everything internally may not be able to benefit from some of the latest developments in the field; while on the other hand, a company that is merely relying on external help either in training or other aspects of HRS cannot develop itself effectively without necessary internal resources.
14. *Planning evolution of the function:* Rushing the introduction of all the aspects of HRS may only result in ritualism rather than effective introduction of the system.
15. *Continuous review and self-renewal:* It is necessary to plan any system in such a way that it has mechanisms of self-renewal, and HRS should provide in it mechanisms for continuous review and change.

Greed and OD

Divisiveness continues. It spreads from caste and religion-based divisiveness, to political divisiveness to linguistic, regional, color, education, health, batch, sector, section, department, etc., everywhere. Some divisiveness for improving the disadvantaged is fine, but divisiveness leading to self-destruction of a large part including that of the Nation is not acceptable.

We seem to live today in a scam-driven world. Every day morning you wake up to hear what new scam has been unearthed: starting with Satyam in the corporate world to CWG, 2G, Adarsh Housing, Mining, and Land grab to everywhere. The nation's time is taken away for protesting MPS, MLAs, and various other agents who should be spending their time in Nation Building.

Are the Indian corporations saved from this? Has OD brought in certain amount of integrative outlook in our corporations? In a world where organizations are built and developed to sell than to build more, has OD done some good? Organizations built by the sweat capital of many people suddenly get sold and the employees discover that they belong to a new management and a new organization that they have not heard before and

(Continued)

(Continued)

they need to learn a new culture. This has become the world order and India is no exception. Sometimes, it appears that we are truly in the business of business and not that of services. In the past organizations used to be set up to serve others and in the process to make some money to sustain themselves and grow. Today, it appears that some organizations are being set up to make money and money alone. Even hospitals' performance is being measured by the numbers in monetary measures rather than patients served and benefits to the society. Government performance is also being measured by the numbers of schools and hospitals set up rather than what they are doing after they are set up.

In other words, short-term orientation emerging out of greed has become the order of the day. The concept of service to the society and working for larger cause is being forgotten except occasional when we talk of corporate social responsibility. Even this CSR gets reduced to projects to take care of the neighborhood rather than getting integrated into everything that we do in our corporations.

I have argued that Udai Pareek's OCTAPACE (Openness, Collaboration, Trust, Authenticity, Proactivity, Autonomy, Confrontation, and experimentation) values should now be expanded to include Extension Value. And all HRD and OD work should be reoriented to this in view of the need to build a strong and healthy world for our future generations.

Source: Excerpts from a key note address at the M. S. University Baroda, February 2011.

Lessons

It is high time that organizations of all forms conduct an HR audit or social audit of themselves, and their programs to renew themselves. This may extent to the political parties and particularly youth wings of political parties and various government agencies and nongovernmental development agencies and agents.

It is also high time we understand and assimilate the OCTAPACE values always so intensely promoted by Dr Pareek. In the end I like to conclude from a quotation from James Allen,

> Man is made or unmade by himself, in the armory of thought he forges the weapons by which he destroys himself; he also fashions the tools with which he build for himself heavenly mansions of joy and strength and peace. By the right choice and true application of thought, man ascends to the *Divine Perfection*; by the abuse and wrong application of thought, he descends below the level of the beast. Between these two extremes are all grades of character, the man is their maker and master.

Let all noble thoughts descend on us, characterize our being with inspiration from the writings of great gurus like Uda*ji*.

References

Mehta, P. (1994). *Social achievement motivation: Needs values and work Organization*. New Delhi: Human Development Series: Concept Publishing.

Mehta, P. (1995). *Education, participation and empowerment: Studies in human development*. New Delhi: Concept Publishing.

Pareek, U. (1967). A motivational paradigm of development. *Journal of Social Issues, 24*(2), 115–122 (and also in *Indian Educational Review*, 1967, 105–111).

Pareek, U. (1968). Motivational patterns and planned social change. *International Social Science Journal, 20*(3), 464–473.

Pareek, U. (1993). *Making organizational roles effective*. New Delhi: Tata McGraw-Hill (original concept in 1980 published in Pfeiffer and Jones).

Pareek, U. (2002). *Effective organizations: Beyond management to institution building; 1981, 1994, 2002*. New Delhi: Oxford & IBH.

Pareek, U., & Rao, T. V. (1998). *Pioneering HRD in India: The L&T system*. Ahmedabad: Academy of HRD.

Rao, T. V. (2010). *Managers who make a difference*. New Delhi: Random House.

16 Institution building: Experiences, learnings, and challenges*

Kavil Ramachandran

What is an institution?

Basically, institutions are organizations with a unique identity and respect. We feel proud to be associated with them. They are unique in a number of ways, and they are not easy to be copied. It is not for a moment in history that they shine, but they continue to shine, spreading light of guidance, inspiration, and leadership across many decades, and sometimes centuries. Three such pillars of global eminence in the educational field are the universities of Oxford, Cambridge, and Harvard. The earliest records on the Oxford University date back to 1036 AD, almost a thousand years ago. Cambridge University, again in Britain, is over 800 years old. Harvard University in the USA is close to 400 years old. These universities have been the destinations of the brightest minds of the world across countries, and they continue to be so. There are many other widely admired institutions, in academics and outside several with shorter histories existing in India and elsewhere. Some of the familiar names are the Indian Institution of Management Ahmedabad (IIMA); the Indian Institute of Science; The Ramakrishna Mission; the Christian Medical College, Vellore; commercial entities such as the General Electric, Tata group, Unilever, and Merck Pharmaceuticals, Germany.

It is worth recalling the tremendous efforts made by the Tata group with a clear purpose of contributing to nation building. Proud of India's heritage of industrialization prior to the industrial revolution, 150 years ago, Jamsetji Tata wanted to provide Indians with an opportunity to rebuild the nation. To quote Mr J. R. D. Tata, "With that object in view, he decided, almost single-handed, to launch India on the path of modern science and industry and

* This chapter is a modified version of the speech delivered at the N. J. Yasaswy Memorial Lecture on October 8, 2013, IFHE Campus, Hyderabad. It is specially modified for the Udai Pareek memorial book. Both Yasaswy and Udai Preek were institution builders in their own ways.

to risk his fortune in the process. That the great projects he conceived and his sons carried through were successful is less important than the motives with which they were launched, than also the sense of social consciousness and trusteeship which Jamsetji inculcated in his two sons and my father, R. D. Tata, and which have continued to this day to inspire and guide his successors and, through them, the management of the various enterprises which they promoted" (Lala, 2004). Such clarity of purpose creates enthusiasm among like-minded people to come together and contribute to the creation of social wealth. People associated with such institutions feel delighted to contribute to their efforts to accomplish the vision. This is what Dr Vikram Sarabhai did when he constituted the founding team of the IIMA. In recent years, the Indian School of Business was the vision comes true of a number of people who wanted to create an institution of global eminence.

Dr Udai Pareek, one of the early day stalwarts associated with building the IIMA, believed the following as the three features that characterize an institution.

1. "Its functions and services are related to society's commonly agreed requirements as tested by its adaptability over time to handle human needs and values.
2. Its internal structures embody and protect commonly held norms and values of the society to which it is related.
3. Its achievements over time include influencing the environment in a positive way." (Pareek, 2002)

In essence, institution building is different, very challenging, but very enjoyable too. Dr Pareek distinguished an institution from an organization and said that while an "organization is both a system of consciously coordinated activities and a rational instrument engineered to do a job, an institution is a responsive, adaptive organism which is the natural product of social needs and pressures" (Pareek, 2002). Dr Pareek was one of the pioneers in India to work on institution building. He defined institution building as, "the process of establishing or transforming an organization into an integrated and organic part of a community in a way that will help the organization play a proactive role in projecting new values and become an agent of change in the community" (Pareek, 2002).

There is sometimes an assumption that institutions are primarily driven by social purpose with very limited economic purpose. I do not believe that economic organizations cannot be institutions. They too do. The extent to which different organizations contribute, directly or indirectly, to the society would vary. All of them have a role to play. The essence of an institution is the constant process of evolution and revolution an organization undergoes while continuing to be an important contributor to the preservation and growth of the society.

We realize that institution building is not easy. It does not come automatically to all. Besides all the qualities discussed earlier, it requires leaders to possess qualities of what I call, detached passion. It sounds like an oxymoron! Detached passion assumes one to be passionate with the ideas and goals, but also have the ability to take a couple of steps back and look at the organization, self and one's own thinking and activities objectively, all done in the interest of the long-term sustenance of the organization.

Learnings

I have attempted to capture some of the key learnings on institution building from the experiences of a number of organizations. The core to all these learnings is the recognition that it is easy to create a document of strategy but it is not always easy to achieve the same in practice. Execution excellence is core to institution building.

- One should enjoy what one does. Work is joy and not a duty to all those involved in institution building. This means that those who are chosen to be members of an institution should share the purpose and process of institution building. S. R. Ganesh and Padmanabh Joshi had in a study of Vikram Sarabhai's leadership noted that "the process of institution building is the energizing of people so that not only they internalize values that transcend narrow self-interests but they also become infused with a sense of mission in their total life" (Ganesh & Joshi, 1985). Jamsetji Tata's selfless efforts at creating high-quality education in India is worth recalling. In 1898, he took the initiative to set up a world class academic institution and contributed a donation worth ₹30 lakhs in those days. It was half his wealth. The other half he left to his sons (Lala, 2004). That was the origin of the Indian Institute of Science. Sacrifice for the society was a joy for him and his successors.
- Institutions not only identify their core values but also preserve them through their practice across the organization. For example, the 102-years-old IBM, the institution that constantly transformed itself and maintained respect and leadership across the world, did something about 10 years ago normally unbelievable for an organization of its size to do. In 2003, it ran interactive exercises with 50,000 employees across the world as part of revisiting and refining its core values. Incidentally, IBM's three core values are:

    - Dedication to every client's success
    - Innovation that matters for our company and for the world
    - Trust and personal responsibility in all relationships

In 2004, they did a similar massive exercise covering 52,000 employees worldwide to share their best practices. IBM leadership did not see such exercises as rituals but serious and dedicated efforts to strengthen their organization.

There is a need to learn from institutional history. All institutions draw energy and confidence from sharing the elements of culture, values, and best practices. They get inspiration from experiences when they withstood tests of their greatness. One of the reasons I consider the IIMA as a great institution was one such incident that a senior faculty member narrated to me soon after I joined there as a young faculty. In the early days of discussing the proposal for government funding for the Center for Management of Agriculture at the IIMA, Professor Matthai had a meeting with a senior cabinet minister in New Delhi. The minister agreed to support, but at the end of the meeting, he apparently asked for a favor— admission to IIMA for one of his close family members. Professor Matthai stood up and politely told the minister, "Sir, we have a very transparent and objective admission process without any interference even by the director. If you wish to link the grant money with the admission of your ward, I am sorry, I will have to politely decline the offer." This gave me the confidence and reassurance that IIMA was beyond any individuals' influence under any circumstances. Also, I could always stand up to my principles and values. I wanted to be associated with it always. Yet another incident I recall is when a very senior professor walked up to the fourth floor office of mine in my very first week at the IIMA to offer me any support in building up our activities on the entrepreneurship front. You should remember that this was something he had been passionately pursuing in the previous years, and I would be taking it away from him. That was Professor Vora for whom institutional continuity was the most important criterion in his decisions.

- I am told that Goldman Sachs, set up almost 150 years back, and which has been a heavy weight in financial services and banking, retain several offices of very senior retired executives for their use whenever they visit the office. They are encouraged to interact and share experiences with younger executives whenever possible. Also, they have several executives who have been with the company for 10, 15, 20, or even 25 years. They pass over the history and experiences of excellence across generations. News about them are shared with the existing staff. They retain an atmosphere of family and constantly build their institutions.
- Well-established family businesses follow exactly the same principle. For instance, Merck KG, the German pharmaceutical giant, was set up in the year 1816 as a small manufacturing company. The Merck family has preserved the remnants of their humble beginnings, for the benefit of retaining their legacy. For the past more than 80 years, the family has been involved only in governance and strategy, leaving overall strategy

formulation and implementation to non-family executives. The family remains the custodians of the wealth, and has created very clear governance mechanisms to preserve and grow the wealth. They are also aware that they are not the best managers to run the company. There are quite a few well-established and respected families that have preserved their families as lasting institutions.

Institution building is like a relay race. The founding leaders who realize this and set norms for the successors to follow do well for the creation of institutions out of organizations. The greatness of Professor Ravi Matthai was in doing this and setting a norm for others to follow in all IIMs. He took over as the director of IIMA without any specific end to his tenure. He could have continued indefinitely, but chose to step down after six years because he believed that the institution required fresh thinking on a regular basis. After stepping down, he set a new norm by becoming a regular faculty member like all his colleagues and worked from a faculty office like anybody else. For him, more than anything else was the continued success of the institution. He was not at all worried about the size of the room or its location. Succeeding generations get inspired by the examples set by leaders like Professor Matthai. Institutions that believe in the relay race principle have clear succession plan as well. They try to build as many towers as possible, and avoid posts in between. The efforts made at the General Electric to find a new chairman are well known. Jack Welsh has detailed how he went about identifying J. Immelt as his successor (Welsh & Byrne, 2003).

- Another key observation is that people trust each other a lot in such institutions. There is always an element of informality, and colleagues can get a lot of things done over phone or across table, all based on mutual trust. You can always follow up many things through an email confirmation, but action is not delayed for want of an email or formal communication. Trust will grow only if it is practiced as a value, top down.
- Team members, particularly division heads and faculty members in academic institutions, need a lot of freedom to operate. While accountability is a critical must, individual team members should have adequate freedom to do things, all for the benefit of the organization. In essence, empowerment should happen regularly and routinely.
- One of the key qualities of a good institution builder is his or her ability to listen to others, encouraging criticism and disagreements. Such leaders know that there are smarter and more capable people around and there are many ways of addressing organizational challenges. All great academic institutions have completely open faculty meetings where leaders encourage dissenting views.

Challenges in institution building

Let us now look at some of the key challenges of institution building that we face today.

- Leadership vacuum: I believe this has two dimensions. One, the pool of candidates to select from, and the other is the process per se. Unfortunately, leadership selection process is often either biased toward job skills or inadequately careful about the emerging needs of the institution. Attitudes are equally if not more important than skills and knowledge. It is also increasingly becoming more difficult to identify leaders with the true spirit of custodianship values. The whole challenge boils down to succession planning. There is a lack of anticipation and homework completed in time. The recent selection of the director of IIMA and the Chairman of the Tata Group took more than a year of search and short listing. Institutions should deliberate and list down the qualities of the leader they look for. The same applies to faculty members of an academic institution. Some of the young academic institutions forget that research and publications alone are not good enough to build institutions. You need faculty members to have dedication to the organization. You need people with ownership passion for the organization.
- Horizontal entry: We often find recruitment of leaders at different levels in the organization without any appreciation for the rich heritage of the organization. This is one reason why respected institutions very often prefer one among them as the best choice to be their leader. Cultural synergy is very important if the prevailing culture is good. Of course, institutions that need some resurrection may benefit from fresh flood provided the person coming in is aware of the context, and is appreciative of the institution's past strengths.
- Shortening life span: Similar to shortening of product life cycles, executives tend to move from job to job frequently. This has become a challenge for anybody to develop a shared culture in such organizations. Unfortunately, educational institutions and hospitals with larger social purpose have also become a prey to this trend for a variety of reasons.
- Vulnerability to pressures: Many institutions are threatened when leaders take decisions to protect others under the influence of colleagues, friends, and influential people. Little do they realize that such decisions, apparently small, affect the health of the organization. Very often the drift from clear and established norms takes place gradually. Such deviations, particularly compromising values and codes of conduct, cut at the very roots of the organizations. Many organizations do not tolerate instances of honor code violations very seriously.
- Organizational politics: Power and politics is present in all organizational entities. The question is only about its intensity and spread. Institutions

constantly resist forces of power and politics from spreading and killing them. A powerful manifestation of the same is lack of appreciation and recognition for people who slog, selflessly.

- Too much of systems and too little human: For a variety of reasons, we are all becoming prisoners of systems and processes. An unfortunate outcome of the over emphasis on systems is the neglect of the importance of values in organizations. This leads to a breakdown in informal communication across departments. Institutions of repute do emphasize the roles of both and work out a way for the coexistence of both, synergistically. In other words, balancing the roles of structure, systems, processes on the one hand, and culture and values on the other is not always easy.
- Arrogance and lack of benchmarking: Success in institution building sometimes becomes the very reason for them becoming arrogant, lethargic, and passive toward dynamic benchmarking. Several organizations, set up with very clear vision and other building blocks of institutions, have fallen under the weight of their past success. Let us remind ourselves that humility is never a weakness but a virtue.
- No life beyond the founder: Founders have a responsibility to ensure that leadership succession takes place in time and that too into safe hands. This does not always happen because the leaders themselves cling onto positions beyond their relevance. Because of their overbearing image and clout, other members of the organization dare not broach the topic. There are plenty of instances of organizations that show characteristics of becoming institutions crumbling under their own weight. Dr Pareek had noted that institution builders voluntarily dispose themselves, as they see it as a means to build the institution further. Unfortunately, we have many leaders of lasting organizations refusing to step down from their long possessed positions. They seldom realize that such institutions may crumble in the absence of smooth and prepared leadership succession.

In essence, all stakeholders should proactively nurture the institutions they are associated with. They should remember that they will not be able to build anything if the foundation itself disappears.

Challenges of nation building

I would like to close with some thoughts on the challenges of nation building we face today. This is not limited to India. Most countries face a dearth of leaders with a strong sense of custodianship. The scenario is very alarming in countries such as India that are struggling to be on their own, and yet to build strong economic foundations. Sacrifice of millions of people for our independence seems to have gone in vain when we look at the thoughts and actions of most of our politicians, jailed or otherwise. Two or three generations ago, Indians masses sacrificed their lives with the hope and satisfaction that their efforts would make their future generations happier. They were driven by a purpose and custodianship values. They wanted to transform the country and make it rich and a happy place for everyone to coexist. Unfortunately, down the road, there are many instances of drift, and we have fallen into a morass from which we need to get out, the soonest. We need to improve our governance drastically. I often reflect over what Mahatma Gandhi had said. To quote, "whenever you are in doubt, or when the self becomes too much with you, apply the following test. Recall the face of the poorest and the weakest man (woman) whom you may have seen, and ask

yourself, if the step you contemplate is going to be of any use to him (her). Will he (she) gain anything by it?" (Pyarelal, 1958).

I would urge the younger generation to be realistic of the challenges ahead and their implications, and do whatever is possible by them to build institutions in the society. That alone is the way to make our society an institution. There are sparks of change, on lines with the message from the Gita about the arrival of a messiah to lead us. There are elements of the messiah in all of us. Let us do our bit. I am hopeful because I believe in humanity, and our responsibility to the future generations.

References

Ganesh, S. R., & Joshi, P. (1985). Institution building: Lessons from Vikram Sarabhai's leadership. *Vikalpa*, 10(4), 399–413.

Lala, R. M. (2004). *The creation of wealth: The Tatas from the 19th to the 21st century*. New Delhi: Penguin Books.

Pareek, U. (2002). *Effective organisations—Beyond management to institution building*. New Delhi: Oxford & IBH.

Pyarelal. (1958). *Mahatma Gandhi, last phase* (Vol. II, p. 65). Navajivan Publishing House.

Welsh, J., & Byrne, J. A. (2003). Straight from the gut. *Business Plus*.

17 Managing an institution of excellence: Reflections on my years as director of IIMA*

V. S. Vyas

I am thankful to Dr T. V. Rao and Dr Anil K. Khandelwal for providing me with an opportunity to pay homage to my esteemed friend Dr Udai Pareek. I was fortunate to come in contact with Udai in the early 1970s, and enjoyed his friendship till his death in 2012. Though we both belonged to Rajasthan, it was only when I came to Indian Institute of Management, Ahmedabad (IIMA), that I met him personally. He and Rama Ben were the first to welcome us in IIMA campus the day we reached there. He acted as my friend philosopher and guide throughout my stay there, and played the same role when both of us settled in Jaipur after our retirement and were associated with the Institute of Development Studies (IDS). I learnt a lot from him on management of the academic institutions during my tenure as Director of IIMA from 1977 to 1982, and later as a colleague on the Governing Board of the IDS, Jaipur. I think it appropriate on this occasion to share my experience in institutional building and organization management, a subject closer to the heart of Udai, taking my experience as the Director of IIMA as an example.

At the outset let me point out two limitations of this presentation. First, this chapter is not based on serious research on institution building or organization development. It is largely based on my personal experiences and reflections. Second, my experience as director of IIMA pertains to the late 1970s and the early 1980s, and therefore may be considered outdated. However, on the latter score I am less diffident. If one agrees that there are certain principles of good management of organizations, which are of lasting relevance, then, I think that the experiences that I am sharing in this chapter have some value. I also

* An enlarged and revised version of a lecture delivered at Institute of Rural Management, Anand (IRMA).

had the good fortune to be involved in the governance of some other institutions in this country and abroad. Those experiences have further emboldened me to say what I am going to present in this chapter.

Let me begin by giving some background about my professional career. Before I came to IIMA I did not have any formal training in management. I was trained as an economist and specialized in agricultural economics. After doing a PhD in economics I started my career as a lecturer in the School of Economics at Bombay University. Within a couple of years I got an appointment as a reader in economics in Sardar Patel University, which was then known as Sardar Vallabhbhai Patel Vidyapeeth. Soon after my joining the Department of Economics in that university, the Agro-Economic Research Centre (AERC) covering two states, Rajasthan and Gujarat, was established in Vallabh Vidhyanagar. Since I belonged to Rajasthan and was working in Gujarat and earned some reputation as an agricultural economist, the authorities thought that I was the right person to head the Center. At AERC I could assemble a very good team to work with me. We started from scratch, but within a short period the AERC Vidhyanagar was recognized throughout the country as a good institute in the area of agricultural economics and rural studies.

It was during that time when Ravi Mathai, director of IIMA, sent D. K. Desai who was the chairman of the Center for Management in Agriculture (CMA) in IIMA, to invite me to join the institute as a professor in CMA. By that time S. V. Vidyapeeth had already promoted me as professor in the Department of Economics. I declined Ravi's offer and told D. K. that I was very happy where I was. A couple of years after that I went to Delhi as a Member of the Agricultural Prices Commission (APC) but kept my lien on the job in the University and came back to Vidyanagar after two years. When I was in Delhi, Samuel Paul, the then director of IIMA, repeated the offer that Ravi had made, that is, that of the senior professor in CMA. Although I did not say no to him but I was postponing the decision to leave Vidyanagar. I came back to Vidhyanagar and took over as a professor and the head of Department of Economics, and was also elected Dean, Humanities and Social Sciences. By that time my children were growing up, and my wife and I decided that Ahmedabad was a better place for us, and offer from IIMA needs to be taken seriously.

The main attraction for me was CMA in IIMA where a group of well-qualified and enthusiastic scholars were doing excellent work in the area of agricultural policies and management. Of course, the reputation of IIMA as an institute of excellence was also an important consideration. Finally, I told the University authorities that I would be resigning my post and joining IIMA. This is how I came to Ahmedabad.

I was invited to join CMA as a senior professor, and was happy with the thought that I would be able to contribute to the research and teaching in that area. Soon after I came to Ahmedabad, the Gujarat State Fertilizers and Chemicals Limited (GSFC) established a chair in Agriculture Policy in CMA. The board, on the recommendation of the director, made me the GSFC Chair

Professor. I was made chairman of the Research and Publication Committee of the Institute and was invited on several occasions to serve on the Selection Committees for recruitment of senior faculty. Professor Paul, then director, Udai, and other senior colleagues in the Institute made me feel that I was a valued colleague. It was very nice to get recognition from persons for whom I had great respect.

After a few years of my joining IIMA, Professor Samuel Paul decided to step down and offered to continue as a professor, a tradition that the first Director of IIMA, Ravi Mathai, had established. The Institute was in search of a director. My name was also included among the prospective candidates. Two of my senior colleagues, Udai Pareek and S. K. Bhattacharya, asked me whether they could propose my name for the director's position, I told them that I may not be the right person for the post of the director as my background, qualifications, and experience were not in the area of management education. However, I was told that neither Ravi nor Samuel Paul, the two former directors, had a formal degree in management. According to these friends formal qualification in management was not a precondition for leading IIMA. What was important was to have a director whose profile met the then existing requirements of the Institute. It was felt by several faculty colleagues, and later by the board that I met those requirements. They thought I had leadership qualities and I enjoyed respect among academic and policy-making circles. In short, leadership qualities and respect in the wider world were considered important for a person whom they would like to select as the director.

There is a lesson, which I feel is worth underlining. At different points of time decision-makers in the institutions have to be clear as to what kind of person they would need as a director. Quality of leadership is of course important, but the type of leadership needed in different circumstances will differ. At that time the IIMA board, and the faculty were looking for a director who was known in the policy-making circles in the state and the Center, enjoyed respect among academics, including of course the IIMA faculty, would give due emphasis to the areas other than private sector in management research and education, and further would strengthen the Institute's relations with its various stakeholders. They thought that I met these requirements.

I will not like to dwell on the developments during the tenure of my directorship. Suffice is to say that there were certain objectives that I could achieve, and there were others, which I could not. There were successes and failures. In mentioning these, I believe, I will be pointing out some facets of the management of an institute of excellence.

At the outset let me emphasize that the Post Graduate Program (PGP) of IIMA was excellent at that time, as it is now. My job in that regard was to ensure that we adhere to the high standards for which IIMA was just famous, and do not allow any slippage. The individuals who worked as Academic Deans during my time, particularly Mohan Kaul, Raghavachari, Narayan, made this task easy for me. They were stalwarts in their subjects and uncompromising in maintaining standards. Because of these friends I could spend more time in non-PGP type of activities.

I could establish much greater contacts with the noncorporate sectors. With the corporate world the Institute had very good contacts right from the very beginning. But the relationship with central and state governments, with other academicians, with the civil society institutions needed to be strengthened. I could break invisible barriers. I tried to get more faculties with diverse expertise though related to the core subjects in our curriculum, and invited people from different walks of life who had significant achievements to their credit to share with us their insights and interact with the faculty.

Another area where I took initiative was the development of management faculty in different institutions. I was convinced that even with their Fellow program the IIMs would not be able to meet even a fraction of the demand for teachers required by a growing number of management institutions. Management departments of the universities were, generally, very weak. Most of these departments were erstwhile Commerce Departments, they had only changed their signboards—instead of Department of Commerce they were designated as the Department of Management! I thought we should do something about this.

We had a short-term faculty development program, of the duration of ten days or so. It was more of an orientation course, by whatever name it was called, and was not taken seriously either by us or by the institutions sponsoring the trainees. With the help of senior and experienced faculty led by Professor A. R. Kulkarni, we designed a regular faculty development course (FDC) of 9-month duration, oriented to the needs of small universities and colleges. We launched this program successfully in India and later took it to Nepal, where we established collaboration with the Tribhuvan University in Kathmandu.

An initiative worth noting is the arrangements for summer training for students who had been given admission from ST/SC quota, most of whom had difficulty in English and mathematics. We organized a residential course for them in English and Mathematics during the summer vacation. It was made obligatory for these students to attend the course. All of them were provided fellowship during the period of their stay in IIMA campus to attend the course. The course was of seven to eight-week duration. Apart from improving their knowledge of the subjects, participating students developed courage and confidence.

There was significant expansion in infrastructure. We could get plot of land more or less adjacent to main campus, which now is the second campus. We built management development center and named it after Kasturbhai Lalbhai, one of the founding fathers of IIMA. Instead of organizing management development programs mostly in hotels as we used to, I thought that there should be a facility in the campus itself where we could organize most of our management development programs. There were other infrastructural facilities, transit houses, more faculty housing, etc., which were developed during my time. I was keen to forge international linkages. So far our relationships with the institutions of the Western world were one sided. Every year a few of our students or faculty used to go to these institutions for higher learning, mainly to pursue PhD programs. I wanted to make a two-way traffic. We forged that sort of relationship with INSEAD one of the leading management institutes of Europe. As I had worked with Food and Agriculture Organization of the United Nations (FAO) and World Bank, I could persuade them to locate some training or research projects in IIMA or award consultancies to the faculty.

These are some of the areas where I can claim some success, some achievements. But there were a few failings as well. For one thing, despite my best efforts I could not encourage much research, and as we all know if we do not have a good research output the standards of teaching also suffer. All my efforts were frustrated by endless discussion on what could be considered *relevant* research in management, rather than doing something about it. It was a serious disappointment for me because before becoming the director I had acted as the chairman of the Research and Publication Committee. There were few individuals who did good research in their areas; many more were engaged in case research; but I could not create an atmosphere conducive for research.

A major failing was the inability to manage the discontent among the administrative staff. To some extent, the conflict with staff is inevitable. We in the faculty believe that academic institutes are faculty-centered institutes, but other employees also have their aspirations, which they genuinely feel are neglected. This is what happened in IIMA. We neglected research and administrative staff. Anyway discontent was brewing. It got aggravated due to external factors; some of those were largely beyond our control. Some individuals had captured unions of the administrative workers of some of the most prestigious institutions in Ahmedabad, for example, IIMA, National Institute of Design, Physical Research Laboratory, etc. They were nihilistic in their approach and were bent upon wrecking these institutions. They were not prepared to have any constructive dialogue with management. Our failing was that we did not anticipate those troubles and did not take necessary steps. I am narrating these because one can learn lessons from successes as well as failures. I will like to underline these as I proceed. There was a strike on the strategically chosen convocation day but all the faculty came alive shared the work and got the convocations process go on. The only disappointing part was it went on without the staff and the Chief Guest J. R. D. Tata. Looking back I am not sure if it could have been avoided. Institutions do mature with such occasional crises. Even today IIMA's smooth functioning is clearly attributable to loyal and committed administrative staff who share the burden of faculty.

Some reflections on what makes a good institution

Now let me leave aside this autobiographical part and see what makes a good academic institution, an institute of excellence. People have given thoughts on it. Udai has made seminal contribution in this area. Others, including Pradeep Khandwala, T. V. Rao, Ranjit Gupta, and Tushar Moulik, have made excellent contributions in this area. I must admit that I have not done any systematic study. What I am suggesting below are reflections based on my experience of managing an Institute of Excellence. And I consider IIMA as an example of the institute of excellence.

The first thing that impresses one about IIMA is that it is an institute, which could maintain the reputation for quality education for nearly 50 years or so. How could it maintain this reputation consistently for all these years? There are several institutions in our country known for the quality of their teaching or research for some years, even for a decade or two. There are numerous examples of the institutions enjoying their hour of glory, but what factors have sustained the reputation of IIMA, for all these long years? I believe instituting certain procedures and practices on the one hand ensures excellence, and the role assigned to, and played by key individuals. In this chapter, I will discuss both the aspects. On the former, I will suggest Ten Commandments (!) taking IIMA as an example.

First, there should be clarity about the mission of the institution, especially among the founding members, and their commitment to achieve the goals. This should be reflected in the preparatory work, in the people hired for meeting the objectives, and resources provided to accomplish the tasks in an adequate manner. People like Vikram Sarabhai, Prakash Tandon, and Kasturbhai Lalbhai were clear about the type of management education they wanted to provide, they sought the right persons such as Ravi Mathai and Kamala Choudhary to accomplish this task, and they gave them necessary resources and full backing to organize the programs.

Second, the institutes have several constituencies. It is important to have a sense of discrimination, that is, a sense to decide which are important constituencies and how much weight should be accorded to them in a particular situation. Successive directors nurtured the relevant constituencies, the industry, government, academics, without being subservient to any of them. In time of need the Institute could received support from one or the other, sometimes from all, of these constituencies.

Third, there should be overwhelming concern for quality. Reputation of an institution, even of an "institute of excellence," cannot be taken for granted. It can easily evaporate. There should be built-in systems to ensure that high standards are maintained in teaching and research. A corollary to that is constant review and adaptation of the teaching material, and exacting review of the research done by a faculty. IIMA had such mechanisms in place, especially in postgraduate teaching.

Fourth, the Institute should evolve continuously. For an institute of excellence, say in the area of management, training mangers for the corporate sector is not enough. Adding to its core competence it should extend to new but relevant areas. Thus, IIMA, a management institute, started with a CMA, decided to work with Public Systems, took up newer challenges, e.g., a separate course for the management faculty in different institutions, a three to four month Management Education Program (MEP), a 3-tier course for top bureaucrats, and so on. If an institute does not evolve, and just continues to do more of the same, it stagnates and then decays. An institute of excellence makes constant efforts to extend its horizons, without losing sense of its mission. In IIMA periodically high-power Committees on Future Directions were appointed, and the reports of these committees were always taken very seriously.

Fifth, another prominent feature of such institutes is a very strong sense of autonomy. One can see this in the case of IIMA. During last few years when efforts were made to bring greater government control and interference in the functioning of institute, it is not only the Board that resisted such move, the faculty was equally agitated. It is not of much consequence if only a section prizes autonomy. It should be shared as a value by all concerned.

A minimum, but by no means a sufficient condition, to ensure autonomy is the financial self-reliance. We know that one who pays the piper calls for the tune. If the institute is fully dependent on the government, or on a particular section of the society, autonomy can be easily compromised. When I was the director the institute's relations with the government, at both the state and the central levels, were excellent. It was easy to get additional grants from the government. We were, however, careful not to have more than 49 percent of our budget from government grants, because at that time if the institute was earning 51 percent of its budget from its own sources, it was not subjected to CAG, that is, central government audit. With government audit comes the government culture, which is not good for an academic institution. The moral of the story is that even when IIMA was a favored institute, it opted to earn a large part of needed resources from its own efforts. This was a contributory factor in maintaining autonomy of the institute. (I am painfully aware that financial independence is no longer a guarantee for institutional autonomy.)

Sixth, the institutes of excellence have the ability to attract and retain good faculty. Nowadays, it is not easy because there are competing offers, a qualified faculty may opt for some other organization or leave at the slightest pretext. In IIMA the proportion of those who are staying on for a number of years is very high. Of course, at the margin there will always be some movements, and for good reasons. Sometimes such movement is desirable; otherwise the environment will become stale. There should be some movement, but it should not be a large-scale exodus. When some faculty leaves the institute should have the capacity to attract equally capable person. Even today with substantially higher salaries offered by private sector, IIMA is able to attract very good people. There are good students to teach; there is good environment to work; there are good opportunities for consultancy; and, of course, there is lot of prestige attached for being associated with IIMA. It is difficult but not impossible to emulate these conditions, especially if the institute is accommodative in meeting the genuine demands of the prospective faculty.

Seventh, another important feature of an institute of excellence is the insistence on accountability. In good institutions there are in-built systems to ensure accountability. It is not that every one has to pass a litmus test all the time. But there are various ways by which people know who is doing what, and the rewards and punishments are distributed accordingly. And both the rewards and the punishments can be substantial. For example, promotions should not be automatic. In my view in our university system incalculable harm was done when the "career advancement system" based on the number of years a faculty was engaged, was introduced. In IIMA a committee of very senior professors assessed performance of every faculty, and the Director would normally follow their recommendations in giving promotions or rewards. It is equally important to have deterrent for the people who are not able to meet the requirements expected of them. I had to ask a very bright faculty to tender his resignation as he was prevaricating in submitting his teaching plan for the following term, which was essential for proper planning of the academic activities.

Eighth, the most important feature of an institute of excellence, in my view, is the peer culture. It is the faculty, which takes upon itself the task of maintaining high standards. During my directorship IIMA of nearly five years, I never issued a memo to anyone. This is because if there was something not proper, immediately there will be corridor talks. The tradition of peer pressure has to be consciously encouraged. For example, we used to have monthly meeting of faculty where all faculty members would be present. In these meetings, a newly recruited associate professor could challenge the director, and the director without being haughty or upset had to explain why certain decisions were taken. This type of respect for the peers is extremely important, so also peers taking their responsibility seriously.

Ninth, such institutions are always under public gaze. Therefore, as a matter of policy all the decisions should be transparent and nondiscriminatory. It is difficult to adhere to these norms on all occasions. For example, personal files of the employees have to be confidential. Similarly, some people can always cite the promotion or rewards to an individual as an act of favoritism. If there are clear norms on what is confidential and what is not, and public knowledge of the standards expected for promotion and rewards, there are better chances of the decisions of the authorities being accepted by different constituencies. In IIMA we tried to adhere to these practices.

Finally, one thing, which we did not have at my time and which I now realize extremely important, is an internal mechanism for conflict resolution. Conflicts are bound to arise even in the good institutions. But systems have to be designed which ensures that the conflict is resolved expeditiously and on the principle of subsidiary, that is, you try to resolve the conflict at the level it arises and do not allow it to fester and go on to progressively higher level.

Qualities of an institutional leader

These are the important features, which characterize an institute of excellence. These distinguish such institutions from the pack. If we agree that these are the prominent features of an institute of excellence, then we have to ask ourselves as to who determines that they are not compromised or tampered. To manage an institute of excellence and keep it on the right course, I believe three entities play a decisive role: the Director, the Faculty, and the Chairman of the board. I will narrate my experience in IIMA and draw lessons from that.

To be acceptable and respected, a *director* of an academic institution should have a reputation as a scholar in his own discipline. His/her peers, within the institute and outside, should recognize him/her as a serious scholar. If one has not got a reputation as an academic, then however good a manager the person may be, he/she will not get respect of the faculty in the type of institution I am talking about.

The second very important characteristic of a good director, in my view, is that he should have the capacity to bask in the reflected glory of his colleagues. Sometimes this becomes a problem when a very young bright person is invited to become a director. If he is not able to appreciate the successes of the colleagues, then things fall apart. A director should be proud to say that his faculty has been able to achieve this or that. He should be able to admire the achievements of the colleagues, certainly not be jealous. I have seen in many institutions in India and abroad the directors feel insecure and they start behaving as competitors rather than mentors of their colleagues.

A director should also build a reputation for fairness. It is not enough to be fair but should have a reputation of being fair. If one fails on that count then the faculty, students, and staff will discover it very soon. If there is an element of unfairness in the dealing, whichever way it manifests, the respect of a director gets diminished. The director should have an image of someone who is even handed and who does not discriminate.

The other important characteristic, which I find lacking in many heads of institutions in our country, is the courage. Courage is very important. There will always be people who would like to dominate or ask for undue favor. On such occasion the director has to take courage in both hands and when there is need to say "no" he/she has to say no. Let me give an example. In IIM Ahmedabad, the most important thing then, as it is now, was to get admission. We did not have the problem of someone pressurizing us for admission from the government of Gujarat, or the industry because IIMA had already built a reputation that nothing could be done to facilitate someone's admission without due process. But once I got a call from Prime Minister's office and a senior officer from Prime Minister's Office (PMO) said that they would like to have the son of the Speaker of the Assembly of a friendly country admitted to PGP in IIM Ahemdabad. I said that to get admission, foreigner or Indian, everyone had to take the examination. The officer was holding high position in PMO and was close to prime minister. He said that what he was asking was in our national interest. I told him, "Sir, there are several ways by which we can serve our national interests. Giving admission in postgraduate program of IIMA to someone who has not completed the process is not the best way to serve the national interest." I was quite firm but he continued to argue. Then I said, "Alright, give me in writing what you would like me to do and I will put it before my board. I will request the Chairman of the Board to call a special meeting of the board and discuss this matter." Of course, nobody gives such instruction in writing. And that was the end of the story. The director should have courage to stand firm however influential the person is.

The other important role of a director, which I admit is a little bit controversial, is that a director has to act as a "gate-keeper." As Ravi Mathai explained to me, the director has to be a gatekeeper on two counts. First, he acts as a shield for his colleagues to protect them from any onslaught from outside. Second, he has also to see that some matters, which are strictly internal to the institute, do not get leaked out or used for giving bad name to the institute. In that respect his/her role is also that of a monitor.

The role of the *faculty* is critical in several ways. After all, the reputation of an institution is largely built on the quality of the faculty. It is not enough to attract good faculty. There should be constant pressure, by the students, by the director, and by the peers, to improve capability. Reasonable opportunities should be provided by the institute, and should be availed by the faculty, to interact with the leaders in the profession. An important way to do this is through publications. I do not believe in "publish or perish" and can also concede that an exceptionally good teacher may not be able to publish much. However, this cannot be permitted as an alibi by all and sundry for not publishing a chapter or a book, which can stand the test of the renowned scholars in the area.

Apart from qualities as a teacher and a researcher a positive attitude is equally desirable. Faculty should have pride in the institution. A cynical faculty can do as much harm as an incompetent one. The personal traits as much as academic achievements should guide the entry and subsequent progress of the faculty if the institution has to maintain excellence. Each faculty has to take responsibility to maintain high standards, not only in academics but also in his/her personal conduct. Earlier, I had underlined the importance of the peer culture. The peer culture also places important responsibility on the faculty.

The chairman of the Governing Board of the Institute has an equally important role. In several purposes he is the "face" of an institute. The status of the chairman in his own field should be that of a senior and respected leader. It adds to the reputation of the institute. At the same time, he/she should be easily approachable and should have genuine interest in the activities of the institute. There are institutions where very high-ranking people are chairpersons, but they are unapproachable, nor are they really interested. It is difficult for the director to get an appointment to meet them. Such chairmen are more a liability than an asset to the institute.

Easily approachability should not mean interference in the day-to-day functioning. Let me recount my own experience in this regard. When I took over as the director, the first thing I did was to go to Bombay and meet Keshub Mahendra who was then chairman of the IIMA board. I told him that I had taken over as the director, and I would need his guidance and advice. He said "Vijay, don't expect any guidance or advice from me. If you have any problem you are most welcome to come and discuss with me. You can see me at any time." He called his secretary and said that "Whenever Prof. Vyas comes to meet me give him the priority." Then he said, "It is for you to decide when you would like to meet me and ask for help. I am not going to tell you how you should run the institute. It is your job, you have been hired for that." That is the type of chairman, any director will look forward. One who has high standing, approachable, but non-interfering.

The other important thing for a board chairman, which sometimes one does not find in otherwise well-qualified persons is that a chairman should be in control of the board. They should really shepherd the board. The chairmen who are otherwise good in themselves but cannot take their board with them really do not serve the institute well. I had good fortune of working with some very good chairmen, with H. M. Patel in Vidhyanagar, with Keshub Mahendra in IIMA, with M. M. K. Wali at the Institute of Development Studies, Jaipur. I also had the privilege of working on the Board of Trustees of the International Food Policy Research Institute, when Sir John Crawford was the chairman of the board, and on the Board of Governors of the IDS, Sussex, when Gordon Conway was the chair. The way in which they used to conduct the meetings of the Board was a pleasure to observe. In large institutions people with different backgrounds, with different interests, with different qualifications serve on the board. The competent chairman handles such boards in a manner that members are brought together, consensus is reached on key issues, and no factionalism is allowed. These are the things, which one should expect from a chairman of an institute of excellence.

I realize that what I have narrated above are some stray thoughts, I will not call these considered or well-thought out principles in the area of organizational development. It would need a genius of Udai Pareek to weave the type of observations I have made into a coherent body of knowledge. It is one more reason that we miss Udai.

18 An ideology-based institution: Some values and dilemmas—An ISABS experience

*Somnath Chattopadhyay**,†

The prevalence of ideology based institutions in societies requires a close look into the origin, development, and the mode of sustenance. The expression "ideology based institution" has two parts: "ideology" and "institution." "Ideology" defines itself with the existence of mind, ideas and ideals, and the system that binds it; it derives meaning of the objective reality as it exists in space, time, and physical processes. Ideology has its antagonists and protagonists. "Institution" is a reputed organization that concerns itself with the promotion of science, education, or any other social purpose, particularly of caring, helping, and bringing positive change in people's lives and living.

People form an ideology based organization with voluntary association, not for earning money, its equivalent, or profit. They essentially derive satisfaction in finding a scope of application of their knowledge, skill, and expertise, and above all, in being a part of a mission that emerges out of the idealism, thereby contributing meaningfully to a social purpose. The governance of an ideology-based institution is different from those obtained in usual industrial, commercial, or governmental organizations. The aim is elimination of coercion. Individualism and holism coexist. Individuals and the institution add value to each other. Ideology based institutions emerge from a movement, in the context of a society, and obtain their sustenance from that movement, which, in turn, invigorates the movement further. The vision and values create the movement, and the ideology provides the spirit and motivation for the movement.

* This chapter is dedicated to Udai Pareek. When I shared some of the ideas and the structure, Udai insisted on a write-up.

† I express my gratitude to Jude K. K. for his association while writing and the various help that he rendered, and to Suryamani Singh. Thanks also to Deepankar Roy and Sudha Shankar for editing the manuscript.

Many movements emerge and flow. Many stop. Many institutions have a vigorous existence; however, if the inner dynamics are not confronted, discontinuity sets in and the institution perishes.

For a comprehensive understanding of these aspects of an ideology based institution, the example of Indian Society of Applied Behavioural Science (ISABS) is presented here, as a case in point.

The experience rests on:

1. Generating a movement: Institution of Laboratory Education, as a new method of education, in India
2. First-hand experiences of Learning Laboratories: Training Laboratories and L-Groups
3. Stabilizing of the L-Group and Learning Laboratories
4. Development of Laboratory Educators: Faculty Development/ Facilitators' Development; preparation for scope of their development, identifying them, etc.
5. Formation of a center: In this case, it is "ISABS"
6. Re-emergence of ISABS and its stability
7. Continuance and reinforcement of the movement

With these in view, the highlights of the historical background are indicated here.

1. *The background to development of training laboratories in India:* No movement is autogenous. Laboratory education in India was also not a sudden innovation. The background includes a process that flowed from Tagore's establishment of Sriniketan; Gandhiji's visit to Santiniketan and Sriniketan; Tagore's efforts for new methods of education; Gandhiji's establishment of the Wardha Ashram; and other ashrams established at different places in India.

   The effort, at that time, was aimed at reaching out to people in villages, with an idea of all-round development. The methods had to be different. The old modality of "teacher-taught subject matter in a classroom" had to yield to something new that touches adult life.

   Gandhiji's *Nayee Talim*, the Sarvodaya Movement, Sarvodaya School, Sewagram–Jharkhand, Gandhi Gram (Institute of Rural Development), basic education, and multipurpose education are some of the reminders of that turbulent period. One wonders what happened to these institutions. Several institutes such as National Physical Laboratory (NPL), NCLB, research institutes, agricultural universities, universities, and deemed universities were established, and were supposed to energize different walks of national life.

   But academic settings do not automatically translate to seats of new methods of learning. These had the infrastructure but they lacked vision, strategies, concepts and approaches, methods, and technologies of education, to meet the emergent need of social relevance. The past mode of education seeped into the present and the preoccupation remained subject-matter teaching.

2. *Emergent needs:* A cursory glance at the end of the 1950s and the beginning of the 1960s reveals that India had about 5,000 Community Development (CD) blocks (the first one being established in 1952) and National Extension Service (NES) blocks. This required one chief for each block like the Block Development Officer (BDO); under him about 10 specialists worked in different areas such as agriculture, veterinary, industry, cooperatives, and *panchayat.* Under each specialist staff were officers, besides village-level workers/*gramsevaks* and *gramsevikas*.

   What kind of persons would they have to be? How should they behave? How could they be acceptable, relevant, and valuable to the people in the villages? What would be their training? Merely knowing the subject matter in which they were educated was not enough. They had to work in an unknown setting. There was not sufficient clarity about what their work was and how they should perform it? Above all, who would educate them and how?

   Universities and other institutes of higher learning, research, and development were carrying on their usual teaching. The only university (Calcutta, now Kolkata) had Education, Social Anthropology, and Applied Psychology. Applied Psychology restricted its attention to Clinical, Industrial, and Educational Psychology. Later, other universities also began offering applied subjects. However, almost nothing focused on application, behavior, science, or society. The only exception possibly was the work of Stephen Max Corey at National Institute of Basic Education (NIBE). He emphasized Action Research, psycho-dynamic teaching, workshops for teachers, and the formation of facilitators' groups. The development of laboratory education saw, briefly, Stephen Corey conduct L-Group sessions in the evenings at his home at NIBE, New Delhi, with Udai Pareek, Prayag Mehta, and four others. This background provides a glimpse into the nature of the forces operating for changing and change.

   To say India is an old country is a cliché; yet, anything of substance—all major religions, precepts of life, norms, mores, customs, rituals and values—is at least a million years old. Endless are its diversities, paradoxes, and polarities. So is its craving for unity. The old existed, the new had to emerge. Present in everything remained one fact—the abysmal poverty. This fact with its multiple implications contributed to the core of social relevance in India.

   Amidst all these, the urge for nation-building and the idealism of a newly emergent nation molded the zeitgeist of the time. Tagore, Gandhi, and Marx provided their individual thrust in the design of the action lines within the minds of those responsible for social change. Through the contradictions, confusions, paradoxes, and polarities, democracy moved on with its activities in diverse fields. Very necessary, therefore, is the need to understand the wind that was scattering the seeds of change.

3. *Foundations of training laboratory and L-Group in India:* SIET Institute (now National Institute of Small Industries Extension Training [NISIET]) provided a venue for one of the seeds of change. SIET Institute, with its cluster of cottages, at the outskirts of the city of Hyderabad was an idyllic setting for the labs, surrounded as it was by rocky hills, a large tank, vineyards, and distant palm trees.

   SIET could be credited with at least two major initiatives in training; one of them was Training Laboratories and the other the T-Groups in India. These were the creations of three colleagues and intimate friends—Rolf, Udai, and myself. The principal director, R.N Jai, of SIET institute provided the patronage and administrative support to us for our work. This was the initiation of the lab movement (L-Group and Laboratories education) in India.

   Rolf Lynton founded Aloka for youth from Asian countries for training future leaders of developing countries. One of the foci was to stimulate awareness about the problems of individual behavior in a group (Lynton, 1960). Rolf, Udai, and I worked together almost serendipitously, and evolved and founded the Training Laboratory and Learning Group (L-Group synonymous with T-Group) in India. Rolf and Udai mention it in "Here and Now" and elsewhere. SIET work had a character of its own; it was not an extension of National Training Laboratories (NTL) or Tavistock.

   From the beginning, there was a need to increase the resource base—the original three were joined by Manohar S. Nadkarni, K. J. Christopher, and, subsequently, Sujit Bhattacharjee and SG Raghu from SIET institute. Occasionally, Abad Ahmad and Prayag Mehta from Delhi came to SIET for taking some part in this new venture.

4. *Founding beliefs:* The foundation of the new endeavor was based on the beliefs that:

   (a) Learning, to be useful to life, should have cognitive, emotive, and conative foci. These were observed in the emergence of the psycho-dynamic processes in the actualities that one experienced. These experiences were in the "here and now" within self or as a member of a group.
   (b) Knowledge and generating new knowledge through research have their own importance in the laboratories as in the department of sciences. The moot point was to discover, recognize, and apply it in being and doing. It was an innovation in many senses.
   (c) Earlier, subjects were taught in a teacher-taught environment, with one-way communication. Rolf and I organized a seminar for professors of a university. The discussion was on the need for two-way communication between a professor and her/his students. Despite

some resistance, a different method of education was introduced, wherein participants, based on their own experience, were learning to behave differently with free and open communication.

Laboratory education in the context of adult re-education ushered in learning by discovery in a group setting. This brought in two innovations in education—Training Laboratories (in short, lab) and Learning Group (L-Group, synonymous with T-Group). In 1964, a question was raised during a meeting of facilitators by Manohar Nadkarni about the use of the name T-Group because the main focus was learning and not training. Therefore, L-Group would be a more appropriate name for the learning group. From then on, in India, the term L-Group is being used.

The difference between a lab and an L-Group is that an L-Group is a basic constituent of a lab, which might have a "C" Group (Consulting Group), a "D" Group (Development Group), an "N" Group (New Group) befitting the objectives for which the lab was organized. Other types of groups may be organized, for example, problem-solving groups, be it, social, organizational, or educational problems. A lab may not, in some programs, have L-Groups as its integral concept. The intensity that could be attained in L-Group work may get reduced by other activities of the program. A lab may function as a crucible in which theory could be understood better through the experiences generated by practice, praxis, and processes of change. The theory could emerge from psychology, sociology, anthropology, politics, mathematics, etc., so long as the focus is on learning of human life and behavior. The facilitator's choice of preference and the participants' choice of direction of exploration together determine the orientation of the lab and the modalities used in mobilizing energy and its direction.

A lab provides wide choices in the range of orientation of its work. The diversity of potential thus available in a lab is not only in terms of humans forming the group but also in the spatio-temporal variations in group dynamics. Some of these potentials that surface in a given moment are attended to.

A lab, therefore, requires a lot of planning, central understanding, and agreements on assumptions and values.

5. *Spread of L-Groups in India:* The first lab was organized in SIET in 1962. A series of labs followed that year. In 1965, to create a nucleus of 10 to 12 faculty members, who may develop into laboratory trainers, an inter-institutional Faculty Development Program was organized by Rolf and Udai in Dalhousie (Lynton, 1965). In this program (April 14–May 2, 1965), the first part was a lab facilitated by Warren Bennis and Rolf. Some of the participants were from IIM Calcutta (Nitish De, Gouranga Chattopadhyay), University of Delhi (Abad Ahmad), SIET Institute (R. P. Lynton, Adhikari, A. V. Nagaraj, Ghiara, and myself), Vohra

Foundation Bombay (Shipchandler). Most of these people carried forward laboratory education in India.

A few members from different institutes such as IIM (Ahmedabad and Calcutta), IRHFP (Gandhigram), and others got their selected training at NTL. In 1966, Abad Ahmad, Iswar Dayal, Raja Deolalikar, Francis Menezes, Suresh Srivastava, and myself met at the conference center at NTL, Bethel. We met to consider whether we could build an institution like NTL in India. Suresh was not coming back to India. Iswar Dayal excused himself from such a proposal. At the end of the meeting, I told Abad Ahmad, "Abad, we will build our institute in India. With Udai and Rolf, we will be able to make it. Let us resolve it here and now. Our people do not have to come this far to get trained." We grasped each other's hands and Abad agreed. Immediately, I wrote to Udai about it. On our return from NTL, I maintained contact with Abad in Delhi and Francis Menezes (at Tata Management Training Center [TMTC], Mumbai) and spoke about our intent to Sujit Bhattacharjee, K. J. Christopher (SIET Institute), and Nitish De (Kolkata). Fred Massarik and Howard Baumgartel showed keen interest in this development.

From 1967, a regular 14-day lab was begun by me at SIET Institute with the focus on Personal Development and Group Dynamics. These were attended by many people from industrial organizations, the Indian Administrative Services, universities, and development institutes. A series of two-week labs, conducted from 1967 to 1969, was specially designed for chief general managers, general managers, and chief managers of State Bank of India. Many of them later became chairmen and managing directors of different organizations.

Around this time (1969), I was with Nitish De, Gouranga Chattopadhyay, Dharni Sinha, Dilip Lahiri, Mritunjaya Atreya, Rege Mascarenhas from IIM (Calcutta), who organized events that, held multiple labs in Barrackpur. At about the same time, I joined Udai at National Institute of Health Administration and Education (NIHAE), New Delhi, and organized a Faculty Development Program, a lab in which all the faculty of NIHAE participated. We invited Suresh Srivastava to join Udai and me to facilitate this lab. At NIHAE, Group Experience Labs (GEL) were very popular. NIHAE had two kinds of major programs (i) staff colleges in the pattern of Administrative Staff College at Henley and Hyderabad and (ii) hospital administration. The faculty thought that GEL was very effective for both the programs—facilitating the unfreezing, freezing, and creating a culture of learning and transfer.

A significant development at the end of this phase was Kenneth (Ken) D. Benne's (who worked with Kurt Lewin and was a cofounder of NTL) stay with us in Delhi. In earlier years, Udai and I had interned with Ken Benne at NTL. Our relationship was reinforced by lengthy discussions on lab processes and learning. A facilitators' lab was organized in Shimla and the faculty comprised Ken Benne, Abad, Udai, and me, with about 25 participants in two groups. J. M. Ojha was the administrator of the program.

There was no culminating event to indicate the termination of this phase. This phase could be characterized as the time when a seed germinated and became a young sapling, growing with vigor, and bearing some flowers and fruit. None can demarcate where the sapling ended and the woody trunk evolved. Individuals and institutes continued to offer programs. A new development was the emergence of behavioral science-based consultancy, utilizing laboratory training.

Significantly, the advent of laboratory learning was due to the work of people with an academic base, working in national institutes and some universities.

6. *Distinction between L-Group and training laboratories*

(a) The L-Group

In the L-group, the objective was to create a learning environment, in which:

(i) learning is enhanced from the data generated,
(ii) the perspective of each individual behaving alone and as a participant is explored in the presence of other members,
(iii) the "here and now" is adhered to, with the least amount of structure, and
(iv) the facilitators display non-authoritarian behavior.

Some of the points mentioned such as learning, data, "here and now," the role of the facilitator(s) and that of the participants, and the minimal structure need to be noted. With the learning comes the realization of cause and effect. The dependent variables are one's behavior and the group's behavior. The causes are numerous, beginning with the behavior of the participant, that of other participants and facilitator(s) in the group, forming a miniature society. The objective, therefore, is a member's learning and re-education.

A participant may be helped to:

- Identify and collect data about her/his transactions
- Explore the causes in own self
- Become aware of the impact of any other member's behavior on self
- Perceive the impact of other members, in plurality

The participants experience the impact of these on self, on others, and on the perceived group. They observe, directly or indirectly, through inferences, the intentions and effects, the motives, feelings, strategies, and actions in the setting. These are the data that are available for analyses, patterning, and condensing. Participants have to note what emerged from the data and to reflect on these. The reflection may lead to cyclic brooding or some concepts, and to, probably, a new theory of their being. Whether participants experience this or not, they have to continue as active members of the group.

A participant learns more about himself by unfreezing, re-examining, modifying, altering, or adopting new behavior (a) in a spirit of scientific enquiry (b) helping others who are also going through similar processes at the same time. They learn how relatively or totally unknown individuals that assemble together become (a) a group and (b) from a group to an intimate group, a mature group or productive unit, a viable organization, a miniature society. The inalienable constituents of an L-Group, therefore, are person–inter-person–group–micro-society, and the processes. The processes that make the flow of learning energy of different aspects, through changes in cognition, emotion, and conation, ultimately result in transferability in the being, becoming, and doing in life outside the lab. If any one of these components is missing, for example, group dynamics, cognition, or transferability, that learning may be labeled as anything but an L-Group.

(b) A training laboratory

A training laboratory (lab), which is a temporary group (preferably residential), is designed for the learning requirements of its members. The learning within L-Groups was expanded later to include testing, experimenting, simulation of roles and action, and change. This occurs in a climate of support, protection, and help before the planning of change, before undertaking action in organization or society, outside the laboratory. A lab could include skill practice sessions, paired interviews, pairing or smaller group interactions, tape listening, community sessions, psychodrama, socio-drama, alter ego, games and exercises, alone time, diary writing, and concept or theory sessions.

Designing a lab is a very important activity for the faculty. The design must address issues of unfreezing, movement, and freezing, as well as pay attention to pretraining, training, and post-training activities (for elaboration, see Lynton & Pareek, 1967/1990/2000 part 2, pp. 47–57). The design of a lab may, therefore, have different types of time allocation, over and above its basic L-Groups. The intricacy of the design increases as one plans a lab or labs, or as part of larger training programs, or part of organization development, social development, change or special purpose programs. The illustration of how a social development project with 11 faculty members can be designed and directed has been described in a case (Chattopadhyay & Pareek, 1982). In those years, the scope of use and application of labs was enormous, in conjunction with on-going training programs offered by training institutes in the country, generating, and sustaining a "movement."

The labs were of central importance in faculty development for one or more institutes, in the entire faculty of an institute attending the same lab, in Professional Member Development, in developing Internal Facilitators (IF) in OD programs, and in Internal Change Agents Program. As an illustration of a special-purpose program, in the 1960s, I offered a 10-day lab for members of an inter-institute, four-country research program. The research program was for 26 weeks, and members were research scientists from varied backgrounds. According to the Research Director in that program, mutual acceptance and trust were essential and the lab provided that momentum and acceleration.

7. *The basic values:* The participants of a lab require opportunities not only for experiential learning but also for exposure to theory and concepts of research and studies. Despite the diverse variety of material, making for unlimited choices in what to include in a lab, the program has to be based on some cardinal principles.

The cardinal principles arise from the description of L-Group and labs mentioned earlier. More specifically, some values and norms provide the framework for laboratory learning. These basic values are:

(a) *Democracy:* This primarily denotes the individual's free choice in being one's own—in beliefs and actions, processes, and practices. This includes non-authoritarian decision-making and relationships. It requires upholding the autonomy of a member. This could be a very challenging proposition for individuals, who have been brought up in societal feudalism and steeped in an organizational culture characterized by autocracy, power, control, and coercion, which they may not be even aware of.

(b) *Science:* The learning is based on scientific enquiry, in which one values collection of data, analyses, acceptance, or rejection of hypotheses, and formulation of conclusions. Data in a lab are facts about participant behavior, feelings, and motivations. Personal and collective potential of persons and groups are also part of the data. Modal processes of participants' lives, their perceived realities of situations and environments, and the paradoxes in their lives are also subject to scientific treatment. Facing facts, bringing objectivity to the extent possible, moving away from subjective perceptions, biases, prejudices and opinions, exploration, and confrontation are also constituents of scientific enquiry.

(c) *Helping:* This means building a relationship of mutuality— "one for all and all for one," "live and let live." No group, no micro-society can be developed without concern for building helping relationships. Cooperation or collaboration, and partnership of a superordinate goal pursuance, are the functions emerging from the concern for helping. The daily activities of self-presentation, giving and receiving feedback, interpersonal relationships, etc., depend on this concern for helping. (For these three basic values and further elaboration, see Benne et al., 1964, pp. 8–12).

Some other concerns were added to these three basic values.

(a) *"Here and Now" focus:* "Here and Now" means the data generated by the behavior can be seen and interpreted by the entire group. At times, clarity is required in the meaning of "here and now," space and time, proximity, and recency. One dominant view of the "here and now" is that it concerns the behavior that occurs during the life of a group. Some insist on immediacy, that is, when episodes occur with one or more members in succession. In other words, it might be here and now, in relation to one person or many persons, or in relation to time between one significant episode and another. Significance is determined when a participant in a group can do something about the issue at hand. Life history, or the memory content, would be "there and then"—the reported description of an event, and its cognitive, emotive interpretation, or impact could be of importance in therapy. It could be of importance to an individual for greater understanding of her/his behavior or the motives and intentions behind a behavior or a tendency in the patterns of a particular kind of behavior; nothing, however, could be done about that event then. Rewriting history will always be an impossible task.
(b) *Openness:* This comprises expressing whatever goes on inside a participant and her/his readiness to receive feedback from outside, allowing for the processes of "giving and taking." This may take place at different levels within a person, that is, through sensation, perception, emotion, cognition, conation, evaluation, intentions, motives, and so on. In addition, a participant may focus attention on structures, events, processes, and persons. They may not retreat into a shell or build a boundary around the self; instead, she/he may contribute at appropriate times.
(c) *Truth:* This refers to the search for truth in life and in relationship with others, in the "here and now" existence. This does not mean some abstract conceptualization of truth. Increasing acceptance of self and others requires identification and awareness of (i) true self and (ii) also of falsities, illusions, and deceptions in self, which need to be minimized. The reduction of defenses, fears, and greed is essential to the learning process.
(d) *Trust:* This was conceived as an interpersonal phenomenon between one person and others, wherein one can predict the behavior of the other person, with the assurance that no harm will be done to the

focal person by the other person. When the fears from the self and the other person(s) are reduced, one does not have to resort to defensive behavior. The person may feel the acceptance of self by self and others.

(e) *Creative self:* This is to reinforce a participant's vision and energy for personal growth and development. Usually, they seek containment of joy and distress, past and present, and seeks to be at peace with themselves. This, however, may not be enough when they need to surpass themselves. They need a cognitive redefinition of themselves, to rediscover themselves, at ever higher fullness.

(f) *Unlimited potential:* This exists in a person so that she/he can actualize not only their potential, in order to enhance their own self or their work but also their contribution to the world. This is a plane where equality among people prevails. Everyone has the potential to be and to contribute.

(g) *Group:* The group is as important for growth as an individual is in re-education. Interpersonal processes, intra-group, and inter-group processes are fundamental to generate enormous opportunities for growth that is of special significance.

(h) *Place for all learners:* All persons intending to participate will be accepted without any reference to age, social or organizational status, or hierarchy of any kind as long as they are healthy. Quiet, silent people can learn as much as others; introverts as much as extraverts.

Values at the verbal level are easy to accept; these, however, need to be part of the action ideology, practice, and processes. The operation of these can be very delicate and the extent endless. The interpretations and practice, and living up to these values, however, may have variations with individual orientation.

8. *Center for laboratory education—Need:* The need for a center for these labs became imperative. At first it was just an idea; soon many labs were being conducted and the requirement for facilitators became acute. To add vigor and consolidate the laboratory movement required the establishment of a center.

A. Practitioners' needs

a. Creating a greater scope of mutuality of learning in laboratory education.
b. Sharing wider experiences, both from own and others' practice, inside and outside India.
c. Creating a forum for conceptual understanding, learning, and refinement of practices.
d. Creating a scope for sharing research on laboratory learning, reported in research journals and periodicals, in published

material from other centers of similar pursuits, and from individuals around the world.

e. Creating a center for giving and receiving help in professional activities.
f. Evolving modalities on operational values and preventing the possibility of extreme variations.
g. Creating a group resource for undertaking developmental work in specified directions.
h. Forming a group dedicated to a profession of social science-based knowledge and skills, and its application to help individuals and society.
i. Fulfilling the needs for togetherness, cohesion, affiliation, belonging and collaboration, and movement away from professional isolation.
j. Generating collective group strength.

B. Participants' needs

Participants include those currently engaged, future participants, and organizations or institutions interested in lab learning. They will need:

a. Opportunities for re-education, L-Group exposure, lab learning to pursue growth and development
b. Scope for experiential learning for group dynamics, building groups, and teams
c. A place where a normal person can explore and confront her/his own dysfunctionalities in behavior
d. A resource center enabling planned change in a chosen domain—be it in an organization, a community, or a domain of society
e. A center that could facilitate development of the entire group of faculty of an institute
f. A resource center for Organization Development (OD)
g. An institute devoted to professional development that will enable and help an individual become skilled in process learning, conducting L-Groups and labs, and being a facilitator of Community Development or OD.

9. *Formation of a center of training laboratories in India:* The lab movement, founded in the 1960s, gained momentum. L-Groups were being offered in different settings and the formation of a center was inevitable. The center emerged in 1971 and was named "Indian Society for Applied Behavioral Science" (ISABS), comprising a body of laboratory educators, who wanted to further the cause of laboratory education in India.

It was registered as a Society. It had a constitution and an organizational structure that comprised an executive board and an executive committee, with a minimum of seven members. There was an Executive Director (ED), a Dean (Education), and a Treasurer. The first four EDs were: Francis Menzes (1971–1973), Dharni Sinha (1974–1975), Pulin Garg (1976–1977), and Purnima Sinha (1978–1979).

The prime activity of ISABS was conducting a two-week summer event annually, comprising a few laboratory programs run simultaneously, with emphasis on L-Group work. Concurrently, an Internship Development Program for a select group of people was held. Aroon Joshi, Gopal Valecha, and WG Naidu were among the first batch of interns who completed the internship program in 1975.

In 1978, at the summer event at Masobra, Shimla, ISABS split. While the labs (including the internship programs in three phases) were going on, Pulin Garg, Indira Parikh, Sushanta Banerjee, Jose Kunankal, and a large part of the faculty left ISABS for good. They formed another organization, which subsequently stopped functioning, and yet another group emerged. The ED tried in vain to resuscitate ISABS by conducting one-day seminars. Many members of ISABS abandoned it. That seemed to be the end of ISABS.

Why did an organization built to enhance a movement meet its closure? The causes often are very close to the phenomenon itself. A critical analysis of these events may reveal some interesting dynamics that may be applicable to many such organizations. Some of the following ideas may indicate the pitfalls of an organization, which if not attended to at an appropriate time may lead to its collapse.

10. *Dilemmas ad some dysfunctional aspects of ideology-based institutions:* An ideology based institution may be beset with conflicting ideas of individualism and holism; Gesellschaft and Gemeinschaft, that produce two opposing orientations—the contractual social impersonal relationship governed by legalities, rules and sanctions, intellect and reason, and the other is bound by relations, governed by likes and dislikes, love and concern, emotions and feelings, and the community bonds. These two orientations operate at the same time within the same organization and can be observed by their manifestation in the following dilemmas. They are themselves not the causes, but they are still there.

    (a) Ideological orientation
    An institution needs an ideology. The ideas and, particularly, the ideals serving as its foundation of beliefs, action ideologies, and praxis are essential for the definition and the operation of the institution and the movement it generates. It is the raison d'etre of the institution. Its high value, by itself, does not guarantee acceptance by all. The differences may start at the level of ideas, its constructs,

in the lines of action, or in its methodology. Any part of these has the potential of becoming the symbol of the whole. The deviation may come from trivia and can consume the totality. This is seen in religious, political, and social institutions, time and again. ISABS had an ideological base. Often, what emerges is a comparative evaluation of functionalities and utilities, and expressed thus: "Mine is more important than yours"; "Yours is dangerous"; "Mine is the only way"; "Pursue the only way I see it or perish." Such fissiparous tendencies may lead to actual fission and the very existence of the organization may be threatened.

(b) Valuation in ideology

All ideologies are subjected to valuation. One may say, "This is scientific, rational" and that, by itself, may constitute valuation. It evokes the issues of ethics, endlessly debated by theoreticians. Ultimately, what the members in an institution think is better, prevails. A similar valuation, cognitively and/or emotively, brings in support and strength; from there, power generates more power and the growing power refuses to be contained, spreading rapidly. It tends to obscure the vision of the idealism, of superordinate-ness of the totality. Almost forgotten become the romantic bonds of coming together, and the cohesion and the dreaming together during the initiating days. The emerging power finds a reverse direction, infighting starts, and hostilities may go as far as killing a person or persons in large numbers and destroying homes and hearths, work, and institutions.

(c) Genealogical purity

The institution after being in existence for some time may engage in a search for its real ancestry, to determine the purest from among the cousins. The purest is, as if, the true inheritor of the original ideation. The linkage with ancestry makes the pursuit genuine, real, and true; simultaneously, others' ideas are seen as impure, gilded, fake, mixed, and adulterated. This ancestral linkage serves as a bond to bring "the true believers" together. It may lead to adherence among each other and maintain a boundary for the adherents, to save and protect it from external "impurities, adulteration, and conversion." When it acquires power and becomes strong socially, politically, or economically, it can release its aggression on the others, who they perceive are heathen, non-believers, lack culture or the moral principles required of a true professional of the institution.

(d) Disruption of the institutional ideology

The ideology with which an organization becomes an institution soon reaches a point when the forces that render it dynamic may face challenges that shake up the very existence of the institution. The institutional ideology comprises (a) visional properties—the vision of the future, the purpose, and the strategic goals;

(b) intellectual properties—the theories, rationalities, and cognitive mapping of the state for which the institution was built. The analyses and data covering the past, present and future, the hypotheses, postulates, assumptions, reasoning, policies, plans, etc., also form part of the intellectual property; (c) emotional properties—the dreams, commitments, involvement, feelings, interpersonal attractions, attachments, detachments, distances, likes, dislikes, satisfactions, joys, hesitations, fears, anxieties, losses, aloneness, anger, hostilities, disillusionment with people, and the other pro- or contra-elements; (d) action properties—the lines of action, processes, praxis, practices, systems, acts of changing, performing of tasks, moving, freezing, stabilizing, generalizing, etc.

These properties and their constituents concatenate to uphold the ideology that is the mainstay of the institution. When this is challenged, a disruption takes place. In history, illustrations abound about interventions and their consequences. One of the consequences is indeed the split of the organization, with a loss of vigor and vitality in the original institution. A splinter group may have to go through the processes of building afresh.

The human costs are many. Some persons may be considered irrelevant in the new environment, some may be lost, and some may try to be in more than one splinter organization and be rejected by one or the other organization. The positions of significance, importance, and leadership may go through a sea change. The new order takes over, and newness prevails in all the properties mentioned earlier.

(e) Commitment and voluntary contribution

At the initial stages, an ideology based organization gets built by the commitment and voluntary contributions of the members of the organization. These members are, in a sense, the builders of the organization. The sense of voluntarism and commitment may prevail at any period of time in the existence of an institution. Commitment may be at least of two kinds (a) personal commitment, which originates from within the person, from his own values, attitudes, motivation, needs for application, abilities, competencies, fulfilling the urges of his creative, actualizing self, and (b) institutional commitment, which originates from outside him that requires him to act devotedly to fulfill what the institution requires as determined by the role-holders of the institution. This voluntarily committed contribution requires, explicitly or implicitly, appreciation, recognition, and valuation, and is abhorrent to comparison. But the "hierarchization" of voluntary commitment does take place in the valuation process, directly or indirectly, overtly or covertly. Its consequential impact may bring about disintegration in the collective cohesion that was the mainstay of the institution.

Voluntary commitment, without expectation of any return, has been considered a supreme value in many cultures. It is eulogized in literature and history. Great epics are woven around this theme. The pursuance of this principle can make heroes of men and women but many become nameless, faceless flowers that are "born to blush unseen and waste its sweetness in the desert air" (Gray's Elegy).

Is the contribution of some person superior to that of others? Is it possible to build an institution wherein the contribution of all will be valued equally? Norms of feudalism, bureaucracy, and modern industrialism influencing organizational culture require the emergence of these two opposite forces: valuing and devaluing voluntary commitment. As an organization develops, it sows the seeds of its own destruction, leaving those who volunteer aside. It may not be infrequent to come across claims of superior contribution, quantitatively and qualitatively. In the absence of any neutral standard of measurement, the clamor cannot be quietened or disputes settled. The claims may be silent or soft murmurs of "Have I not contributed more than what that fellow ever did?" These thoughts in the self are like termites that silently eat up the roots of a tree, which withers and falls.

There are also examples of the tendency of a volunteer sacrificing almost like a martyr at the altar of a cherished cause. It is the giving up of what the volunteer possesses, including his position, influence, and even life. Sacrifice is an exalted value in society. But when it turns into competition and from there to a conflict, it enters the life-blood of an organization and engenders turbulence and turmoil. Consequently, the institution will lose its vitality and sense of direction. The simplicity and purity of sacrifice now become matters of reinterpretation and doubt.

(f) Power need and the institution

The main function of the institution is "change" and more particularly "changing." The action properties mentioned earlier are all dealing with change in behavior. It involves two parties: the change inducer and the one to be changed. The change inducers are the facilitators and/or the group. The job is that of influencing, of persuasion in a fundamentally emotive group culture. The major issue is the power motive of the actors. This may be high or low in a person. The issue is whether she/he is aware of it and whether it is congruent and authentic with her/his motivation. If she/he has high need of power, how does she/he apply it to the work she/he does wherein a major value premise is that of non-authoritarianism. How does a person with a high need for power function as a member of an institution? If she/he has a low need, how does she/he perform her/his influencing function? It is easier to assume that each member of this influencing community understands, and has explored power motivation and its effects on her/his mode of application. She/he knows fairly well the reasons for joining this particular fraternity. The whole world, since the mid-1940s, examined authoritarianism and fascism and their effects on individual and society. But more often than not she/he is unaware of her/his own power motivation. She/he satisfies her/his need "naturally." Power begets power. More people join and the ones, who are already powerful, exercise more and more power. Soon, a war breaks out. The ideology is shattered, the institution destroyed. The hunger for control over others, the group, the role, and the organization increases steadily and is seldom mitigated.

In an environment dominated by this type of control power, other kinds of power that enable others beat a retreat. Consequently, democracy in action is devalued. The people, who derive fulfillment from generating "enabling" in the other person, become less in number and the surrounding environment becomes less supportive of their mode of working with others. These "other-centric" persons may have control on the means of influence on persons. The means are their wisdom, knowledge, emotional richness, and mode of engagement in their work—in short, their whole personality. They help the person move of her/his free will from one state of being to another, and release more of her/his creative potential in becoming. When these people do not volunteer actively with the organization, decadence sets in, in every sphere.

(g) Deification

The ideology based institution exhibits, at times, a tendency to deify persons. This cult of deification, accompanied by rituals and worship, may be a function of control power and can be fashionable in the organization. Many members derive a sense of satisfaction in deification. A member, who is instrumental in deifying, may either have her/his own need of dependence fulfilled or may have a covert wish to be deified in turn. The institution may face a tough time, however, when there is more than one person waiting to be deified. What does it do? The members fight and begin a process of elimination of one or the other.

(h) Transference and counter transference

An ideology based organization may eventually become an abode of gods, an ashram of *Babaji*s and *Mataji*s, and their devotees, leaving aside the unconscious erotic redirection of transference and counter transference, as outlined in psychoanalysis or by Jung. The aspects that get affected immediately are the organizational processes and practices. A new entrant gets entry to the organization only with the blessings of a *Babaji* or a *Mataji* and it continues at every stage of the onward journey of that entrant. A reference of a venerable *Baba* or *Mata*, a recommendation from them becomes a memorable event for the person.

She/he, in turn, learns the technology of obeisance and joins with others and together goad their respective *Babaji* or *Mataji* to start a new ashram on the banks of the Ganga or Narmada. *Babaji* or *Mataji* do not need any goading; they gloat over their success and count the number of devotees they have been able to convert with their magic wand. This process between the *Baba* and *chela* is mutually reinforced and can go on and on merrily till disillusionment takes place, if at all it does. Neither feels the need to recognize, explore, or confront her/his own deep-seated psychological factors that exert a compelling effect on her/his behavior.

A person may unknowingly redirect her/his deep-seated feelings—anger and love—on to the guru and ultimately finds a new parent-figure. What does a *guru* do to enable the disciple to understand and deal with her/his inner dynamics. Just like the disciple engages in transference, the *guru* participates in counter-transference. The two processes go on unabated.

If this happened and remained contained within the confines of any two persons, it could be left to their freedom of choice. When this happens in an ideology based organization, however, it becomes an issue of professional ethics. For ideology based institutions, the emergent culture is vitiated, and the microcosm created by the group perpetuates the practices of the old culture, which it had wanted to change. When transference and counter transference prevail in an institution, it ceases to become change inducing, purveying old wine in new bottles, with new labels and a new language.

(i) Affection and affiliation

An ideology based institution, unlike other institutions, makes "emotion" one of its prime concerns. As intellectual orientation and action orientation predominate in society, and emotional orientation is, by and large, neglected or left to the creative world of literature and fine arts, the acculturation processes in an ideology based institution cultivate the neglected field of emotions.

It may appear to be paradoxical that people, who champion the cause of influence, are at once engaged in expressions of affection. The explorations in interpersonal relationships and the expressions of feelings, mutuality, anger, hostility, love, and hate charge the air that the institution breathes. There is seldom a coherent conceptualization of these activities within the main goals of the institution and its members. Take, for example, "catharsis" in the Aristotelian sense of (a) lessening the emotional tensions by expressing them in aesthetic experience or (b) refining them by sharing in emotions universalized and artistically portrayed, or (c) in the Freudian sense of reliving (here and now) the events of the past (there and then), especially those that have been repressed (there and then till here and now), and explorations and insights from these events. How is this phenomenon of catharsis understood and brought under concepts that promote the institution? The same question arises for influence and affiliation. Danger sets in when these expressions are ends by themselves and are not cathected to any goal that an institution stands for. The "feel-good" factor may be good for a hedonist or a political leader; for a lab, however, its utility is doubtful.

One can ponder the genuineness and depth of meaning, the authenticity of the users, and the tendencies of overemphasis in affiliative expressions. These are the words one may think that together constitute esoteric passwords for the shibboleth of institution faith. Once this happens, the institution is set on a path of doom.

(j) Conceptual incoherence

Hundreds of academics are generating knowledge in every branch of science and their application over decades. It is not possible to integrate them. An analogy from medicine may be cited. Innumerable research is being conducted every day worldwide, to break new ground and enlarge the frontiers of knowledge. Medical practitioners, specialists and super-specialists, attempt to keep abreast with the advancement of knowledge and skills because there are patients to heal. They operate from coherent concepts although the concepts may not be comprehensive. An institute can engage with all of them. Similarly, an ideology based institution requires coherent concepts, generative of practices, and maintains harmony among its properties. Intellectual properties must be coherent with the other three properties mentioned earlier, namely, visional, emotional, and action. Without coherent concepts, what may happen is that too many concepts from too many sources are collected and referred to, and none is explored in depth.

The elimination of cultivation of concepts finds no rational basis. Laboratory education, being a serious business, requires alertness in data generated in experiencing, and their analyses need to be coherent. The elements in the learning climate are the individual and the group, with their theories-in-action and espoused theories. Groups and the transient actualities and the dynamic processes, covering both the members, as individuals, and the group, holistically—all these elements require coherence and relevance. This will not only be satisfying but will help build strength among thoughts, emotions, and action, eventually reinforcing the vision and the organization. If this nexus is devalued, the organization tends to disintegrate. This paucity need not be covered up by pseudo-theories and folk tales. If the thoughts, emotions, and actions are discrete and disjointed in segmental preoccupation, unitary conceptual coherence may be the casualty. This coherence is needed to build solidarity. Diversity, exploration, experiments, and expansion are necessary, provided there is unity.

(k) Adherence to professional competence
Profession requires a body of knowledge and consequent practices at the least. It should be well founded on science, or evolved on well-tested epistemology, on a focal area that generates, maintains, and guards its application. A set of normative behavior, like the Hippocratic oath in medicine, with a body for reference should be constituted by experienced persons. Practices have to be exoteric. If the practitioner uses unprincipled freedom, that freedom is liberty, likely to be abused. A highly skilled ability to do a job and the performance of jobs done on strictly contractual basis, irrespective of what happens as a consequence, are not enough in a profession. The institute has to have a clearly defined area of work and profession. There is otherwise a danger of quackery.

(l) Goal clarity
An ideology based organization needs to have a fairly clear idea shared by its members of the change that it is seeking. It may happen that all the members in the organization may not have the same depth of understanding of the ideas and values embedded in an ideology, and also about the one-on-one connection between ideology and action. For example, the Indian National Army's slogan, "Give me blood, I'll give you freedom" or Mao Zedong's "We shall overcome someday" were not only emotionally arousing but were supported strongly by practical operational efficiency. In real life, therefore, the ideology and action require continuous review and adjustment. The process that it employs for changing should be clearly delineated. The focus of work may be an individual person, the group, the macrocosm, or a part of society. Change in a part does not change the whole if the change is not accepted by the whole. It is a debatable assumption that if one changes a part, when its science and application are known and the required competence is assured, the whole will be healthy. A reputed heart surgeon once said, "I can operate on a patient but would I be able to see him walk back home and lead a normal life? If I were not sure, I would not take up my scalpel." This sense of responsibility and concern is essential among members of an organization that professes "changing" as an objective. If this is missing, it makes not only the individual but the organization irresponsible and weak.

These 12 aspects are needed in any ideology based organization. If any of these features malfunctions, the institution loses its vitality. The dysfunctionality that developed slowly during the

1970s brought about the closure of ISABS in 1978–1979. It was not the action of one or the other person. Creative processes accompany destructive processes. In an organization when the destructive processes are not heeded to and the creative urge muffles the murmurs of malfunction, the dissonance gains in strength silently over a period of time, leading to the demise of the organization. The critical reexamination and analyses of the functioning of the institution, its state of health, and periodic rejuvenation are often ignored as unnecessary by the powers that be. This blindness brought ISABS to disaster.

11. *Institution building and stabilizing training laboratories*: The lab movement had to continue. People felt the void. The need to have a center that generates and sustains Laboratory education, be it ISABS or some other organization, by some other name, was felt again. A few people resolved that the institution had to be built anew.

At the formative stage, the initiation came from Deepankar Roy, Kirpal Singh, Ramesh Galohda, Somnath Chattopadhyay (the author), Sukhvarsha Roy, and Syed Akhtar. These six formed the first nucleus. Syed Akhtar did not continue for long because he had to leave for work abroad.

A special mention must be made about this first nucleus, which operated as a team intensely, devotedly, and efficiently, particularly Deepankar Roy and Sukhvarsha Roy, who shouldered much of the work.

This group then contacted Abad Ahmad, Manohar Nadkarni, MMP Akhouri, Prayag Mehta, Rolf Lynton, and Udai Pareek, who agreed to join the nucleus group. C. N. Kumar, Dilip Lahiry, Mahaveer Jain, Partha Sarathi, Sujit Bhattacharjee, and VK Jain constituted the third wave.

All agreed to this new formation of ISABS. These were the members of the new ISABS.

There are many others who helped: Vasudevan, who went through all the old files and took relevant extracts; Col. Sheshadri, Wing Cdr. Ranga, Madhukar Shukla, Geeta Saxena, Poonam Sehgal, K.K. Anand, Alexandra Merryl, and Michael Merryl joined ISABS at that time.

No payment was made to the members for their contribution. Deepankar and Sukhvarsha's home functioned as the office of ISABS for many years. Everything was done manually. There were no computers, no Internet; they relied on telephones and the services of Posts and Telegraph.

Some of the problems faced initially were:

- The old books of accounts, money accrued to the accounts, and fixed deposits were all frozen; therefore, the new ISABS started without a single paisa to its credit. The accounts had to be regenerated.

- The operating files, including the constitution of old ISABS and its legal documents, were not available.
- There was an acute shortage of facilitators for the labs.

A series of activities began simultaneously.

The re-negotiation between Pulin Garg and Dharni Sinha was undertaken by Alan Batchelor and later by Manohar Nadkarni. The efforts failed to revive the organization. All that was obtained were some files. At that stage G. D. Dutta (Group Manager, BHEL) provided space for ISABS office in a part of his own office, and was personally involved.

The institution had to review the dysfunctionalities of the old mode of operation and the new organization had to protect itself from these.

The decision was to further the dreams of 1964 and to continue with L-Groups and learning laboratories. The inner strength of experience based learning Laboratory prevailed and sustained the venture.

What was required in the new venture was affirmation of the vision afresh and its values and processes upheld.

A logo to symbolize the new institution was created by Professor Bilas Bose of the College of Art, Lalit Kala Akademi, with inputs from Somnath Chattopadhyay. The institution had to build its structure, systems, and procedures anew. For this, the following were done:

- New accounts were created. One member put his own money to create the initial fund.
- New memberships: Two categories were created—Professional Members (Life Membership and Annual Membership) and Associate Members.
- It was decided that those who had left the organization would not be taken back. Also, those who did not show any involvement in the organization or its processes, after seeking their choice, were dropped from membership. It was decided at that time that only those persons who completed the Professional Development Program satisfactorily will be accepted as Professional Members; among them, those who subscribed for the stipulated fee (at that time ₹500) were absorbed as Life Members. Co-option of members or Presidential nomination was disfavored; however, a person with outstanding professional competence in experiential learning could be considered by the Board for membership.
- The designation of executive director was dropped. The new designation, in conformity with the nature of the institution, was president. The first president of ISABS was Manohar Nadkarni.
- L-Group and training laboratories were the organization's main work area. Other laboratories could also operate but only on the sidelines.

- Group work, group processes, group norms, group values, building the microcosm, experiences generated in the "here and now," understanding, cognitive awareness, and, above all, learning should constitute the main elements of the work of the lab.
- Learning should be built on self, the other person, and the impact of one on the other, between each other, on the group, on the organization and on society, feelings and their effective expression, understanding the other person not only through words but also through expressive interactions, feelings in the other person and what makes her/him behave the way she/he does, behavior in the group, skills of dealing with self, other people and group, should be part of the institution's activities.
- Learning should aim at enhancing insights into humans and their relations wherever they are, in groups, organizations, societies.
- The systematization of labs was done in two streams Basic Human Process (BHP) Laboratories (Bradford, Gibb, & Benne, 1964) and Advanced Human Process (AHP) Laboratories (Benne, Bradford, Gibb, & Lippitt, 1975)
- The development of people who will be facilitators and future professional members of the institution needed very careful work, and a special committee was created for this purpose.
- In the Professional Development stream, it was decided to reduce the requirement of the period of candidates' stay on an ISABS campus. Sponsoring a candidate event after event, year after year was a major burden for organizations. The new system required candidates, after their BHP and AHP, to stay for two events (Phase I and Phase II) and carry out non-campus, inter-phase work.

The idea behind formulating these programs was to introduce a graded degree of sophistication, for example, the AHP would be more advanced than BHP. Similarly, the progression through the Professional Development Program (PDP), that is, from Phase I to the Inter-phase work and on to Phase II and then to Internship provided a graded degree of progress. Participants start from being a collection of strangers to becoming persons (each one in a group), confident in their competence in effecting change and development. Their domain of work may not remain contained within the walls of personal growth and development, and may extend to society through their competence in groups. The challenge was to create the required opportunities for such a graded education.

Although the new institution experienced paucity of facilitators, it did not want to increase the pool of professional members "somehow." Quality and not quantity was pursued.

Log writing, which emphasized the candidate's ability to write and to form a habit of writing, was introduced. This was believed to sharpen awareness of data, process sensitivity, and provide for intellectual reinforcement. Review of literature was part of this process.

A book of relevant references for further studies on human process learning, *Basic Books*, was published by ISABS.

It was thought very strongly that the same group of people should not be on the Executive Board, term after term. If a person were to continue indefinitely, new centers of power were likely to develop.

The institution would welcome senior people for its events. They would be invited as resource persons.

12. *Regionalization of ISABS:* The all-India character of the organization, the new ISABS, was envisaged. The country was divided into regions. Along with the central body of the Executive Board and the Executive Committee, there were regional bodies in Ahmedabad, Bengaluru, Chennai, Delhi, Kolkata, and Mumbai. Professional members of ISABS were present in each region; the coordinator of a region was also the representative member to the Executive Board. Each region was empowered to conduct BHPs and AHPs.

Conclusion

ISABS has now sustained its growth over the last 30 years. From 1980, it stabilized as an institution. ISABS may like to review its functioning, to ensure continued vitality and longevity. The number of persons being trained may not be the only indicator of the institution's growth. As indicated earlier, the trends of dysfunctionalities lie within the functional area of an institution. It becomes almost imperative, therefore, for an ideology based institution to review and rejuvenate itself. An overview may indicate:

a. *Reduction in the presence of academics:* The early phase of the laboratory movement saw the involvement of academics, engaging in the transfer of learning to different domains of life in society. Gradually, that flow has reduced.
b. *Reduction in intellectual pursuits:* A person's readiness to be involved with knowledge for its (a) generation (b) cultivation and broadening awareness of contributions, both in the world of academics and the world of action (c) sharing (d) dissemination (e) storing (f) application and utilization. There are many ways of dealing with each of these aspects. This, one may think, is a matter of individual preference, but a culture requires emphasis on intellectual pursuits as one of its strong components and demands its prevalence at a normative level. Any culture that neglects this, suffers.
c. *Emphasis on group and group processes:* This seems to have reduced in the institution. If this premise is even partially true, it throws up a set of questions regarding laboratory education, its goals, methods employed, the level of competence of facilitators, their choices and motivations, the

level of safety, and the potency of transferring the learning beyond the boundary of the life of the group. The institute was founded to enhance learning about effective group processes to impact areas of social needs. By now, there are specialized institutions to provide professional care in various fields. ISABS, being one of its own kind, may like to preserve its own identity. The plurality in "applied behavioral sciences" indicates the diverse areas from which resources can be brought, provided these are adapted, integrated, and internalized after value addition to contribute to more effective utilization by society. Preference seems to be to confined oneself to the individual and her/his feelings, and stay within the streams of the person–interperson nexus.

d. *Personal and interpersonal effectiveness:* Both are very important and necessary. ISABS has already done tremendous work on these. The entire preoccupation of ISABS may not remain confined to this alone. Group processes and group development may be the focal areas of energy mobilization. The locus seems to be: Individual person–other persons–inter-person–small groups–large groups–organizations–communities–society.

Before the 1960s, there was newness and uncertainty in working on groups in a variety of settings. That phase of novelty and uncertainty has been traversed by competent ISABS facilitators. Their experiences in working with different groups need to be captured, with greater focus on learning about group processes and group development.

The more the attention to group development, the more will be the availability of expertise to work in groups beyond the laboratory environment. As a consequence, more cases and concepts, models and theories will be developed. It may be relevant here to note what Udai Pareek said about, "moving from individual to the larger group and the community, and developing insight into larger group and community processes must become the agenda for ISABS" (Pareek, 1999).

In *T-Group Theory & Laboratory Method: Innovation in Re-education,* Bradford, Gibb, and Benne (1964) in the last chapter "A look to the future" where they point out their concerns about the professionalization of T-Group training, expansion of resources and theory development, extensions and modifications of T-Group methodology, and extended use of T-Groups and laboratory methods in 'non-laboratory' settings. They hoped that answers to the future "will be found in responsible efforts to solve the problems on which continuing development depends" (p. 486). Ten years later, the authors (1975) after examining the same four dimensions of concerns reiterated their hope in the same language.

These four areas of concern are also relevant to ISABS. In addition, ISABS may undertake:

- The promotion of applied behavioral science manifested in a line of activities. It needs theory development, research, and studies. Conferences, discourses, seminars, writings, and a culture of discussions of concepts are associative elements to this promotive venture.

- Enhancement of professional development of facilitators. ISABS has helped develop many effective L-Group facilitators of national and international repute; however, it still needs to pay careful attention to the development process. ISABS has found its potential and is today a viable organization. With nurturance and removal of obstacles, it can continue to grow as a socially relevant and significant institution in the years to come.

More than two thousand years ago, Gautama Buddha found the strength of the group and Buddhists chanted *Sangham Sharanam Gachchami*—"To the group I be beholden." Jesus established the group of 12 Apostles. In the last century, Kurt Lewin explained the science of groups. During the last half century, the sapling of a lab has become a mighty tree, spreading its branches in a thousand directions.

One can imagine the emergence of new persons, who, after acquiring the expertise of laboratory learning, will transcend technologies and the language of the lab, and with confidence in self, unfold their immense creative faculties, to make the world better, braver, and more beautiful. This venerable path contains the footprints of those gone by and will be traversed by the young, decade after decade in the future. Let that journey for mankind be ever vigorous.

I do not say much of the relationship I had with Udai as my teacher, mentor, colleague, and friend. It does not convey the quality of the relationship. Whenever we were together in the country or abroad we had intense discussions. Many a time, I had the privilege to elucidate and elaborate in lectures the basic points of thoughts that Udai used to provide. This complementarity continued breaking new grounds. The relationship of 55 years made the two families intimate. His soft care, his concern, his helping, and his tenderness used to reach at the center of the heart. Thinking about this breeds intense sadness.

References

Benne, K. D., Bradford, L. P., Gibb, J. R., & Lippitt, R. O. (1975). *The laboratory method of changing and learning theory and application.* Palo Alto: Science and Behavior Books.

Bradford, L. P., Gibb, J. R., & Benne, K. D. (1964). *T-Group theory & laboratory method. Innovation in Re-education.* New York: John Wiley.

Chattopadhyay, S., & Pareek, U. (1982). *Managing organizational change* (pp. 275–319). New Delhi: Oxford & IBH Publishers.

Lynton, R. P. (1960). *The tide of learning: The Aloka experience.* London: Routledge and Kegan Paul.

———. (1965). *Inter-institutional faculty development programme.* Hyderabad, SIET Institute: (Mimeo).

Lynton, R. P., & Pareek, U. (1990). *Training for development* (2nd ed.) New Delhi: Vistaar Publications.

———. (2000). *Training for organizational transformation.* New Delhi: SAGE Publications.

Pareek, U. (1999). *T-Group journey: A sunrise talk.* New Delhi: ISABS.

19 Institution building: Case study of development of the University of Delhi's South Campus

Abad Ahmad

Dr Udai Pareek was an intellectual prodigy, a prolific writer, and a highly creative mind. Above all, one of the finest human beings I have known in my life. He was a friend, philosopher, guide, and mentor for all who came in contact with him. For me he was an ideal role model as a scholar, writer, teacher, and colleague. I had the privilege of working with him in many major organizational development (OD) consultancy assignments. He was a highly "appreciative" inspirer. Even short informal conversations over lunch or meeting him on other occasions gave me new insights, ideas for research topics for PhD scholars, and research projects. Working with him as a colleague and listening to his lectures was the most enriching and enlightening experience. Reading his books opened new vistas of knowledge, clarified complex concepts, and theories which were always backed or illustrated by empirical studies, case studies, and research.

He was truly a doyen of applied behavioral science. More than all his scholarly attributes, his broad world view and open mind, humane values, his openness and humility, and his quest for knowledge and deep concern to develop his colleagues and students were amazing. Once when I was working with him and Rolf Lynton in a consultancy project with Health Ministry officers in Indonesia, with whom we conducted T-Groups also, the participants invited me to join them in Friday prayers. Dr Udai Pareek said he would also like to join if they had no objection. They welcomed his desire, and thus I and Dr Pareek stood together for prayers in the mosque. Similarly, when he read my article that was published in the *Journal of Human Values*, based on my keynote address at the Center for Human Values at IIM, Calcutta, he complimented me, but questioned why I had based it only on Srimad Bhagavad Gita and not on other scriptures also such as Holy Koran.

I owe my inspiration for learning about Institution Building to Dr Udai Pareek's book: *Beyond Management—Processes in Institution Building*, published in 1981. Introducing the concept of "Institution" as different from "Organization," he has stated in the Introduction to this book that "*the main characteristic of an Institution is its normative character and role—it is characterized by a mission, a set of values, and by its role in making impact on a smaller or a larger part of the society*" (emphasis added). Quoting other authors such as Blaise on the subject, he has clarified that institutions are "organizations which embody, foster, and protect normative relationship and action patterns, and perform functions and services which are valued in the environment." He has also quoted Selznick who has defined it as "a change-inducing and change-protecting formal organization."

Dr Pareek has described "institution building" as "the process of establishing or transforming an organization into an integrated and organic part of a community in a way that will help the organization to play a proactive role in projecting new values and become an agent of change in the community." As stated by him, institution building has also been used for the process of the internal development of an organization. He has further specifically stated that "institution building" has been used with two meanings: one meaning refers to the process of development of some institutions, by an outside expert institution, as total systems, including the development of values and norms that are relevant for the type of work the institutions have been doing. The other meaning relates to the internal development of an institution to be able to play its role effectively. From this perspective, as stated by him, "institution building" can be seen as a process of developing institutions (organizations imbued with a mission to make impact on the society) through special efforts to develop capabilities of the organization to grow into "institutions." It is this second meaning of internal development and transformation that I have chosen as the theme of this chapter, which has been illustrated by a case study based on my personal experience.

Institution building can, therefore, be described as a process of creating, transforming, and developing organizations with a normative purpose and values that can impact a smaller or a larger part of the society. It is the process of embedding certain core values, motivation, and inspiration that leads to high level of commitment to the purpose, values, and identity of the organization. Through the process of institution building, an organization lasts way beyond the life of its promoters and tenure of managers. It acquires a distinct identity and character that distinguishes it, and commands support and respect of all stakeholders, as it makes higher values-based contribution to the society.

In essence, therefore, I am utilizing the concept of institution building as the process that transforms an organization into a purposeful, values-driven, innovative entity that serves a larger purpose through which it benefits all its stakeholders in a "win–win" relationship, achieves high levels of excellence, becomes sustainable, and acquires a distinct identity and stature that is valued by all stakeholders and the society.

In this chapter, I propose to share one of my several personal experiences of institution building with which I was privileged to be associated.

The experience I would like to share is of building a new campus of the University of Delhi which is called South Campus. The University of Delhi is a premier university of the country that was established in 1922 on the model of Oxford and Cambridge Universities with constituent (not affiliated) colleges. It gradually developed postgraduate departments. It continued to grow rapidly with the establishment of many new colleges by private trusts, the State Government of Delhi, and the University itself. Because of increasing demand for quality undergraduate education, and turbulence in other north Indian universities where examinations were not held regularly, the pressure on Delhi University kept increasing. The number of colleges in the University of Delhi continued to increase, including opening of evening colleges, non-collegiate courses, and correspondence courses. The number of colleges became more than 60 and number of students became more than 100,000 during the 1970s. The number of postgraduate departments also grew to more than 50, and number of postgraduate students increased substantially. The other university in Delhi that was started by the Central Government was the Jawaharlal Nehru University that had confined itself to a residential unitary university with specialized courses mostly at the postgraduate level, with no affiliated or constituent undergraduate colleges. It, therefore, did not share any load of expanding numbers of undergraduate and postgraduate students and colleges.

In this scenario, in order to maintain a manageable size and ensure good quality education, the University approached the government to start another university to share the load of the large number of colleges as well as the of postgraduate departments. The government did not agree to start another university in Delhi, but agreed for the development of a new campus in South Delhi. As a first step, in the year 1970, some postgraduate classes were started in South Delhi, and in the year 1973 the South Delhi Campus was established.

To start the new campus, a number of residential buildings were rented in South Extension, in which different departments, administrative offices, and library were located. The buildings were scattered on both sides of the main road. The 24 colleges located in South Delhi (18 day and 6 evening colleges) were placed under the jurisdiction of South Campus. Postgraduate classes in humanities, languages, and social sciences such as philosophy, history, English, Hindi, Sanskrit, political science, mathematics, economics, and commerce were held in South Campus. In due course, a good library, a health center, and administrative offices to administer the postgraduate teaching as well as the work related to the 24 colleges were developed.

The concerned departments and their heads continued to be at the North Campus, and teachers in-charge managed the classes in the South Campus. The campus was headed by director of South Campus, who had the rank of a pro-vice chancellor, but could not be so designated because of limitation of only one such position in the Act of the University of Delhi.

Professor Amrik Singh, the well-known educationist, was the first director, and Professor K. B. Rohatgi, professor and dean of the Faculty of Law, and an excellent administrator, was the second director of this new campus. They had laid strong foundation of the South Campus by recruiting high-quality teaching staff and very efficient administrative staff. Most of the faculty members and supporting staff members were young, energetic, and cooperative.

The earlier directors and vice-chancellor had got about 70 acres plot of land allotted near Dhaula Kuan for the new campus. The master plan for the construction of the new campus was developed in an open competition of architects, and also some grant in the five year plan for building the new campus was secured. The foundation stone was laid by the then president of India. However, for some reasons the construction work could not be started for several years, the grant had lapsed, and even the foundation stone was lost. The hilly plot of land with thorny bushes remained barren, leading to some encroachment by a public sector organization that built a storage godown there, and by its workers and stray people who built *jhuggis* on it.

I was invited by the newly appointed vice chancellor (VC) Professor Gurbakhsh Singh to take charge of the South Campus as its director in the year 1980. I was then Professor of Management and had completed my term as the dean of the Faculty of Management Studies of Delhi University. In the Delhi University system, professors may be given university's administrative positions for a limited period of three or five years depending on the terms for the position, after which they revert to their academic position.

On joining as director of the South Campus, I noted that since the parent departments in all subjects were in the North Campus, and most of the senior teachers were based there, many students preferred to attend classes there. The physical ambience of the residential houses being used for holding classes in the South Campus was also not satisfactory. The location of offices, teachers' rooms, and other support facilities in residential houses were not conducive for an academic institution of a university's stature. The morale of teachers, students, and office staff was low, especially because the campus was like an extension of main departments, as the teachers of South Campus had little or no role in the academic policy matters. Similarly, the Administration of the South Campus had only notional role in managing the colleges, as most of the problems had to be referred to the dean of Colleges and the VC located in the North Campus for decisions. Some people derogatorily called it a "post-box campus."

Keeping in view the above-stated state of affairs, it was evident that the South Campus did not have a well-defined purpose, character, and identity. At the same time, it could be seen that there were a large number of young and bright teachers in various subjects. The library and administrative staff was also quite competent. The location of campus in South Delhi, which is the more developed part of New Delhi, was another advantage, especially for the students. The challenge, therefore, was to develop it into an institution that would have a distinctive character, a well-defined educationally and socially important purpose, a participative and democratic culture that would create a sense of belonging and ownership in all its members, and an identity of which the teachers, students, and staff would be proud.

The first meeting I had was with all the teachers of the campus. A series of meetings with the teachers were held, and a system of regular monthly meetings with them was started. An attempt was made in these meetings to discuss the present situation, problems, their solutions, and the future possibilities for the development of the South Campus. In order to familiarize myself with all the members of the teaching and support staff, I visited each teaching department, all offices, the library, the health center, etc. During such visits, attempt was made to meet most people in small groups, and many of them individually. The purpose was to understand their roles, their problems, what could be done to facilitate their work, and their expectations for themselves and for the development of the campus. After this, a meeting of all nonteaching staff was held in which I tried to share with them my perception of the strengths and potential of the campus, and future possibilities of its development. In order to understand working of the colleges, the principals of colleges were invited for meetings in small groups. Further, all colleges were visited personally, as and when suitable opportunities arose. A system of quarterly meetings with the principals was initiated. (This had to be stopped after about one year, as it was noted that some principals began to use it as a pressure group for influencing the university authorities for their benefits.)

During the meetings with teachers, I could sense their pressing need to have connectivity with North Campus, and some facility to commute between the North and South Campus. Many of them were living in North Campus, and those living in South Delhi wanted to come for academic work to use library and other facilities that were located in North Campus that was about 20 km away. They had to travel by public transport or their own vehicles for about one to one and a half hour each way through narrow congested roads for this purpose. Therefore, immediate steps were taken to start a mini-bus commuting services between the two campuses.

During this period there were some embarrassing events because of the location of the campus in the residential buildings with narrow lanes for access. An ambassador of an important country, who wanted his daughter to be admitted to the campus, came to see me. He was highly exasperated as his car got stuck in narrow lanes, and he could reach my office with considerable difficulty. He decided not to get his daughter admitted to the South Campus. This incident strongly precipitated the feeling in my mind that the first priority for developing the campus should be to build its new building. It was obvious that the new building was also essential to start new courses that would attract good students and teachers to the campus, and would give it a distinct character and identity.

I could sense a great opportunity for the new campus because of its location in the more developed South Delhi, proximity to airport, and location of some distinguished colleges such as Lady Shri Ram College, Sri Venkateswara College, Jesus & Mary College, Kamla Nehru, and Gargi College. Besides, being a new campus, there were enormous possibilities of starting new courses and other academic initiatives.

The University of Delhi was passing through a very turbulent time. The agitations by students, teaching, and nonteaching staff were very frequent. Being a part of the VC's team (consisting of the VC, pro-VC, dean of colleges, and director of South Campus), I had to be present in all the meetings and during crises in the main North Campus. In fact, we were "*gheraoed*" several times, ranging from a few hours to 14 hours. Some agitations were so violent that there was almost threat to life of the team members. However, Professor Gurbakhsh Singh was a very seasoned and patient administrator, who handled these tough situations without being tense and vindictive.

When the situation eased a little, I discussed with him the need for the construction of the building of the South Campus. Since he had the experience of developing a totally new campus as VC of the Central University in Hyderabad before joining the University of Delhi, he was enthusiastic about it, and agreed to visit the plot of land allotted for South Campus. When he visited the site and found that it was a semi-hilly plot with lots of rocks, uneven surface, and thorny bushes, he was very disappointed, and remarked: "Abad, you have the worst plot of land I have ever seen, and it will be a tough job to construct building here." However, he was a pragmatic and optimistic person. He advised me to immediately start building a boundary wall around the plot to stop further encroachments and enable construction of new buildings. He also extended support of funds from the university for this purpose.

In order to inspire the teachers and supporting staff with a vision for the future of the campus, I asked the South Campus engineer, Mr Uttamchandani, to organize a special function at the site by clearing some space at the site, and to display the master plan and model of the proposed new campus. All teaching and supporting staff was invited and taken to the site in buses, and a picnic-like atmosphere was created. Some teachers were skeptic about it when they saw the tough terrain and distance from the existing location. However, it gave them some hope of the possibility of a new campus being built there. Despite the fact that we had no approved grant left for the new campus, the work was started. When the boundary wall was complete, the second step was taken to develop a broad landscape according to the master plan, and to plant trees. For this purpose, another function was organized on the site, and each teacher was asked to plant a tree.

In order to inspire all the stakeholders to support the new campus, the idea of multicampus system for the University of Delhi was proposed on the model of well-known universities such as the University of California, where I had gone several times as a visiting professor. It was stated in different university body meetings and other relevant platforms that the multicampus system fosters the development of a unique character and role for each campus within the framework of one university. It makes for efficient management of a large number of institutions through decentralized administration. It ensures maintenance of uniformly high academic standard through overall common philosophy and objectives, and central coordination mechanisms. It enhances productive time and satisfaction of students and teachers by cutting down commuting time, and by provision of satisfactory library and other academic facilities on each campus. It offers opportunities for greater interaction between the teachers and the taught, thus enabling closer relationship between them and preventing alienation of students—which was one of the root causes of student unrest and dissatisfaction. In addition, it offers an opportunity to start new courses that may be more in line with the aspirations of students and developments in the fields of knowledge.

Before starting the construction of the main buildings for the new campus, the major challenge was to remove encroachments, for which discussions were held with the managers of the public sector organization, who agreed to vacate, but the challenge was the other unauthorized encroachments. When they were asked to leave, they approached the local MLA, who tried to intervene. However, Professor Gurbakhsh Singh took a tough stand and advised us to take firm action to remove all encroachments to ensure proper development of the new campus.

Detailed discussions were held with the architects to ensure a unique architectural character and pleasant learning ambience. It was decided to build the entire new campus with "Delhi Blue" stones (gray in color) that were quarried in Delhi, and octagonal form was chosen for large halls in the building and for the library for giving a distinctive architectural character to the campus.

Since the funds provided in the five-year plan for the building had lapsed, it was a real challenge to start construction. However, we decided to start construction of the first teaching block (that was called the Arts Faculty building in the Master Plan) in the hope of approval from the University Grant Commission (UGC) and with the assurance of support from VC from the university. To obtain funds from the UGC which is an extremely difficult task, a number of visits were made along with the VC to meet the secretary and chairman of the UGC. In order to get their support, the vice chairman and chairman of UGC were invited to the site and progress of work was shown to them. To get support of thought leaders, eminent persons such as Dr D. S. Kothari were taken to the site and the concept of the new campus was shared with them. The VC knew from his experience that lots of UGC funds remain unutilized by the end of the financial year. Therefore, the VC and director used to go to the chairperson of UGC, Ms Madhuri Ben Shah, in the month of March with our building plans and requested her to approve them from such funds, which she graciously did directly on our submitted proposals.

The Bennett Coleman & Co. (Times of India Group) was at that time interested in starting S. P. Jain Centre in the University of Delhi and was willing to offer funds for this purpose. This came as a golden opportunity for the South Campus where we could offer space to build such a center. The matter was vigorously pursued by the Director and VC who visited some key members of the propriety family several times at their residence. An exclusive architectural plan for this center was prepared, which got an international award in an exhibition abroad. Bennett Coleman & Co. gave a sizable amount at that time without any strings attached. When the center's building was ready, in addition to the MBA Evening Course that was started in it, a proposal was made to the UGC and Ministry of HRD for starting a new MBA (Public Systems Management) Course at this center. The course was approved, and 12 teaching positions (two professors, four readers, and eight lecturers) were sanctioned for it. We were able to recruit some of the best teachers and researchers on these positions for the S. P. Jain Centre that became the southern wing of the Faculty of Management Studies.

The entire campus was carefully planned with the help of well-known architects to ensure aesthetic and functional ambience in academic and administrative buildings, library, sports complex, hostels, students' center, health center, residential quarters, etc. The underlying thought in developing the buildings and landscape of the campus was to provide pleasant learning ambience in the classrooms, teachers' rooms, meeting rooms, conference halls, and green garden-like ambience with wide open spaces across the campus that would attract best students and teachers and make them feel proud of their campus. The landscape was carefully planned and a large number of trees in consultation with horticulturists were planted across the campus before the buildings were constructed. Attempt was made to retain as many rocks as possible as these were geologically the oldest rocks and also aesthetically quite artistic.

Special attention was given to make classrooms interactive with good acoustics, and the furniture and fittings in classrooms were specially designed to make them functional and comfortable. All furniture was chosen to be made in teak wood, and not steel. An amphitheater design conference auditorium, and a hall for large meetings, were specially designed to organize seminars, conferences, and meetings with teachers and staff. The guiding principle in all this work was "Attention to Details" by the director and the other key members of teaching and administrative staff.

In order to ensure quick progress and quality of work, frequent personal site visits were made by the director, and the VC was also invited from time to time. To arouse interest of the government, community, and students in the new campus, press reporters were invited and shown the new building under construction, and the concept of new campus was shared with them. Based on such a visit of reporters, the *Statesman* newspaper published a report entitled, "Campus on the Rocks" with a nice write-up and photo of the upcoming building.

In order to expedite working of the campus in the new building, and to ensure quality of work, the office of the director of South Campus was shifted to the new building, even though there were problems of water and electricity because of lack of proper infrastructure. Gradually, all the teaching departments and offices were shifted to the new building. However, the teachers of some departments insisted on continuing in the South Extension premises, as they felt that the existing location was more convenient for students and teachers. They were not forced to move, and in due course they themselves chose to move. By the year 1983, most of the teaching departments and administrative offices were shifted to the new building, and a number of new courses and departments were started.

In order to develop a distinctive identity and character for the new campus, the strategy of making it "campus with a difference" was followed. Therefore, keeping in view the fact that the North Campus had great strength in basic disciplines in sciences such as physics, chemistry, biology, and botany, and in almost all disciplines of social sciences and humanities, it was decided to develop new courses in the South Campus that would be "inter-disciplinary and applied." Therefore, new departments and courses were started in bio-chemistry, micro-biology, genetics, bio-physics, electronic science, plant molecular biology, business economics, applied psychology, applied operations research, modern European languages, etc.

The guiding principles for starting new departments and courses at the South Campus were: "Relevance, Innovation, and Excellence." This message was repeated in all meetings with the teachers and other university authorities. Besides, meritocracy in selection of students and teachers, and for promotions of teachers and staff, was rigorously followed. A culture of open communications, participation in decision-making, listening to needs and grievances of teachers, staff, and students, and addressing them as quickly as possible was followed. Close and personal contact was developed with the Union of South Campus nonteaching staff, and elected union of the South Campus postgraduate students. On almost all major festivals, functions were organized by the staff in which teachers and director participated. Regular monthly meetings with all the teachers of the campus were held. In addition to elected students, the top two meritorious students from each class were invited for informal get-together with the director and some professors. As a result, a distinct culture began to develop in the South Campus, as contrasted with the main campus. There were no students', teachers', or staff agitations in the South Campus. So much so that once when the representatives of Karamchari Union (non-teaching staff union—DUCKU) of main campus came to persuade the South Campus nonteaching staff to join them in some agitation, they politely refused and requested them to leave them alone to do their work. However to show solidarity with them, they agreed to send a few representatives, but did not stop the work in South Campus offices.

In September, 1984, a grand inaugural function for the new South Campus was organized in which the Minister of Education, Ms Sheila Kaul, was invited as the chief guest. A detailed brochure giving a full coverage of history, philosophy, and objectives of the new campus, its new departments, the courses offered by it, and details about its colleges, their main activities, courses offered by them, number of teachers, students, etc., were described. This brochure was released by the minister. The function was attended by a large gathering of all university authorities, a large number of deans, heads of departments, professors, principals of colleges, members of teaching and supporting staff, and all students of the South Campus. Wide press coverage was given to this function that brought the South Campus in limelight. In his address, the pro-VC, Professor P. C. Mukharji said: "Dr Abad has passion for the South Campus."

It is not that everything was smooth in the development of the campus. Enormous problems were encountered at every stage. Getting approval of the Academic and Executive Councils of the University for new courses, and then obtaining approval of the UGC, were often time-consuming and tough. Getting funds from the UGC, clearances from various civic bodies, getting water supply and electricity connections, resistance of some departments to shift to the new campus, threats from some owners of rented premises in South Extension pressing to vacate their buildings, and attempts of Delhi University Students Union to hold boisterous functions in South Campus were some of these challenges. These challenges were handled persuasively, and a few of them—like DUSU's program to hold a boisterous jam session on the campus—were handled very firmly, sometimes at great risk.

There were also problems in construction work, such as delays in giving necessary drawings by the architects, in few cases running away of some contractors leaving buildings incomplete, and getting clearances from government authorities. These obstacles required enormous time and persistent efforts. However, the support of faculty colleagues, administrative staff, and above all the VC and university authorities made it possible to steer through these tough challenges. There were also tense situations when because of mishandling of situations by principals of some colleges the agitated teachers of some South Campus colleges would come in large numbers and let their anger out on the director of South Campus. Other than few such cases, most such matters would go to the main campus office of the VC, where the director and other team members of the VC had to face such tough situations.

In 1985, the VC Professor Gurbakhsh Singh retired. After some gap, because of the sad demise of the pro-VC, Professor P. C. Mukharji, I had to officiate as VC. It was a very turbulent phase because of students' unrest. We had to face highly agitated students at the main campus. However, such problems remained confined to the main campus, and the South Campus remained peaceful. After about six months' gap, Professor Moonis Raza, who was at that time Rector of the JNU, was appointed as the VC. He had developed the JNU with a very distinct multidisciplinary character, and highly innovative courses. He was a distinguished institution builder, and had embedded strongly intellectual, liberal, democratic, egalitarian, and socialist values in the JNU. He was all for a multicampus system, and developed a grand vision for the University of Delhi with four campuses, that is, North, South, East, and West Campus, in which in addition to the present two campuses he wanted East campus to focus on studies on human brain and medical sciences, and West campus on Engineering and Technology. Since he fully supported the concept of interdisciplinary and applied courses, we got a lot of support from him for the development of the South Campus.

The most important support that we got from new VC was his help in developing probably the first of their kind new faculties: The "Faculty of

Inter-disciplinary and Applied Sciences," and the Faculty of "Applied Social Sciences and Humanities." These were the first such faculties in the country. He also helped us in persuading some of the most well-known scientists to join the South Campus, among whom I would like to specially mention the name of Professor B. K. Bachchawat. He was one of the most respected senior professors and researchers in life sciences at that time, and was working in Kolkata. His joining the South Campus made all the difference in developing life science departments of the campus. He was able to attract finest scientists from the country and overseas. In order to attract and retain such scientists, we offered them houses on the campus, and in some cases their spouses who were also good scientists were offered research positions. And the most important attraction was our arrangement with the Springdale school that was located just adjacent to the campus, where their children could get quality education.

Surprisingly, there was no provision for girls' hostel in the master plan, but there was provision for a boys' hostel. Keeping in view that most of the students in science departments were girls who had to stay for late hours in the laboratories, the planned hostel was built on priority and allotted to girls. The mess of the hostel was outsourced and its management was entrusted to elected student members. Thus, they could control quality of food, and most of the hostel management problems were avoided. This boosted the morale of students, many of whom came from outside Delhi.

Under the leadership and guidance of Professor Bachchawat, the young scientists got big research projects from government agencies such as DST, DBT, and INSA, and built state-of-the-art laboratories. Within a few years, the South Campus came to be known for its outstanding research and teaching programs in fields such as bio-chemistry, micro-biology, and genetics. The work of teams under guidance of Professor Bachchawat on drug targeting was internationally acknowledged. Other distinguished scientists, such as Professor S. C. Maheshwari and Professor G. P. Srivastava, who were at that time at the North Campus, desired to move to the South Campus, and build their new laboratories. We also built exclusive buildings for some of them and provided space for other new departments in the new science blocks. This came as a further boost to the Life Sciences and Electronics Science wings of the South Campus. The work on Rice Genome and Plant Molecular Biology was also internationally known. Thus, world-class researches were being done in most science departments of the South Campus. In due course, advanced courses in computer sciences and information technology were also started.

Later, in my role as pro-VC of the Delhi University, when I went to sign a Memorandum of Understanding with the Heidelberg University, I visited their life science departments. When I mentioned to their scientists about the work at the South Campus, they said that they were not only fully aware of the work being done there, they were in touch with the scientists there and were already doing some collaborative researches with them. The other departments of the South Campus, such as the Department of Mathematics, Department of English, Department of Applied Psychology, Department of Modern

European Languages, Department of Hindi, and Department of Commerce, distinguished themselves by offering innovative new courses, high-quality teaching, and research, and by organizing national and international seminars.

Some of the innovative courses in social sciences were Applied Psychology, Business Economics, and Master of Finance and Control. These courses attracted bright students and they got good placements in leading organizations. In addition to new departments and courses, a number of new centers for advanced studies and researches were also started at the South Campus, such as the Center for Himalayan Geology and Center for Applied Behavioural Science and Action Research. Some of these centers could not be sustained because their financial sustainability was not carefully thought of and provided for. The strong lesson I learnt from this experience was that unless financial viability of an institution is ensured, it cannot become sustainable.

Any institution-building work cannot be done by an individual alone. It is a team work with involvement and participation of each and every member of the organization. I was fortunate to get a lot of support from all the members of the teaching and administrative staff, with of course a few dissensions and critics also. There is always a small team that steers the process with direct involvement and support to the leader. In addition to the enormous support that I got from the vice chancellors and other university authorities, I would like to specially mention the support and contribution of Professor B. K. Bachhawat, Professor Nirmala Jain, Professor O. P. Gupta, Professor G. P. Srivastava, and Professor Lallan Prasad among faculty members, and highly efficient administrative support of Deputy Registrar Mr P. L. Kadalbaju, and Deputy Librarian Mr H. C. Jain.

By the year 1990 when I completed my two five-year terms as director and reverted back to my role as Professor of Management at the FMS, the South Campus was a vibrant, distinctive, well-known, and beautiful campus. We had some of the best, dedicated, and highly motivated teachers and researchers at the campus. World-class researches were being done in its life sciences departments. Its postgraduate courses that were interdisciplinary and applied or professional were highly sought and were enabling students to get excellent placements. There was a palpable sense of pride among the teachers, students, and staff members for being a part of the South Campus. When visiting teams from overseas universities came to Delhi University seeking collaboration or joint research, etc., they were taken to the South Campus to see the high-quality research work being done there.

I would now like to summarize some of the insights and learning from this exciting and gratifying experience.

First, I would like to emphasize that Institution Building goes "Beyond Management," as suggested by the title of Dr Udai Pareek's book referred to earlier. As pointed out by him, "Administration" was concerned more with the successful maintenance of an organization, and running it according to the laid-down rules and regulations. "Management" has brought about a change in the emphasis, and has been primarily concerned with efficiency and task accomplishment. In addition to these concerns, now there is need for two more aspects in an organization—its continued growth and development (self-renewal) and its impact on a segment of the society or the whole society. These concerns are reflected in a new function—the "Institution Building" function. The above-stated case study of developing South Campus reflects how these new concerns were addressed through the goals, values, and processes chosen and practiced for this purpose.

Based on my experience and insights from the literature, I believe that institutions can and should be built on the foundation of their strengths, potential, and opportunities. I followed the same approach in the institution-building process for the South Campus. The quality of faculty and supporting staff, the backing and support of the reputed University of Delhi, the location in the rapidly developing part of New Delhi, the possibility of starting new state-of-the-art courses and research, already allotted piece of land, a good library and health center, and the support of top leadership of the university provided immense potential for developing the new campus. All that was needed was to inspire all concerned with a shared vision for the future of building a distinctive and unique institution with excellence of which all members and stakeholders would be proud. It was necessary to stretch the imagination of all concerned to think beyond the existing state. The theme chosen for such thinking was that all academic activities of the new campus will be guided by three main principles: relevance, innovation, and excellence. The new courses and research should respond to the needs of the society, students, and teachers in the emerging areas of knowledge. To impart a distinctive character to the new campus, it was thought that it should develop interdisciplinary and applied courses and research.

The guiding values were: openness in communication; democratic and participative process of leadership and decision-making; meritocracy in selection of students, teachers, and staff; pro-active search for talent for leadership in teaching and research in emerging new fields of knowledge; maximum autonomy and support to departments for innovative academic initiatives within the broad framework of the university system; professionalism in management (i.e., honoring commitments, fair dealings, integrity, and transparent functioning); and students and teachers-friendly administration.

The processes followed were: sharing vision and guiding principles in all important meetings, seminars, conferences, get-togethers, and other platforms; practicing participative style by institutionalizing faculty meetings and meetings with other relevant groups; opening channels of communication with teachers, staff, and students by following an open-door policy for meeting the Director and other administrators; empathy to listen and understand the problems of teachers, staff, and students, and quick and fair redress of grievances.

Further processes were: public appreciation and recognition of good work and meritorious performance; collaborative relationship with Staff and Students' Unions; relationship building with university authorities, UGC, press, and other government agencies; resource mobilization with government agencies and reputed private organizations; attention to details in all implementation processes; and strong and firm handling of situations that may hurt the interest of the campus, damage its reputation, or create indiscipline on the campus. All support was provided for new relevant and innovative courses and research, faculty development initiatives, collaborative research, national and international conferences and seminars organized by the faculty, etc. Deliberate policy of avoiding personality cult was followed by not publishing the photographs of director and university authorities, and other government authorities in brochures and other publicity material, which is a common practice in most other institutions.

The significant learning from this experience is that people have enormous creative potential, talent, and motivation for personal development and good performance. Right inspiration for serving a larger cause; right direction by shared, clear and concise vision, mission and goals; participative and consultative processes in decision-making; empowering and encouraging them to take initiative for innovation, improvement, and getting results are critical for mobilizing their energies and talent for institution building. Further, fair dealings; good working conditions; ensuring support in meeting their genuine needs for development and career advancement, appreciation, and recognition can take the institution to great heights of excellence. At the same time, it is important for leadership of the institution to ensure that it serves a larger purpose for the society, creates genuine value for stakeholders, is guided by sound values such as meritocracy, professionalism, openness, integrity, and transparency, becomes sustainable by mobilizing and generating resources for its maintenance and development, and earns the respect and goodwill of the society. For achieving such goals, the values and processes of Institution Building have to be followed, in which "Passion for Excellence" is most important. It has been rightly said that "nothing great was ever achieved without passion."

Based on the above-stated case study, and several other experiences of doing such work in organizations in educational and other sectors such as health and industry, I would like to propose the following Action Template for Institution Building (IB) process:

Action template for institution building/OD

1. Relate to people in the organization and key stakeholders
   Personally engage with people at all levels within the organization, and other key external stakeholders to develop relationship and rapport. Understand their needs, concerns, and expectations. Involve them intensely and sincerely in the shared process of change. Be accessible to all. *Make it their project.*
2. Build on strengths and opportunities
   Identify and explore strengths of people and organization, positive-thinking persons who will lend support to the change process, identify competencies and positive points, and highlight the strong points and contribution

of all who matter. Build on hope, optimism, opportunities, and confidence for change and development.

3. Develop a shared vision and mission
   Involve the core staff, influencers, opinion leaders, formal heads, and decision makers in developing a shared, inspiring, bold, but achievable and tangible, "vision"—that is, the desired future state of the organization within a foreseeable time frame. Develop a consensus regarding the main purpose—the "mission" of the organization. State the vision and mission in most simple, easily understandable fewest possible words, and in one or two short sentences. The statements should be easy to remember and express by people at *all* levels in the organization.

   Vision and mission must be directly linked with the existing strengths and potentially developable strengths in the organization; and clearly related to the opportunities in the environment and needs and expectations of the key stakeholders. Follow up with formulation of clearly articulated plan of goals and strategy to achieve them.
4. Inspire, educate, and engage with all stakeholders
   Inspire members of the organization and stakeholders by constantly interacting with them individually and in groups about the main theme of change, development, and transformation. Inform them of the progress being achieved from time to time privately and publicly.

   Invest heavily in developing people by offering opportunities to them for developing their competence, exposure to best organizations, institutions, seminars conferences, and best practices to broaden their thinking and improve personal performance.
5. Publicly acknowledge and appreciate good work and contribution to change
   Ensure immediate personal and public recognition of a good initiative, notable achievement in individual or team performance that helps in attaining the goals of organization, advancing the change process, effective innovations, and any other achievements that improve the image and performance of the organization.
6. Expect high performance, extend high care
   Motivate staff to set high and challenging goals. At the same time ensure that you sensitively take care of their genuine personal and professional needs and aspirations that affect their motivation, morale, and performance. Most important, trust them, but keep your eyes open and be fully vigilant about negative and damaging practices and behavior.
7. Embed sound values and culture in the organization
   The essence of Institution Building is to embed strong sound values and create the right culture. The key elements in values and culture are: meritocracy and professionalism; openness and transparency; integrity and fair dealings; egalitarianism and respect for individuals; honoring commitments and delivering full value to the customer; sensitivity to social and environmental issues; equity and inclusiveness; gender sensitivity, etc.

8. Make big ambitious plans, but give full attention to details
   Implementation of plans, programs, and strategy is the key. Effective implementation requires very focused attention to details. Either do it yourself, or ensure that it is effectively done by some key persons in the organization.
9. Keep raising the bar
   Having achieved a desired level of performance and excellence, do not become complacent. Raise the bar of goals and standards of performance to achieve the next level. Institution building is a dynamic, continuous process for achieving higher and higher levels of excellence.

20 The story of a management institute: Institution building in retrospect

D. Nagabrahmam

Introduction

T. A. Pai Management Institute (TAPMI) stands out as one of the reputed institutes of management in the country. While the institute remained lesser known and mostly a regional outfit in the first decade, the next decade created altogether a different image as one of the top-rated and a pan-Indian institute. Today, it is one of the three institutes in the country that has the distinction of the most coveted international accreditation of Association to Advance Collegiate Schools of Business (AACSB), a highly acclaimed accreditation body in the world.

It has a large, enviable residential campus located in the pristine surroundings on a small hilltop with the best of state-of-the-art facilities. All this and more managed with self-finances and with greater executive initiatives and commitment. The story of the institute is unusual as it has no endowments given and had very low-resource backup. More so, it is located in a small town away from the main stream of business or industry.

This is an account of how the institute was turned around from a small lesser known institute to a leading management school in the country. The chapter presents an account of how all its disadvantages and constraints were managed? What all went into it to become what it is today? Importantly, what are its distinctions? What has been the process of institution building? These major questions are answered through a brief narration of the story over a period of 16 years (1992–2008).

Background of the institute

A brief background of the institute will help understand the context. TAPMI was started in 1984 as an autonomous institution offering a 2-year post graduate program (PGDM). A senior person with professional qualifications and wide experience was recruited as the Director in 1983 with the assistance of a consultant. Soon after his joining, he started looking for avenues for starting the program. However, it took almost a year to start the program in the Engineering College premises and later on the institute was moved to a separate place. The first batch of students were admitted mostly from local areas and from the neighboring state. There were bigger plans including a residential campus with the munificence of state government for about 30 acres of land and the support of a rich industrialist, who was a beneficiary of the founder of the institute. However, these plans never materialized. Thanks to the indifference and negligence of the trustees in general that the director could not help matters any way. It all fell on the new director to initiate a campus plan in 2002 on an alternate site of 40 plus acres of land and he made it possible subsequently.

The Trustees are, of course, well-intentioned people, most of them belong to the extended family of the founder, and few others are his friends, admirers, and well-wishers including the rich industrialists who promised a huge amount for the campus. It was very unfortunate that the founder passed away suddenly rather prematurely. Possibly, it would have been a different scenario had he lived for some years to come. However, his younger brother took the lead in registering the Trust and initiated action for setting up the institute. The institute started in 1983 and the first batch of about 45 students admitted in 1984.

The Trustees in general had experience of running arts and science colleges, schools, a medical, and an engineering college. They seemed to believe that a management institute could also run on similar lines. Hence, many of their assumptions and possibly their understanding of institutions could not take the matter forward. For example, faculty is only necessary for teaching a number of courses, but nothing much else. The fee cannot be higher than that collected in the other colleges. However, they aspired that the institute to become a premier center in management education. The salaries paid to faculty were also similar to the local colleges. Given these beliefs and practices, the director could not do much nor intervene in any other way. Since the time the foundation stone laid for the campus on the land promised by the government, the matter only dragged on with a court case but without any positive result.

There were seven faculty members and most of them were young recruited locally. There was one with a doctoral qualification and one younger faculty had submitted his thesis. One of the senior members was Registrar of the institute taking care of administration, accounts, and finance. However, the daunting aspect was utterly low level of finances and almost hand-to-mouth existence. There were deficits year after year and somehow activities were managed by seeking help from one or the other donors. (To the extent of even two lakhs of rupees). Because of such a situation, practically, everything in the institute looked poor—the classrooms, general maintenance of the place, and the education itself. However, the program continued year after year without attracting students. Thus, the number of applications for admission to the program gradually declined and there were only 200 applications for 60 seats available. Since the number of applications reduced, the fee collected was inadequate to improve any facilities. It was only ₹20,000 for two-year program and the total annual budget remained at 10 lakhs of rupees. The Governing Council and trustees had no way of raising revenues and allowed in effect this situation to continue. The director earlier could not do anything more to improve finances on his own except asking the trustees to cover small deficits. It continued for about nine years until a new director appointed in 1992.

First few steps: Taking important decisions and commensurate action

On the taking over by the author of the chapter as the new director a case study of the institute was prepared and presented in an international conference. The case study mainly covered the activities before 1992 and brought about the need for thorough review of all activities. The review suggested both short and long-term plans for raising quality of the program, with better resources including faculty, new programs, raising overall resources with involvement of faculty, and seeking support of the Governing Council and the Trust of the institute.

Simultaneously and quickly, improvement of facilities was taken up including cleaning of the place, better class rooms, and hostels. The staff was oriented and encouraged to use computers with internally available resources and faculty support. Since there was adequate staff strength, proper deployment was made so that they work with faculty as secretaries. Computers were leased to replace typewriters. This itself raised the morale of staff immensely. However, their compensation was pitiably low and it took a year or so to raise it to a reasonably decent level.

A new vision

As mentioned, a case study prepared by the new director clarified his thoughts and put across a cogent plan to move ahead. This was the beginning of initiating several steps of action. The case study also provided a vision to develop TAPMI as a national level institution. It also argued that better faculty attracts the best of students and investments in such faculty should be the priority. We wanted TAPMI to provide quality management education and attract the best students and best faculty

A compelling vision about the future growth of the Institute helped us to think big and remove the constraints.

Some enablers

A framework of management education, its characteristics, the kind of institute sought, and a strategic plan of such an institute not only enabled seeking an identity conceptually, but also allowed its practice and implementation successfully.

Subsequently, three detailed notes were prepared—one about how to raise finances, faculty, and physical resources, the second on the status of the program, its quality, and delivery, and the third a detailed account of how the program quality be raised including internal processes and its conduct. It was also planned to take some immediate steps to raise quality of the program with improved curriculum and taught by better faculty. Curriculum was revised and few courses were added particularly in the area of finance. It was also decided that the conduct of the program and its organization improved by preparing schedule of classes more regularly with a time table prepared for a month to begin with (Earlier it was for every week and even then a number gaps in the schedules). The program office was streamlined with an assistant taking the responsibilities for day-to-day administration with the support of faculty as PGP coordinator.

Because of irregular schedules and lack of proper coordination, conduct of classes, students took classes for granted and their attendance was erratic. Classes used to be conducted throughout the day with gaps in the schedule, leave alone their cancellation at the drop of a hat. All this and more resulted in students' lack of preparation for the classes and their indifference and in general their casual behavior. Since they had enough time on their hands, many of them used to go to their native places at will. It also leads to some unfortunate incidents including death of two students when they went for a late night party outside in a car that met with an accident. It also created distraction from studies.

Implementation of plans

Creating a learning culture and discipline: Soon, the students were informed about the sessions and the intent for regular classes from 8.30 am to 1.30 pm. A few visiting faculty from IIMB and a young PhD in finance added to the overall quality of the program. More surprisingly for students, 100 percent attendance was made mandatory and absence of students monitored. The combined effect of these steps improved attendance and brought semblance of order in the system. The faculty was also oriented to the methodology of preparing course outline, developing teaching material and coaching of students. It was clear by then the entire education process is serious and no let up allowed. All this facilitation enabled along with the library support, timings, much better utilization of afternoons by students for preparation, completion of both individual, and group assignments enforced. The system of continuous evaluation put into practice with prompt feedback given to the students. Letter grading was introduced. It was clearer for faculty that they cannot miss classes once scheduled. With many of these practices, the general outlook of students toward faculty and the institute changed for better. Their preparations improved and they became regular and disciplined. Simultaneously, hostel facilities were also improved substantially with reasonable boarding arrangement (earlier no such arrangement) as much.

Faculty enablers: A process of enabling faculty to take up responsibilities of teaching and distribution of academic and institutional responsibilities has been put into practice. The director spent enormous amount of time with faculty drawing upon their abilities and oriented them toward issues like developing effectiveness in the program, improvement in the methodology. The distribution of academic administration such as PGP, admissions, placements, students' activities, and an advisory system along small group of faculty members supporting each of the coordinators was also implemented. With the joining of young PhD and that too research oriented, the responsibility of research and publications was entrusted to him. He introduced Working Papers series. The placement for students showed marked improvement with companies such as L&T and EICHER visiting the institute.

Attracting students all over India: Having put in basic processes in place, involving faculty with shared responsibilities and understanding, streamlining procedures and conduct of the program, improving physical facilities and environment with several improvisations, the beginning of 1993 academic year looked more promising. All this happened during 1992–1993 and while these changes made, action was initiated on two major fronts: one on improvised admission process and methods including better and large advertisements in the leading newspapers, the second, preparation of a comprehensive curriculum from semesters to terms and the recruitment of few additional faculty. The advertisement brought about 10 times more applications and with planned admission process, there was all-round improvement in the composition of students admitted with greater diversity and from many regions, including places such Delhi, Calcutta, and Bombay.

The institute was much better prepared to receive a large batch of 90 students with adequate processes of induction, senior students taking responsibilities to welcome juniors. A creative socialization event called EPISODE was organized in which both of the seniors and juniors worked for several cultural events. On the program front, there is a well-prepared curriculum with a new structure of terms and with the addition of three more faculty. The process for implementing the new curriculum and the discipline improved the rigor of the program with adequate faculty support. Composition of the batch itself enabled such a process. There was constant attention to minute details with frequent meetings among faculty and the director, students, and faculty.

Attention was paid to feedback, discipline, and conduct of the program rather systematically. Slippages were largely minimized.

Importance of governance

The year 1993 also saw the streamlined Governing Council and Trust meetings with greater regularity. Director could place before the governing board, all the activities of the institute, information on additions to faculty, financial position, and in general progress made by the institute. First time, there were four meetings of these bodies in a year given their importance which added to the quality of governance of the institute. These meetings were conducted regularly with fixed months (July, October, January, and March) year after year during 1992–2008. Executive initiatives in several respects and their deliberations created a symbiotic process. Gradually and slowly, the trustees realized the value of such initiatives in a scenario of low resources and good performance. However, a few of them were always reticent and even reserved, but others could see such a value and prevailed on those few. They could see how the functional autonomy and freedom resulted in the institute gaining ground as much reputation. Many such initiatives included raising compensation to the staff and faculty as much as equitable distribution of surpluses.

The campus project initiative is a classic case of executive decision but supported by the trustees with some hesitations. It is also a large project involving a huge amount of ₹50 crores and all of it was sought as a soft loan from a commercial bank. The entire operation of campus project and its completion became largely the responsibility of the director with the support of faculty. However, he had to go through very hard and uncertain times bearing the brunt including twists and turns in executing the project. It took almost seven years to complete the campus.

TAPMI has another distinction of really a self-financing institute and its credibility made it easy to seek such a huge loan and be accepted by the banks.

Board of studies

The board of studies meeting was revived with the first meeting held in September 1993 and provided a great opportunity to look at the new curriculum. The composition of the board consisted of senior executives from the industry and a number of them from premier management institutes such as IIMA, IIMB, and MDI. The lively discussions, helpful suggestions, ideas from the very first meeting helped the institute in the conduct of the program, addition of courses, modules. There was no looking back and the meetings continued annually for the next 15 years. The institute benefited by the counsel, caution, and advice of the members. The networking proved helpful, added new sources for support, and enabled the institute to learn continuously.

These meetings also provided a significant arm for useful reviews of all academic matters. Being an emerging institute with its limitations, these meetings served immensely in building the institute for future.

The general environment became purposeful and full of activities. It also brought about students' initiatives in several of the extracurricular activities. One such remarkable one is Brand Scan—a market research fair that has become popular in this region and has got industry recognition. It also brought the students together in creative exercises, games, events for different segments of population and offered them good deal of entertainment and bonding in all these exercises. The unflinching support of the faculty and their active involvement is a major highlight in making the first Brand Scan a great success. It went on and on to become much bigger event thereafter attracting ever-increasing offers of products and services for Brand Scan from a number of organizations. An event turned out to be tradition almost over the years.

All put together, the word spread in the community that TAPMI as an institution is different. With many initiatives, the foundation for an institution of promise was laid down. The foundation grow only stronger with continuous urge to do well, aiming at a big picture, constantly remembering for what the institute meant for resulted in series of plans and activities. Within the limitations, new methods of work, delivery, and teaching methods were pursued vigorously which added to the quality of program. These also spurred few innovations such as management in practice (MIP) using summer internships for developing cases and teaching materials.

The eco-system for quality in every aspect developed much early along with many interventions. Discipline, regularity and systematic functioning, faculty sharing larger responsibilities, quality and diversity of students, better physical environment, thrust on academics, and extracurricular activities, faculty interaction with students, coaching, advisory system, a rich and well-organized library, informal working system, and flexible approaches were all hallmarks of the changes made.

Some important steps in making the institute

Capacity building: Curriculum development

It was not the number of students alone but focus on building capacity by development of curriculum and strength of faculty and development. Over this long number of years, there are continuous efforts in developing curriculum as comprehensive as possible and current on par with the best of institutes in the country, perhaps even somewhat new and innovative. There are annual reviews, active participation of faculty, systematic collation of feedback of industry and students, alumni, peers, better methods of delivery, adding young PhDs as faculty in strengthening curriculum and its quality. There is continuous search for new approaches, experiments and innovations, and ideas added by the Members of Board of Studies. The members met annually and offered their advice, experience, knowledge as much as caution and care in developing and offering courses. The institutional processes within and with board of studies benefited the institute enormously. The enriched discussions and support were so valuable that the institute continued with this institutional mechanism.

There were many initiatives made in the curriculum. Probably, TAPMI was one of the earliest to introduce Ethics as a required course way back in 1993. It became one of the most popular courses sought after by students. Thanks are given to the faculty member who developed this course imaginatively for both quality of material and lively teaching. No wonder, all this material published as a book, viz., Ethical Choices in Business (Response Books) in just about five years' time and a second edition also brought as there is good demand for this book across business schools. Seminar courses on managerial leadership including requirement of reading of one biography and autobiography of a leader by each student, their presentation, summary, and analysis provided students with an opportunity to develop insights into the problems of business. Projects locally and more so with entrepreneurs, Brand Scan—market research fair, an annual event and joint research of students and faculty, and their publication are other initiatives creating greater capacity of the institute.

Strengthening the standardized curriculum of a business school, introduction of measures for enhancing experiential learning, practice of students, and collaboration of students and faculty made enriched curriculum and its continued development. There is also certain element of evolution of processes and methods, as there is continuous search and improvement of curriculum, of course, with a clear purpose and focus.

The introduction of MIP in 1998 is one such example that emerged out of concern to raise practice component for students. It became not only a meaningful learning exercise with good deal of intensity but also highly productive contribution by both students and faculty. No surprise, it was highly commended as an innovation in management teaching and received an award in 1999. Significantly, about 600 cases authored by TAPMI faculty were registered with ECCH by 2007, at least 15 percent of them were used by many B-Schools across and received royalty.

The approach toward effective learning and better methods did not stop with it. New approaches were sought in making the curriculum differentiated and highly learning oriented. Almost two years of deliberations among faculty, director, dean, a thematic curriculum, first of its kind among business schools, was developed by 2002. The course structure was replaced by well-defined themes with their rationale, faculty-taking responsibility for specific themes its components and modules. There are altogether 10 themes divided equally within the two-year period with a theme coordinator for each of the themes. The themes included: (a) Personal Competency; (b) Individual and Roles; (c) Firms and Markets; (d) MIP Phase1 + Out Bound Program; (e) Firm and Functions;

(f) Firm's Performance; (g) Managerial and Leadership Roles; (h) The Firm's Context; (i) Professional competence; and (j) The Complete Manager.

The importance of formal approaches toward curriculum and its development in raising qualitative capacity equally matched by and reinforced by informal means. The importance of deliberations, discussions within the faculty group, and enabling a process of student evaluation of courses and faculty on a regular basis enhanced the value of both formal and informal approaches. Strengthening and evolving structures such as Faculty Council and Academic Council and the processes involved certainly brought out greater degree of deliberations and concerns of faculty as much as their openness. It was also because of developing a concept and practice of working groups within each specialization and discipline. The idea is borne by the need felt at a time when research and publications became institutional responsibility as much as individual and groups. While teaching should not suffer at any rate, the working groups became useful means of accommodating and working, teaching and research plans for each faculty. For instance, those actively pursuing PhD. program were accommodated and supported to have one term free from teaching responsibilities. It became almost the responsibility of each group to help support their colleagues. In fact, this became a norm for research environment. AACSB accreditation would have remained a dream if such a norm were not in place.

The process of curriculum development became not only continuous, but also part of faculty development as much as raising its quality and standards. Possibly, one could see the effect of such approaches in raising capacity of the institute.

As suggested earlier, curriculum and faculty development are seen as two sides of the same coin rather than separate entities. It is like an organic approach drawing upon the strength of curriculum and respective contributions of faculty in general. However, there are other developments and evolution of plans and programs for faculty to learn and excel with the basic tenet of faculty autonomy and freedom. Reinforcements happened rather by the interest and motivation of faculty centric approaches through many initiatives. Some such are worth recalling and mention.

Faculty development

It is the beginning itself, better faculty and their attraction to the institute became an important consideration. However, given the circumstances, it was not an easy proposition. Nevertheless, relentless efforts in spite of several constraints including resources as much as the location of the institute paid off well.

To start with, a few highly qualified and young faculties joined the institute on their own initiatives. There is certainly the professional network, former colleagues, friends, and well-wishers in some of the premier institutes brought about significant relief in getting needed faculty complement in the initial years. It has led to a word of mouth that the institute is highly faculty driven. There is greater value placed on faculty in the institute. A number of visiting faculty members from IIMB and few others from IIMA, IMI, and MDI readily agreed to teach in the institute. This, a shot in the arm, really made the task of attracting faculty relatively better.

Incessant efforts are others like contacting potential faculty from those who are about to complete their doctoral studies in some of better-placed institutes, inviting them to apply and see that they are adjusted to the place. Asking friends to encourage potential candidates, using visits outside to go to institutes and selected university campuses seeking meetings with the candidates and having a word with them proved to be useful. Besides, former colleagues and students of the Director joined the institute.

In other words, the strength of the faculty was mustered through rather informal means in the early years and continued in later years. The difference, however, is evolution of a system and formalization of faculty recruitment. The important consideration remained the same that the right kind of faculty be identified and selected. There was a turnover of faculty initially, by the year 2000 or so there was greater stability. By 2004, there were 25 faculty members from best of institutes and 80 percent with Doctoral qualifications.

The institute supported faculty who joined the institute without PhD degree to acquire Doctoral qualifications. The institute initiated a PhD. program in collaboration with a University for the formal award of a degree and the registration of candidates. Design of the program, course work, standards, examiners, and so forth, became institute's responsibility. Later, a doctoral qualification became a prerequisite for selection of faculty.

Faculty Development became the hallmark of the institute that brought better faculty year after year. Research and publications activity is one highlight of such development. This has enabled the institute to seek international accreditation to raise standards of learning and institutionalize research activity. One may recall that one of the first initiatives in the early 1990s was to introduce the working paper series, which received further boost in later years. However, other initiatives like MIP and faculty student collaborative research through the field-based projects added to overall climate of the institute. Majority of faculty had doctoral qualifications by then also changed the quality of research and growth of publications.

Vision 2005

During this time and around 1998, a friend of the director and a management professional offered his services to draw up a long-term plan and a vision for the institute. It was a coincidence at a time, the institute was aiming big and it was in fitness of things to have a long-term view of it. However, it was not simply a consulting report but a process of intensive efforts by the entire faculty and director in developing a document called Vision 2005. It took almost a year. The efforts brought together greater understanding of the institute across the faculty and most importantly, their ability to view from a perspective of long-term orientation.

Comprehensive Performance Management System (CPMS)

Coinciding with the process of international accreditation, a thorough and well-designed performance system was introduced called Comprehensive Performance Management System (CPMS). The concept and design was finalized in various meetings of the faculty. There was a trial run before its formal introduction. The basis of performance is comprehensive enough covering teaching, training, research, and publications and administrative responsibilities with due weight and credits besides incentives for outstanding performance and rewards. There is also clear indication of poor or indifference performance and its treatment. By then, faculty annual plans of work were in place. The accreditation process reinforced performance at all levels with a clear focus. It facilitated faculty with self-evaluation, along with peer reviews by two colleagues and student formal feedback. Based on inputs, there are two levels of assessment by the dean and director separately.

The vital ingredients in building the institute are curriculum and faculty development simultaneously, building upon each other and together. The synergy generated several spin offs and as much the influence.

Structures and systems: Formal and informal

Informal structures and systems such as working groups of faculty, annual faculty workshops, industry workshops, MIP, faculty student collaboration, field-based NGO stint for students and research, not to speak of the weekend executive program, first of its kind and successful one in Bengaluru are important spin-offs. They came more naturally and evolved. The importance of process approach cannot be overemphasized. Nevertheless, formal systems and structures were created as the institutes grow such as Academic Council, Faculty Council, Strengthening of Constitutional Bodies, Research and Publications Group, Center for Executive Education, Bengaluru, Alumni Association (Registered), Library Committee besides the existing Chairpersons and Committees for Academic Administration. The functioning of these formal groups strengthened and became effective due to the existence of informal groups and their active involvement and support.

Impacts

The impacts are clearer. These are largely reflected in faculty contributions in terms of publication of books, research papers, strengthening of academic base and quality, norms of behavior, and setting examples of teamwork. In addition, demand for admissions and visit of industry to campus, placement offers, etc., are symbolic of the changed brand and growth. In addition, this also contributed to the enrichment and empowerment of stakeholders of the institute—be they students, faculty, staff, and management. All these initiatives helped give a unique identity to TAPMI and get the prestigious international accreditation.

The first phase of growth of the institute is complete through basic foundations laid and creating "Institutional Capital." However, the next phase and its importance are in initiating renewal processes to take TAPMI forward its sustained growth.

Conclusion

As Ravi Matthai once said: "Institution Building is an act of Faith." Udai Pareek suggested nine criteria for Institution Building: attention to process: significance of goal or uniqueness of the filed-urgent social needs; innovative nature; autonomy; generating new values; impact; multiplication of know-how; linkages; and development of people. This experience indicates how all these aspects were attended to in building TAPMI.

Perhaps, there cannot be a better way of presenting the conclusion of the story than to cite a reflection of the author along with the students, and three faculty members, their understanding of the institute, and experience of going through its pangs of growth, constraints including hardships as much as their growth.

Reflection: Comments from a faculty member

"We acted more naturally, independently and exhibited relatively greater degree of autonomy. Every aspect of our institutional life, its culture, and its ways and means of conduct reflected this behavior. The freedom enjoyed by the faculty, their participation and involvement in the institutional affairs are legion. Their support was beyond their call of duty. The staff exhibited quite a bit of enthusiasm and took active role in the institute's work and welfare. Their levels of motivation and morale were high. They felt good working in the institute and used to think it is different from others in the neighborhood and in the vicinity.

The students realized the value of their studies in the institute in spite of lack of proper facilities and constraints of space and other comforts. They valued more, however, the academic rigor, quality, discipline, and people—both faculty and staff. They also valued their participation in several co- and extra-curricular activities as much as curricular. Formal systems of their collaboration with faculty, colleagues, and other lesser formal or even informal ones were some that are not only innovative but also of educative and learning value. There were several such avenues."

TAPMI experience—A student's comment

I can recall three distinct things that set the platform for the transformation. They are: (a) Discipline with zero tolerance; (b) Boundary-less learning environment of a high standard; (c) Diversity and talent of the students.

Comments from faculty members

A faculty member on leadership: Professor Nagabrahmam's leadership was key to successful turnaround of the Institute. Professor Nagabrahmam's greatest leadership qualities I saw were his hard work, ability to recover from shocks quickly and bounce back with optimism, positive approach to life in general, never say die attitude, perseverance, and networking. I know of cases where some good faculty had left the institute, Professor Nagabrahmam would keep in touch with them and pursue them to return to the institute for several years. For recruiting faculty, Professor Nagabrahmam would make visits to IIMs and other institutions of excellence, he followed up with the leads he got, and persuaded graduating young PhDs to join TAPMI.

A faculty member on placement: "I distinctly remember visiting a corporate house in Mumbai seeking placements for TAPMI students. A meeting was set up with HR head through some acquaintance (we would not have got an appointment if we had cold called!). We were grilled for an hour about the institute from application process, selection, curriculum to placement process. Some of the questions posed to us were whether we followed case pedagogy, whether a particular case was discussed in the program, whether institute paid for the cases used! At the end of it all the HR head said that they will take one student for summer placement on an experimental basis and we have the freedom to select the best student for the job! I distinctly remember sharing this experience with the Director and he used to quote this on several occasions subsequently. What was more interesting was Director's own experience which he shared with us. He was made to wait for several hours by an HR head of one of the leading FMCG companies and then asked him to come the next day. Director was promptly at his door next day at the appointed time! When I asked him if he did not consider this embarrassing to go back the next day, his reply was that he was doing this for the institute and not for himself or his family. Few years later this company was part of the list of companies recruiting from TAPMI. Learning from this incident stays with me even today. If you are doing work for the institute, there is no scope for words like humiliation and embarrassment."

A faculty member on camaraderie: "Within a short span of time TAPMI was able to attract some young, like minded group of faculty—Professors Sivaram Srikandath, Sathosh Pai, Ravi, Obaidullah. We gelled well as a team and shared each other's class experience. The success of team effort was one of the things that gave me high apart from some good feedback from students on my classes. I guess this is one of the things that leaders should learn. Create an environment that attracts talented young people who can form team that enjoys each other's success. Even after almost 20 years of leaving TAPMI, I am still in touch with some of those colleagues, though we all parted ways in search of greener pastures. TAPMI gave me an opportunity to experiment and establish my career in teaching. I used to say TAPMI was my first love as far as teaching career is concerned has remained so after all these years. I would like to thank Professor Nagabrahmam for giving me this opportunity."

21 Academic leadership of universities of tomorrow

Indira J. Parikh

Introduction

All tomorrows begin with a "once upon a time." Now, India is an ancient civilization dating back over seven thousand years. Its rich heritage has meant that many events evolving and revolving around education have emerged over centuries. There has been a "glorious past" but there has also been a "decadent past." The glorious past has witnessed the flourishing of our education and educational institutions, which now have the wisdom of centuries. However, there is also now a downfall and decay being experienced in our educational industry. It is important to note here that now what educationists and educational institutions will create will become the universities of the future. And in order for this to happen India will have to create a system of Academic Leadership and Academic Leaders who need to be innovative and creative in the redesigning and recreation of an academic curriculum which is relevant and meaningful for the generation of today and tomorrow. It is in this academic creation that new knowledge will emerge and the country will contribute to the world. It is time that the country stops taking the role of adapters and adopters, borrowers and followers and instead emerges as contributors of theories and theoretical frameworks relevant for human existence across the world.

Before I begin my chapter, I would like to share an experience and example of Academic Leadership. About 12 years ago, an academic associate of mine was nearing completion of her PhD degree. All work, formalities, protocols, etc., had been completed with only the viva examination remaining. However, two years passed and no examination had taken place. On her behalf I intervened and spoke to her guide requesting him to share with me the reason for the delay. He explained that there was a difficulty in finding examiners and if I could help him out by speaking with Dr Udai Pareek. I immediately called Udai and he instantly agreed, provided the student would travel to Udaipur as he had fractured his leg and was on complete bed rest. The student went to his residence at Udaipur and resting on the bed

Dr Udai Pareek completed the viva and the student was able to complete her PhD. degree. Such is the commitment and dedication of an academician who would go out of his way to accommodate a student in order for her to be able to complete her requirements. However, this is not a unique example or experience involving just one student. Over the years there have been many, many such stories involving many students at all of the academic institutions where Dr Udai Pareek taught. He added immense value to both the institution and its student body through his level of commitment and involvement.

Therefore, the role of Academic Leadership centers around creating an environment for learners to engage with the learning process. The role also involves being able to navigate any "road blocks" that may come on in the way for whatever reason that may prevent/hinder institutes from achieving this objective. A parallel objective of Academic Leadership is also to help students to grow and emerge as good citizens of the country. For instance, Dr Udai Pareek was at all times an academician first, but he was also a stalwart in supporting individuals, institutions, educators, education, and students.

Both these roles are vital because today the industrialized and IT worlds are emerging in diverse and multiple forms, shapes, structures, textures, hues, and colors at a rate never witnessed before in human history. For instance, one innovation in technology today catapults the world and its people into a spiraling effect as there is almost always an immediate avalanche of subsequent accompanying and complimenting applications. The people of today respond and catch onto these innovations and the world is then abuzz with these new innovations. The young of today (who get younger by the day) are almost instinctively familiar with these innovations that the older generations read about in the media and wonder if they are actually fact or fiction. Today, India is known for the technological skills of its people and the contributions made by them in the development of technology. However, it is in the development of new knowledge and the documentation of that research which needs to be strengthened.

The same phenomenon repeats itself in schools and colleges as well as in universities. This makes everyone wonder as to what will be the next big change, how fast it will occur, and what will its intended and unintended consequences be—not only in the specific field or institution, but also on the culture, society, families, the life space, roles of individuals, communities, as well as institutions of learning. Now with all of the transformations occurring in academia, the role of leadership becomes very significant and critical as to the choice of direction institutions can make by providing direction of movement and growth.

Today, educational institutions at one level are seeped in tradition and at another level are making movements in directions that never existed before. Universities have been grounded and rooted in the tradition of teaching and preparing the next and subsequent generations with an academic curriculum and pedagogy time tested and proven for decades, if not centuries. However, in the recent past, a silent revolution has been simmering in the country. This simmering is reaching a critical point where, if appropriate, pressure releases are not created, then there is every possibility of this revolution erupting into a volcano. This, in turn will create chaos which will then be difficult to manage and contain. Simmering has a way of gathering the energy and momentum to a critical juncture where at a point either it can be disruptive or the energy can be channelized into a meaningful direction. If the energy intensified with the simmering is not channelized, then the intensity has a way of flowing like lava and becoming destructive and unruly. But if the energy is channelized and provided a direction, the chances are that innovative forms and shapes will emerge which will propel the growth and movement forward. It is the role of academic leadership in educational institutions and universities of the country to provide a direction in which this new knowledge and wisdom can be translated into creating a learning environment for the generation of today.

Academic leadership

Leadership in today's organizations is markedly different from the leadership in yesterday's organizations. Similarly, the leadership of the universities and colleges or for that matter any academic institutions requires role models, leadership styles, and approaches unlike those employed in earlier times. Leadership literature is full of descriptions of earlier models of leadership of universities. However what I am asking is: what will be the requirements for future academic leaders in these most turbulent, ambiguous, and uncertain times? Similarly, regulatory bodies, faculty, and students search for direction and guidance to manage the most complex and global environment of our times. Education at the present time is perceived to be the most critical and significant process, whereby an individual and/or a group can rise beyond the location of the past and create a new location for the self.

An academic leader of a university requires:

1. *"An ability to visualize" the nonexistent or un-entered landscape of education:* To visualize what this landscape of education ought to be and should be for the coming times and the present. For this landscape will have a different and difficult terrain. However, this terrain has existing people who have experienced discomfort with the earlier spaces of education. These people consist of demanding faculty, unsatisfied faculty, overworked faculty, and inadequately paid faculty. In this terrain are also a few individuals who against all odds continue to bring dedication and commitment to their roles and to their institutions. Faculty today are also grappling with uninterested students, unprepared class students, disrespectful students, belligerent students, anxious parents, and controlling regulatory bodies.

Current academic leadership is grappling with obsolete knowledge, outdated practices, poor infrastructures, and scarcity of resources. In this context, leaders of tomorrow's universities will have to have a vision to transcend all these barriers, as well as to inspire people to join him/her to make new choices and departures. Such departures can be made if the academic leadership defines a target destination where education will be challenging and will create new opportunities as well as an excitement of creating, building, and shaping.

2. *Sensitivity:* That is to say to be sensitive to the new generation of students who are Google addicts, tech-savvy, have the world and the information at their finger tips, but paradoxically are also confused as they have, maybe, too many choices. Parents are confused as they mainly know of traditional streams such as medicine, engineering, IT, and computer courses and the even more traditional choices of arts, science, or commerce. But the world has moved on and there are now hundreds of new career options many of which are also available on the net, which adds to the confusion of both students and parents. Academic leadership must help them to navigate and ensure that meaningful learning is available and that the process of learning continues to challenge and inspire the generation of today.
3. *Pragmatism:* Courses and curriculums will need to be relevant and meaningful for both students and teachers so that they can create a value-adding learning environment. Similarly, the pedagogy and methodology of creating a learning environment will have to be interactive and participative for both to engage with the learning process. In this day and age, learning needs to be centered partly around the students and partly around the faculty, but most importantly it must be relevant to the times. An optimum balance will need to be maintained between the significance of the students' needs, the faculty requirements, and the actual learning process.
4. *Administrative diligence:* Academic leadership today mandates the ability to manage the regulatory bodies of both the government and the accreditation agencies. Today's world is a world of ratings. These ratings influence the quantity and quality of entrants into an institute that in turn influences the level and quality of placements. Each of these can have a potential domino effect wherein one element influences all the other parameters which in turn influences the overall standing of the academic institution. However, academic leaders must decide whether they would run their institutions by the dictates of ratings or whether they would set their own standards of learning and knowledge.
5. *A global perspective:* Universities of today need to be multirace, multicolor, multireligion, and equally importantly to have gender diversity. It is only then that a university will truly be able to impart an education that is anchored in human civilization. The diversity of race, religion, color, and gender will create a unique juxtaposition of rich heritage of various

civilizations and help students to learn from each other as well as to be better positioned to deal with the shattering of myths, beliefs, and stereotypes about each other. It is in this meandering and experiencing that academic leadership will create a body of dynamic and vibrant students engaged in learning about the universe and the rich complex heritage of human civilization, and its history of evolution. This will add to the richness of learning, as well as the growth of students and human perspectives.

6. *Universality:* Today is the time for any individual in a leadership role to respond to the universe. The universe is at the fingertips of all constituencies and as such an academic leader needs to ensure that first himself, then the faculty and the students have an exposure and an experience of the diversity of the universe in regards to language, countries, culture, society, and the life space of people. The transformations across the world are phenomenal and leaders need to ensure that the students and faculty are grounded and rooted in their own culture and then open to exploring and experiencing other cultures, societies, and people. Academic leadership needs to be aware of where the universe is moving and the importance of the collaborative effort without compromising on the unique qualities of their own institutions. Many universities of other countries are already grounding their education in the exposure and experience of other cultures and societies. Traveling and experiencing other countries is a must if real experience and understanding is to emerge.
7. *"A holistic approach" to values-based learning:* Leadership needs to be holistic in its approach to learning and life. It is vital that students are prepared for life and its complexities of both knowledge and the multiple diverse institutions of belonging. Just as preparing for future life and the world of work is very important, so too is knowing how to experience life. Today's education is focused on preparing students for occupation through knowledge and skills. However, it is equally important to provide perspectives on life and the continuities and discontinuities of societies and human existence. Human history has a rich tapestry of heritage and this needs to be part of the approach to education. For this to occur, academic leadership needs to have a perspective on the times of today and the times to come. Education needs to be shaped in such a way that students have a perspective and a spirit of inquiry to explore and discover newer dimensions of content in the context of learning.
8. *Meandering and exploring knowledge:* For centuries, education across the world has frozen itself into narrow streams of knowledge. In India this is much more prevalent. A young student has to make a choice very early on even when the student is not ready or aware of what choices and/or directions are important and/or where his/her true calling lies. Educational institutions do not provide time, space, and freedom to

meander and explore the multiple choices available through experience and or dialogue. Meandering is an important element of learning and the ability to make choices, to enable students to discover, and/or to get in touch with their callings. Academic leadership of tomorrow's universities will need to provide the time, space, and people so that students can take their time to move around in the multitudes of subjects and choices and then discover their calling. This means that the universities will need to introduce the newer fields and the newer occupations just as they are emerging in this ever transforming world.

9. *The ability for experiential learning:* For centuries, education has been predominantly through books, rote, or teacher centric. As a result, the inputs are well defined and it is all about existing knowledge and/or "theorized" knowledge by people who have built knowledge from other sources. Experiential learning is frowned upon in the name of lack of rigor and or being too subjective. This has created students, individuals, and adults who can easily recite already existing knowledge. However, their education has not prepared them to evolve their own perspective, be able to take a stand/position, and/or have a viewpoint of their own. This makes students lose interest in the process of learning. The generation of today has no understanding of how to be reflective, explore their own roles and identities, and have a spirit of self-reflectivity. Objective learning is important but without the subjectivity and/or including themselves, the learning does not generate thinkers or reflectors of the phenomenon of life and living. All education becomes occupation centric with a focus on knowledge and skill sets. Tomorrow's academic leadership will need to create spaces where the generation can grapple with concerns of life and living and the transformations occurring across the world where uncertainty and ambiguity are so much a part of life. These need to be reflected and experienced by the students.

Universities of today

Universities of today stand at both a threshold and a crossroad, which is a paradox that never existed in India before. Beyond this threshold where universities are frozen today, lies a landscape ready to be entered and built as well as shaped and defined. The shaping and defining of this space is very important for the nation as without academia and universities taking this onus, we as a nation would continue to be beggars and borrowers, adapters and adopters and not take charge of generating new knowledge, have an underlying philosophy of education, create a learning and research environment, and finally take leadership roles in the midst of nations. In order to acquire and play a significant contributory role in the global scenario of education, the first step the universities of India need to take is to cross the threshold wherein they are frozen and have remained frozen for centuries.

Let us take a look at the frozenness of the threshold where the universities are caught and explore the reasons why they are so trapped. For this, we need to glimpse into the history of human civilization and the education in India.

Universities of tomorrow

Universities of tomorrow will have to cross the threshold they stand at today. They have been frozen at this threshold for a long time. Beyond this threshold lies a landscape ready to be shaped, designed, and defined. The universities that will take up this challenge and space will be the ones who are going to shape the world and the generation of tomorrow. The shaping, designing, and defining of this world are very important for the nation as without academia and universities taking this responsibility, the nation and its next generation will not grow into thinkers and philosophers. They would continue to depend upon other nations for their academic inputs and in the long run feel insignificant and inadequate. The generation of tomorrow needs to develop prose in the academic contributions by the country. The universities of tomorrow need to take charge of the destiny of education in the country, generate new knowledge, articulate the underlying philosophy of education, create a healthy research and learning environment, as well as take leadership roles in education in midst of nations. The universities of tomorrow need to acquire and play a significant role in the global scenario. This can happen only when the universities of today take a step into the future by crossing the threshold at which they have been frozen for a very long time.

Indian universities once they cross the threshold will then encounter new choices and new directions. It is only then that new designs and templates of education will emerge for generation of new knowledge, research, and theoretical frameworks that will impact the nation and the globe. The universities of tomorrow will become alive to the dramatic transformations and then change the basic structures of education and learning processes. The future of education is very turbulent and across the globe universities are grappling with the urgent need to redefine, redesign, and recreate an education that makes sense to individuals, the society, and the nation. Universities need to discover a new voice, a new language, and different ways having students and faculty relate. What and how this academic environment will unfold is unknown, how the generation is going to enter the world of tomorrow is unknown, but what is certain is that slowly but surely the universities of tomorrow and the education that they impart will have to be different. They have to grow today for tomorrow, and this growth will need change.

The universities of tomorrow will need to have a bigger canvas for the students and many more colors. Their innovations and contributions will need to be dramatic. Incremental transformations will be inadequate to cope with the turbulences that are visualized for tomorrow. Besides knowledge, attitudes, and skills, besides values, specializations, and occupations, what needs to be the underlying philosophy of education for the universities of tomorrow is to create a unique and complimentary configuration of all these components.

Therefore, universities of tomorrow need to:

1. Differentiate truth from untruth and pursue truth.
2. Create a space for dreams, differentiate fantasy from dreams, and translate dreams into reality.
3. Create spaces where learning is experienced as learning for life, living with values and not just for jobs.
4. Expose to students multiple ideologies and perspectives of human civilization and learn to move away from the compulsiveness of ideologies and the oppressiveness of one ideology.
5. Help students to discover their passions and true calling.
6. Be anchored in the process of learning and be grounded in the reality appraisal of their culture, society, and their own educational systems.
7. "Tech" and multimedia savvy.
8. Develop new models and frameworks of teaching and learning.
9. Be complex and competitive so as to foster and inculcate a spirit of excellence, entrepreneurship, leadership, and innovation.
10. Be focused on imparting an education that will prepare the next generation to emerge as writers, poets, philosophers, artists, thinkers, scientists, mathematicians, and healers of new world and generate all the wisdom needed to sustain the environment and human civilization.
11. Have a pioneering spirit. With all this discovery and creation, India needs to emerge as the pioneering nation that contributes to the making of choices and directions. For this, departures will need to be made, paths will have to be created, terrains will have to be shaped into landmarks, and students and faculty will have to grow into leadership and entrepreneurial roles.

Future challenges for the universities of tomorrow

1. The doctoral programs for universities will need to strengthen to have a much greater focus on more original and more rigorous research.
2. Universities will need to have an international panel of advisors who would reflect on the nature of the academia, where it is going, where it needs to go, and what is globally required.
3. Create a think tank to reflect on the future scenarios of the world and where each society is going, as well as how each society should design itself given its context and history. This will enable their policy makers to formulate directions and strategies for the interventions that are relevant for growth. This think tank would facilitate researchers and thinkers to dialogue and debate so as to understand global trends in education and multi-societies.
4. There will come a time when students will take courses from different universities which they will explore and find meaningful. Then they

will choose a university where they will want to integrate their learning through values and a perspective that they would like to live by. This would be the true spirit of learning where students can experience the wisdom of centuries and make meaning of it. They would choose their present calling and then focus on their specialization. This process then would reflect the essence of life and living. Each culture, society, and country would create its own contours of learning. Each educational institution would then invite students who would join for the accreditation as well as for its unique configuration of their curriculum as their graduation. The direction of learning is for growth and to move to a space beyond the horizons, which would be full of discoveries and surprises.

5. It would be important for each educational institution to create a *Navratna* Panel that would ensure the quality, rigor, and excellence of the academic institution, as well as emphasis on the human spirit of engagement.
6. Create centers of innovative research interconnected with the diversity of the various streams of knowledge, as well as focused specializations creating a rich collage of knowledge and wisdom.
7. Create a global research center where researchers across the globe will reflect on the nature of human evolution and civilization. This would provide understanding of the society and facilitate coexistence across diverse cultures.

The main challenge for the universities of tomorrow is to visualize what will come tomorrow. It is to beckon and prepare the young generation to shape tomorrow, to respond to technology, and to retain the essence of human existence; above all, to prepare the world for a harmonious and prosperous life.

22 Values and beliefs shaping transformation in higher education

*Lalitha Iyer**

Dr Udai Pareek clearly recognized the importance of transforming educational institutions, particularly colleges and universities. He pioneered in this area of work in the early 1980s and his contributions have continued since then. Dr Pareek and Dr Rao published their *Handbook for Trainers in Educational Management—With Special Reference to Countries in Asia and the Pacific* in 1981 and the issues they addressed then have continued to be significant and topical over the years. More importantly, the values and beliefs embedded in their work are more relevant today than ever before. I respectfully offer some reflections on the theme of change facilitation in the higher educational system applying the basic approaches pioneered by Dr Udai Pareek.

Value and beliefs

The values and beliefs practiced by Dr Udai Pareek have been a source of inspiration for the field of applied behavioral sciences. A common thread that runs through it all is the belief in human potential for achievement, learning, and growth.

I was working in a public sector institution and our organization had benefited greatly from the organizational development (OD) work of Professor Ishwar Dayal, Dr Udai Pareek, and Dr T. V. Rao. I therefore heard of OCTAPACE (Openness, Collaboration, Trust, Authenticity, Proaction, Autonomy, Confrontation, and Experimentation) values identified by Dr Udai Pareek and his associates in the context of effective HRD function in organizations.

* I would like to thank Dr T. V. Rao and Ms Shanti Yechuri for their comments and suggestions that helped me in revising the first draft of the chapter.

When I moved into the learning and development function the handy volume *Training for Development* (Lynton & Pareek, 1990) became my constant companion. I became aware of the term andragogy and found the approach simultaneously challenging and liberating. Simply put, the principles of andragogy assert that

- Adults are internally motivated and self-directed
- Adults bring life experiences and knowledge to learning experiences
- Adults are goal oriented
- Adults are relevancy oriented
- Adults are practical
- Adult learners like to be respected

Using these ideas in designing training and learning interventions proved to be an adventure that I enjoyed. From OD practice I learnt about AI and its principles and the related ideas on the need to focus on strengths and positivity in education and psychology. It was easy to see that stalwarts such as Dr Udai Pareek had integrated these positivity principles into their approach to people. An understanding of the implications of these principles and discovering effective ways to practice them are particularly relevant for transforming higher education. In my view, these beliefs offer the key to understanding and addressing the challenges faced in higher education in India today.

Higher-education–challenges in India

India has an expansive network of colleges and universities, with a well-defined "brand" hierarchy—central universities and national institutions of excellence such as the IITs and IIMs, state universities that are accrediting colleges and directly offering courses, autonomous colleges, deemed universities, and private universities. There are also polytechnics that offer options for skill development, and Open universities offering distance learning and e-learning options.

The system served nearly 26 million students enrolled in various programs in 2012 and the number is expected to be 35 million in 2017 out of the population of 143 million in the 18–23 age group. The GER in 2012 was around 17 percent. Significant differences are noticed in the gross enrolment ratio (GER) between men and women in rural areas—13.7 percent of men and 8.3percent of women are enrolled as against 30 percent in urban locations. GER is merely 7.7 percent for STs and 11.6 percent for SCs (Source: The Twelfth Five-Year Plan, GOI 2012, p 102 Fig. 21.12). There are significant shifts in the quantum and nature of demands on the higher education system. In terms of numbers, a big increase of 10 million students is anticipated in five years (2020) because of the convergence of several trends. These include the expected growth in the population in the age group 18–23 (India's demographic dividend as it is popularly termed), increase in candidates completing higher secondary education with universal primary school education, and therefore the increase in enrolment in higher education. The expectations in terms of what is provided by the higher education system are also undergoing a deep change. The profile of students is changing with increase in first-generation formal learners from marginalized sections of the society taking up university education. These students bring in anticipation that college education will offer a breakthrough for them and they will gain equal access and opportunity. Their progress is eagerly watched by the families and often the entire community that they hail from. Simultaneously, there are students from elite educational backgrounds who expect the human resources (HR) system to match the best-in-class offerings from across the world. There are also students from less elite backgrounds looking for technical and professional education as they settle down in their lives and begin contributing financially and socially to nation building. Youth displaced from agriculture see higher education as the path to escape from a life of poverty and deprivation. Industry, especially in manufacturing and services, is looking for talent and skills, both technical and managerial and employability has become a buzzword. There is much effort to nurture innovation and entrepreneurship in premier institutions.

Simultaneously, there are many constraints on the supply side. There is a shortage of university teachers which coexists with a growing pool of qualified young people anxiously waiting for appointments in teaching positions in colleges. Facilities need to be upgraded, and funds are very scarce. Entrance tests for the few islands of excellence are very tough and many talented young people face much distress because of their sense of failure when they do not make it into these institutions. This gap between the desired quality and its availability has given rise to many unorganized and informal coping systems. Here are some examples.

An entire industry has come up to coach aspirants for various qualifying examinations. This "coaching industry" is a "cash-cow" beyond the scope of formal regulations. It can coach aspirants for any aptitude test in fields such as management or administrative services besides preparing them to face job interviews, group discussions, and so on. Similarly, private education consultants offer much needed to "help" students to apply for studies abroad which includes "standardized" statements of purpose (SOPs). These phenomena are indicators that students are not ready to move on to apply the skills they are supposed to have gained by successfully completing their academic program. We may identify three outcomes that are sought after by both the graduates and the "user systems" such as campus recruiters

- Professionalism,
- Technical and academic competence in area of study, and the
- Capacity for innovation and research

And here are further examples that prompt us to pause and reflect

- Finishing schools are springing up offering the guarantee of "success" in getting a job by taking you past the first-round interviews and group discussions.
- Campus recruiters are looking for fresh minds, open to creative challenges and competing in campuses to "catch them early." Campus recruitments from prestigious campuses make headlines and the success in placements makes a college or school attractive.
- Colleges are busy advertising their success in placements rather than in academic or research arenas.
- If a family can afford it, children are sent abroad at any cost and reports of fake admissions rackets are an annual seasonal feature.

So where are we missing the bus? How can we strengthen our grip on the core principles that will generate the results we seek, instead of being trapped in a haze of frenetic activity.

Outcomes versus underlying belief systems

There are many debates about the challenge of balancing between the means and the ends or results. In learning–teaching situations, there is much data and research to suggest that the process is as important as the content in determining learning outcomes. It is also acknowledged that the methods used in education are instrumental in influencing the behavior of learners. When the focus is on knowledge transfer the methods are usually about gathering and presenting available knowledge in a manner that is easy to understand and recall. Invariably, a teaching style attuned to the learner's preferred learning style or a plan to balance the different ways that people use to learn helps the learner. When the focus is on skill building the methods used focus on practice, applications, and projects. For innovation or creativity, an environment or ambience for learning is created and the individual learners or innovators are encouraged to do what they want. Therefore, reflecting on beliefs about the methods or processes in education will help a great deal in addressing these core issues. Let us explore them in the light of the values and beliefs embodied by educators such as Dr Udai Pareek.

Fostering professionalism

The online Merriam-Webster Dictionary defines the word professionalism as "the skill, good judgment, and polite behavior that is expected from a person who is trained to do a job well." (http://www.merriam-webster.com/dictionary/professionalism accessed on March 31, 2015). The specifics may vary with the nature of the profession. Professionalism in the medical field is defined by the Hippocratic Oath and a chartered accountant is trained to uphold the principle of "true and fair" reporting of financial transactions. A young person graduating in any field from any school or university and joining the workforce is expected to live up to the code of ethical conduct prevailing in the workplace in letter and spirit. Now this is not covered in any syllabus either in school or in college, neither is it evaluated or assessed in the qualifying examination or assessment.

Yet perceptions prevail in the market place. An Ivy League college or Ox-Bridge education is taken to indicate professionalism and academic excellence thought this is not taught as a subject anywhere. So how can we expect that this esoteric thing will be nurtured in our universities? Is not it the job of primary or secondary school teachers and the family? Perhaps it should be achieved by the HR efforts of employers because they are the ones who benefit by this professionalism! The issue is simple at one level and complex at another. It is simple because it is "taught" by demonstrating rather than lecturing. The behavior of the college teachers can inspire students and influence their behavior profoundly. Thus, the teachers must remain conscious and aware that they are role models. It is also complex because there is no ready or agreed upon set of "professional" standards covering all aspects or lines of study.

It is in this context that the OCTAPACE model becomes very significant. It is a well-researched framework, accepted as a set of behaviors that will enhance the productivity well-being in work places. It can be taught—or negated—in colleges, universities in several ways. Here are some examples:

Openness and authenticity: Effective feedback systems can provide good opportunities to practice these values. In academic settings, assessments and evaluation can therefore become an area for practice of openness. For example, students can be introduced to peer rating, self-rating, and external evaluation as they progress in projects and assignments. Students will learn to be open to receive feedback and to offer objective feedback to others. The ability to assess their own work will also improve. These skills go a long way in the work place.

Collaboration and trust: Opportunities to work in teams on academic or research projects will orient students to collaborate and trust each other. The current system is very much oriented to individual performance and a tacit acceptance of a "winner-takes – all" strategy.

Pro-action and autonomy: Student projects and dissertations are the most readily available opportunities for the practice of these values. Choosing a subject, defining a method to gather required data for analysis are important experiences for a student to develop these. A research guide plays a significant role, balancing between autonomy, free choice, and academic rigor.

Confrontation and experimentation: Blindly conforming to authority or rebelling without a cause are both extremes that teachers notice among their students. Encouraging class room participation and group discussion go a long way in dealing with conformity. Reaching out to the rebels and channelizing their urge to make changes develops their leadership potential for the workplace. Working on these two values could prove most important in developing leadership qualities.

Such approaches will go much further in preparing or orienting a student toward professionalism required. The "brand" or ranking of a college will gradually be built on the way the alumni demonstrate these values wherever they go.

OCTAPACE in education

In the context of a college or university, the focus of leadership is on the learning processes within the institution which are well aligned to the emerging needs in the society at large. Learning happens in a web of relationships and institutional networks. The facilities and infrastructure are often mistaken for the "real thing"; they are necessary but not sufficient. For each teacher to emerge as an effective leader of a learning process in the class room, laboratory, or the field of application, the foundation is laid in the quality of interactions. OCTAPACE offers a ready framework for the administration to influence and encourage the teacher-leaders in their system.

This is well illustrated by a study on the culture within Birla Institute of Management Technology (BIMTECH), Greater Noida, which is ranked well within the top 10 private management Institutes in India and carries the BITS brand name for excellence. The recent study by J. Mathangi, Kanika Chauhan, Jaya Gupta, and Ajoy Kumar Dey (2010) offers evidence of the influence OCTAPACE has on the overall performance of an educational institution. They report their findings in their paper Organisational Culture: OCTAPACE Profiling of BIMTECH (February 14, 2010). They administered Pareek's OCTAPACE instrument to a sample of 70 teaching and non-teaching staff members and analyzed the responses. BIMTECH scored "very good" in all eight values as defined by the instrument. The authors concluded that as per OCTAPACE norms the culture of BIMTECH was "entrepreneurial." Collaboration and Experimentation were strongly correlated with rest of the parameters which the authors deem beneficial for an educational institute. Consistency between individual values and group beliefs was also tested and found to be very good, except for a significant difference in employee perception for the value of autonomy and the belief about pro-action.

Another recent study of private technical colleges in Punjab and Haryana by Dr Deepika Goomer (n.d.) found the situation in the selected private colleges quite different.

Only 4.0 percent employees are sure that participation is invited in the management of organization affairs and 10.3 percent employees are sure that organization welcomes participation of employees in the crucial activities of the organization. A few people within the organization are not sure that constructive ways of doing things are promoted in the organization. The author concludes that the organization does not believe in their work force and works on traditional patterns. Though the organization does not encourage employees to gain enjoyment from their work itself the employees do not generally avoid their responsibilities. In total, 44.0percent agree that experts and creative practitioners are invited to share their ideas, indicating that institutions work on traditional patterns and avoid initiatives. Only 4 percent agree that employees are encouraged to attend external programs. Clearly, the practice of OCTAPACE values differentiates the best in class versus mediocre. Discussing the findings the author emphasizes that an "environment resembling to OCTAPACE culture reflecting Openness, Collaboration, Trust, Authenticity, Proactivity, Autonomy, Confrontation, and Experimentation should be created in the institutions" for improving professionalism among both staff and students.

These findings are presented here to highlight how the practice of OCTAPACE within the institution is making a difference.

Developing technical and academic competence

Higher education has to prepare students to continue to upgrade technical knowledge and skills. There is an information and knowledge overload and "learning how to learn" or double loop learning becomes the key differentiator between expertise and mediocrity in any knowledge domain. The principles of andragogy are extremely relevant in this context. They play a major role in developing the ability among students to take responsibility for own learning and continue to learn in every situation through feeling, seeing, thinking, and action.

In respecting students as adults, it becomes important to accept that they have experience and knowledge and are motivated to learn because they aspire to achieve their purpose. Systems such as mentoring by seniors and teachers, and internships and volunteering activities chosen by the students become highly relevant. These offer opportunities to students to relate what they are learning to the practice in the field and apply their learning. The shift to andragogy is gradually taking places in students seems to notice and experience the difference at every stage in their academic path.

As an example let us take the experience of admission. A college usually goes by the marks scored in the qualifying or entrance examination and a "counseling" that indicates what program you can enroll for your given "rank" in this enrolment process. It is not about the interest or aptitude of an individual. This approach is changing today.

Rani S., an international student from Nepal, who has registered for a master's program in TERI university shared her experience when I requested her to tell me how she found her entry into the Indian education system, and she had a very different story to tell. She was interested in sustainability and environment, even though her bachelor's degree was in business administration from a well-rated college in her country. She was invited for the interview and group discussion even thought her discipline was different. The interviewers respected her choice and found her suitable for the course. She continues to feel very engaged with her program and appreciates the freedom and space she enjoys to choose her own learning path (interviewees name changed to maintain anonymity).

This is much more in line with the process followed in leading global universities. When students begin the process of applying, the first task is to submit a statement of purpose or SOP. With this step begins the transfer of responsibility for learning on to the student. Processes in our universities can be re-engineered using the ideas of andragogy, to enable students to excel in the field of their interest. It sets the path for research, creativity, and innovation. The policing approach to testing the store of knowledge has to give way to a more holistic profiling of the knowledge skills and talent or aptitude of each student. Opportunities to practice the skills being presented to the student must become an intrinsic part of the system. For example, students in fields such as management or social sciences can be gradually inducted to support administrative tasks in colleges through bodies such as the hostel or mess management committees, anti-ragging or diversity and inclusion committees, and so on. These opportunities will enable them to understand the "other side" and the challenges of managing preferences of diverse groups with limited resources. Those with academic orientation can be given opportunities to assist teachers and professors in research grading and similar tasks to deepen and enlarge their understanding of their subjects of study and the intricacies of college teaching. Officially and systematically developing such opportunities will go a long way in helping students grow up! In keeping with the anticipatory principle used in Appreciative Inquiry, students will begin to behave like adults when the system treats them as adults.

Promoting innovation and research

Research and innovation in the higher education lays the foundation for the emergence of a knowledge economy that leverages its creative capacity to get ahead in the global arena. This has been amply demonstrated in the Indian economy in the last few years. Our growth in the ITES and related industries is spearheaded by individuals from our "best in class" institutions such as IITs and IIMs. One factor that helps IITs and IIMs stand tall is the rigorous and competitive entrance and selection system. The applicants are highly talented, motivated, and often coached by experts to "crack" the entrance test. With such a talent pool the institution is bound to do very well. Building further on these strengths, these institutors have managed to foster innovation and entrepreneurial behavior. Some of the factors that have been crucial in this regard are the quality and credentials of the teachers, research projects, links with industry and other user systems, and the facilities in terms of laboratories, libraries, and special funds for incubation of new ideas or technology.

If these factors are replicated in other colleges and universities would this suffice to nurture creativity in a systematic fashion across the Indian Higher education system? It is obviously not that simple because these factors are not so easily replicated. The students enrolling are not in the same league as those who qualify in the prestigious institutions. Yet surely there is a talent pool that is deeper and wider than the intake into the premier institutions. Teachers who adopt an appreciative strengths-based approach have been able to reach out and unlock the available creative potential.

Conclusion

The critical factor that makes a difference is the philosophy or world view of the teacher. A teacher who is inclined toward a positive psychology approach and is willing to focus on the strengths of the students she is working with is more likely to foster good quality research that leads to innovation. Thus, the selection and induction of research project leaders and guides becomes an important route to nurture research and innovation. Universities can support college teachers to build their skills to foster creativity and innovation.

To sum up, the main argument I wish to make is that the beliefs and values practiced by the educators in the higher education system strongly determine the results that are desired but remain elusive. This theme has been discussed by Dr Udai and his associates over the years. More importantly, many of us have been witness to the impact that he could create in influencing teaching and practice over the years. All change begins with changes in beliefs, assumptions, or mental models. Perhaps one step we can all take is to review our personal philosophy about why we engage with learning processes. Attempting to answer for oneself the big questions about teaching identified by Goodyear and Allchin (1998) might help make a good start.

- What motivates me to learn about this subject?
- What do I expect to be the outcomes of my teaching?
- How do I know when I have taught successfully?

Bibliography

Goodyear, G. E., & Allchin, D. (1998). Statements of teaching philosophy, DigitalCommons@ University of Nebraska–Lincoln. Retrieved from http://digitalcommons.unl.edu/cgi/viewcontent.cgi?article=1403&context=podimproveacad (accessed on April 15, 2015)

Goomer, D. (n.d.). Organization learning, culture and quality initiatives in technical education institutions: A case study with reference to Punjab & Haryana State. Retrieved from http://www.niilmuniversity.in/niilmjournal/njmr/research%20paper%20deepikagoomer%20%281%29.doc (accessed on April 15, 2015).

Jayaraman, M., Chauhan, K., Gupta, J., & Dey, A. K. (2010). Organisational culture: OCTAPACE profiling of BIMTECH (February 14, 2010). Retrieved from http://ssrn.com/abstract=15 (accessed on April 15, 2015).

Lynton, R. P., & Pareek, U. (1990). *Training for development* (2nd ed.). New Delhi: SAGE Publications.

Pareek, U., & Rao, T. V. (1981). *Handbook for trainers in educational management—With special reference to countries in Asia and the Pacific.* Paris: UNESCO.

The Planning Commission, Government of India. (2013). *The twelfth five year plan*, (Vol. 3, p.102, Fig 21.12). New Delhi: SAGE Publications. Retrieved from http://planningcommission.gov.in/plans/planrel/12thplan/pdf/12fyp_vol3.pdf (accessed on April 15, 2015)

23 Lending wings: Institution building for specially abled

K. K. Verma

Dr Udai Pareek was a great human being and an outstanding academician. He, along with Dr T. V. Rao, founded several institutions such as the Academy of Human Resource Development and the National HRD Network. He was the first to introduce human resource development (HRD) programs at Indian Institute of Management Ahmedabad, various universities, institutes, and companies. It is in fitness to publish Dr Udai Pareek's memorial book. This chapter is my tribute to revered Udai.

This chapter is an adaptation of my book *Saga of an Emerging Voluntary Organization in Gujarat* (Verma, 2015). It is a biographic view of the famous six plus decades of the Blind People's Association, Ahmedabad. Though an old association, it is modern, professional, and progressive. Dr Vijay Mahajan, the founder and chief executive officer (CEO) of BASIX, in the foreword to the book described the association as "a wonderful institution, established by an even more remarkable person, Jagdish Patel." This chapter describes the long old organization through three eras.

The founder's era (1950–1974)

Jagdish Patel, a blind person, founded the Blind People's Association (BPA) in 1950 with the help of a Tea Club that he had formed for the entertainment of the blind. Though there were blind children's schools, there were no blind men's schools. The members of the club became the first beneficiaries by joining courses. Starting with on-the-job training and earning scheme, Patel added a primary school, secondary and higher secondary school, and a music school. He went on adding services for the adult blind. He realized his dream—a holistic service organization for the blind. The dream had four dimensions:

1. Enhancing capacity building
2. Improving self-confidence

3. Inculcating a sense of independence
4. Creating a strong and committed organization

Enhancing capacity building

Systematic and progressive steps were taken to train the trainers and educate the blind:

- Patel believed in Garbage-in, Garbage-out (GIGO) axiom in the teacher-taught relationship. He believed that a teacher can give what he has. He invested heavily on teachers' enhancement of knowledge and skills. He sent them to the best schools in India and abroad for domain training and special teaching technology courses for the deprived, particularly the blind. Qualified teachers only were recruited. For special areas such as community-based rehabilitation, experienced people were inducted.
- The schools were residential. Hostels were created for 260 students, including 70 women who were studying in local schools and colleges but did not have accommodation.
- Three inclusive schools were set up to encourage experiments. Very intensive and extensive work was done to build the capability of both the groups.
- Patel wanted to break the myth that blind were capable of only singing or grinding flour. He believed that they were able to do a variety of tasks. Skill development was created through vocational training.
- A two-year professional training physiotherapy institute was set up for the blind.
- Counselling and health awareness programs were initiated.
- Students were taken on field visits for varied exposures.
- Audio library, Braille services, and other equipments were provided free.

Improving self-confidence

Improving attitudes of adults was challenging. Class room encouragement, counselling, and programs offered at the BPA made a difference to the blind's self-esteem and self-dignity:

- Education was not limited to academics. Students were involved in group discussions in the class and group games in the field. They were given opportunities to discuss and make presentations.
- Emphasis was placed on morality, values, and good behavior. Stories of Mahatma Gandhi, Vivekananda, and other leaders were discussed. Examples of teachings were highlighted in student behavior, like a student declaring that he did not deserve marks on a question as it was not his answer.
- Good performance in studies, sports, and ideation was appreciated.

- Mistakes were not punished. Confession and promise not to repeat was emphasized. Counselling was resorted to.
- Mobility training was compulsory because it was closely associated with confidence building.
- Job-oriented courses, trade-focused skill programs, and a placement and employment department added confidence.
- Community and family were influenced through community-based rehabilitation intervention for better acceptance of the reality vis-à-vis the disability. They promoted among the people with disabilities rapport with and confidence in the village environment.

Inculcating a sense of independence

- One requirement for a sense of independence was to maintain health, as stated in a UNESCO paper. BPA arranged regular eye care, audiometric checks, hearing aids, eye surgery, ENT consultations, etc. Exercise/yoga was part of the schedule.
- Schedule was kept tight. Hard work was inculcated so that students could bear the work demands of the outside world.
- Job fairs were organized. Companies came to select the students.
- Artificial limbs were provided free.

Creating a strong and committed organization

To achieve his objectives, Patel wanted the Dream Organization to have an appropriate structure, staffing, and systems.

- BPA maintained lean and thin structure throughout to enable faster delivery of services.
- BPA recruited also people with disability. It was expected that they would be extra-committed to the cause of the people with disability. It had 40 such staff at all levels, including the Board and the Executive Committee.
- People with professional qualifications were recruited. Managers were given opportunities to participate in training and conferences.
- Patel gave up any work he was busy with if a blind person came to discuss his/her problem. This attitude trickled down the organization.

These were the four components of Patel's BPA. Majority of its students belonged to the low-vision category and about 30 percent to the totally blind. Because the group came from poor background, all rehabilitation services were provided free.

Patel entered into the challenging and complicated services to mentally sick, mentally handicapped, and emotionally disturbed children. This endeavor was a big shift from adult to children and from eye treatment to mental cases. These people needed treatment that was not readily available near home or within the district. The poor could not afford visiting a government mental hospital at far off places or a psychiatrist. In 1990, Patel set up a Mental Hygiene Clinic and recruited and trained the staff. The clinic was open to adults and children. Children with cerebral palsy, retardation, and learning problems were admitted. He also set up the Center for Children with Visual Impairment and the Center for Children with Deafblindness.

Another significant area in which Patel involved himself was to seek a membership of the Lions Club. Joining the Club was natural as Lions Club also aided the poor and down-trodden. This was another route to serve. He did hard work at the Club, gained recognition, and won prizes. He was elected president of the Lions Club of Vastrapur. Later nine of his associates were also given positions. They held discussions with the governments at the center and the state and other authorities. Thus the blind men's movement was activated. Patel played an active role at the National Association of the Blind.

This was a significant phase of BPA. Patel was not only BPA's founder but also its general secretary all the years till he died.

The turnaround era (1974–1999)

The beginning of this era was characterized by tension, conflicts, and bad name. Environment played its part. In 1974, some states faced agitations. Student unrest was going on against economic crisis and corruption in Gujarat. It was led by the social reformer Jayaprakash Narayan under Nav Nirman Movement. Many schools were closed down. The student movement may have added fuel to the fire in case of a strike at BPA. Patel, though a capable leader, could not control the situation and he himself became a part of the problem.

Patel was overall in-charge. He had created BPA and his acceptance as the chief was total. He was fond of keeping contact with the staff and giving them decisions on the spot whenever he met them. Such direct contact with the staff was so important to Patel that he instructed the staff even on minor issues. Once Punani, the number two of the BPA, suggested to Patel that he could stop dealing with trivial matters such as earned leave, sick leave, and late coming to the department heads, and confine to policy matters. A few more suggestions of Punani were sharing of the Braille books with other blind schools, job rotation, re-structuring of staff, etc. Patel reacted strongly by issuing a circular that henceforth even casual leave applications, which ordinarily were being sanctioned by the departmental heads, would be sanctioned by him.

Principles of management suggest that the CEO should take care of the macro management—forming policies, interaction with the external environment, and reviewing the strategic progress and performance of the organization. But micro management of implementation of the policy, plans, and keeping the morale and motivation is best left to all other managers.

Such a division was never acceptable to Patel. Decision-making was not separable from him or sharable by him. Mahesh Thakar, ex-principal of the Technical School, said: "Jagdishbhai told me several times that power must be grabbed, only one should know when and how to grab it. An administrator cannot be tender hearted" (Punani & Rawal, 1997). Patel's taking over the casual leave sanctioning authority from the departmental heads demoralized the managers. He did not listen; he ignored all suggestions.

More than 20 years after BPA was founded, a stir reflecting dissatisfaction by teachers and the principal of Higher Secondary School started in the form of a strike and demonstration in April 1974. The agitators involved the students and some of their parents. The people who went on agitation daily ranged between 70 and 100, followed by indefinite fast by a few. The agitators criticized the general behavior of Patel, quality of food, and low wages. The Executive Committee, which included trustees and others, stood firmly by Patel. But they could not resolve the grievances and declared a lock-out.

Ultimately, the state government took over the administration of BPA. The 3–4-month-long agitation waned when the students got tired and started leaving for their native places. Principal AP Trivedi, the leader of agitation, and many of his followers ultimately left BPA.

The lock-out and the government take-over dented BPA's reputation and image. Normally, the staff of BPA had been under obligation of Patel as he had recruited them on reference by people known to him. Why were the agitators disgruntled? Why did they agitate? An ex-employee shared in confidence to me that "both were hot-headed and even an ordinary matter *mann-mutave ho gaya* (difference of opinion) led to angers from both sides." Trivedi demanded better interpersonal relations, wages, and change in Patel's behavior. Punani and Rawal, who did a detailed study in 1997, noted: "The major cause of the agitation was the interpersonal conflict between this group (Trivedi and his followers) on one side and Jagdishbhai on the other side" (Punani & Rawal, 1997, p. 35).

Patel was always inquisitive about information and his favorites perhaps gave him exaggerated reports. Subhash Datrange, Executive Director of BPA, Bombay, observed: "Due to his nature, he has won friends as well as created enemies who would credit or discredit him with many things" (Punani & Rawal, 1997, p.233). Datrange's observation also goes to confirm that Patel had inter-personal problem with certain people, perhaps those who had independent views and held to them. Both such people and Patel were strong headed and hence a problem to each other.

All at BPA had a sigh of relief when the strikers started leaving for native places. The Managing Committee, after discussing for days, decided to appoint some senior-level professional managers and specialists to share Patel's load.

It was a blessing in disguise that a management graduate with PGP qualification from country's best business school, the Indian Institute of Management, Ahmedabad, Bhushan Punani was recruited. Punani shaped well and retired as an executive director of BPA after 34 years of service. Two more senior professionals were recruited, Ms Nandini Rawal, a management graduate from the B. K. School of Management, and a qualified engineer, Harish Panchal. These recruitments revealed that Patel had a foresight to judge the potentials of the candidates. All three fitted well with the culture and rose to the position of directors. Punani, after retirement as an executive director, was appointed as an executive secretary. Rawal later filled the post of executive director. The young professionals' team of directors took their jobs very seriously. In a couple of years, BPA again was back in limelight.

In the 1980s, many specialists joined BPA in areas such as social work, marketing, finance and fund raising, economics, law, and computerization. Patel took keen interest and gave them induction and training himself. He was concerned that the three seniors were not affected by the old guys. He personally inducted them sharing the history and his experience. He helped them to adjust to the new environment. He often kept them around to give the impression that they were important. They hardly faced resistance from the old guys. Patel's contribution in their long inning with BPA was creditable. Punani was given many opportunities. Many times, Patel took him abroad with himself to give him exposure. All the three I talked with were happy with their career in BPA. Punani did not feel offended that his proposals were rejected as Patel was a senior and founder. All the three grew with rich knowledge, skills, leadership, and pro-action. And they smoothly took the BPA to new heights when they took over after March 1999 when Patel expired.

Professionals era (from 1999 till now)

Patel breathed his last completing five decades of running the BPA his way. He was a capable leader, most effective, but somewhat overbearing and user of authority. Perhaps the decades of 1950s to 1970s were a tough period to raise funds and manage people and Patel filled the slot successfully. With his absence it looked the sun had set for BPA. However, the moon dawned soon.

From day one thoughts of new management were evident. The professionals' era was different. Punani, executive director, was given the additional role of CEO. Punani improved his contact with all staff and shared his participation style. The contact meetings clarified that all had stake in the organization and all should be active and responsive. Thus, the command and control approach began replacement by an participative approach. The flock started responding to the new style and systems. It was thought that easy substitute of the general secretary would not be available. Nevertheless, a substitute in Patel's wife, Bhadra Patel, was a fit case and was welcomed.

1. Systems change

I. Decentralization

BPA undertook new projects such as a rehabilitation center at semi-urban Kutch and an eye hospital in rural Bareja. For the first time establishments were set up outside Ahmedabad. The rehabilitation center was rushed up as a consequence of a severe earthquake in 2001. The eye hospital came up soon in 2003. The beneficiaries were local poor. But the surrounding areas were resourceful and they tapped the local contacts. These centers were controlled and directed from the campus at Ahmedabad. Many more similar centers were contemplated. Decision-making therefore was wisely delegated to the local heads.

Decentralization was the first time experience. Informal working became BPA's management practice. There was no scope for written communications, rules (except service conditions), and instructions. The concerned managers/heads were briefed that they could take decisions concerning their respective centers, except the financial matters and staff recruitment.

I found two instances of management's commitment to decentralization. One, whenever sitting with directors, I found that if a manager came with an issue, he was questioned for his personal views. Then discussion took place for solving the issue. Further while discussing with directors the past 10-year activities, I observed that they referred to "we" rather than "I." This was true of all the three directors. This participatory way showed respect for the manager's thoughts and involvement. Second, in my discussions, senior officers stated invariably that they enjoyed full freedom to plan and execute their assigned work. Such incidents confirmed management's prevailing style of functioning which was positive for decentralization.

II. Leadership focus

Another change in the system was to focus on developing leaders. Punani said that for succession planning they were looking at departmental heads to take up the posts of director. He cited examples of two departmental heads being considered for senior positions.

I went into the details of available records and made my assessment after detailed discussions with them and the three directors. Their identification was no doubt right. Their potentials and performance clearly stood out for senior management positions. However both had one limitation. They were in their present position too long and had little other experience. They needed wider exposure in more critical areas. Management should have provided them with opportunities to expand their experience base. There was another limitation. One of them had a short time left to retire. The leadership strategy was no doubt sound, yet succession remained to be seen.

2. Innovative initiatives

Several innovative initiatives took place. Seven of them are highlighted as follows:

I. Two high growth centers

The Kuchchh Comprehensive Rehabilitation Centre (KCRC) and Navalbhai Hiraba Eye Hospital (NHEH) were unique. The rehabilitation center, at a far off corner of Gujarat, was set up on an urgent basis as the earthquake of 2001 had devastated the town, leaving many people crippled. They needed treatment, support services, and appliances. People already disabled had damaged their appliances and needed replacements. A temporary center was set up at Bachau, the epicenter of the earthquake, to extend the relief on an urgent basis.

KCRC tied up with Indian Medical Association, Bhuj, and with Spandan Medical Relief and Rehabilitation Trust to extend physiotherapy treatment and establish make-shift general hospital to render medical services. The Gujarat Government, the financial supporter, suggested the center be made permanent. The government sanctioned a piece of land of 2,000 sq. yrds. BPA and its partner Spandan decided to make the center comprehensive, that is, extend services for employment, farm sector, skill development, etc. Technical help for assembling of equipments and money were no issues as local corporate and foreign funding were easily available in moment of crisis. In the farm sector and skill development, inclusive approach was adopted. People with disability and others from the main stream were jointly involved. KCRC emerged as a BPA within BPA; it became a center for all assistance.

Likewise, the Navalbhai and Hiraba Eye Hospital was set up in Bareja, a rural center, for all eye diseases including retina on donated land by a trust and financed by donors. It offered medical help for all eye diseases including retina, physiotherapy, ENT, and women's diseases. It became a general hospital in less than a decade and performed over 35,000 surgeries. The hospital also took up the challenge under the Australian plan for inclusive help to the villagers. The hospital provided free surgery, medicines, preoperative and postoperative expenses, food, lodging/boarding, etc. Support from the local community came for food items, and the staff was trained by CBM, Germany. Corporations, banks, voluntary organizations from India, US, and UK extended help.

II. Sensory path

People with disability needed help in mobility so that they do not hit against objects. Therefore, exclusive pathway was built across the walls of the buildings in BPA campus for the visionary impaired and the wheel chair drivers to manoeuvre independently and safely.

III. Dan utsav

Dan Utsav, *Festival of Giving*, was celebrated every year since 2009. The assumption was that people want to give; they did not feel happy to take; they would give to the deserving and trusted. The festival brought together people who want to give money or labor or talent or employment or entertainment, etc. Fifteen lakhs people participated in 700 plus events.

BPA celebrated the *Joy of Sight* in 2011. Four hundred eye surgeries were performed in a week to celebrate the Father of the Nation's birthday. In 2012 Joy of Employment was celebrated. Thirty-seven companies participated and selected 170 people with disabilities who came from different sides of Gujarat and nearby states. The approach was pitched with positive sentiments.

IV. Helping other voluntary organizations

BPA's leadership assisted smaller and new NGOs even at their own cost. For example, Stitching Shared Vision of Netherland asked for guidance for not only setting a unit in Rajasthan but also proposing a director to head it. BPA recommended their one promising project manager annoying a senior management personnel as BPA itself was short of project managers. BPA has been helping developing countries such as Mauritius and Bangladesh and many domestic voluntary organizations for setting up physiotherapy institutes and other services. BPA was helping not only other institutions but also themselves by growing their own internal resources.

V. Online fund raising

BPA undertook online fund raising. Both domestic and international platforms made sure that the applicant enjoyed transparency and reputation. BPA not only garnered money but also international exposure. Funds raised internationally were entirely from different donors than from the normal ones.

VI. Cataract-free Ahmedabad

On Louis Braille day, January 4, 2013, BPA took up "Cataract-free Ahmedabad" project. It was estimated that 20,000 surgeries were needed. The project started from the slums and then moved to other areas. The patients were invited to the campus or the eye hospital for surgeries.

VII. Experiment with passion

Punani selected six professionals from within BPA based on two criteria: capability and willingness to make sacrifices. He invited and encouraged them to participate in meetings with him and other directors. They were sent for select seminars and conferences in India and abroad. "We interacted frequently with them," stated Punani. The organization provided training and development opportunities by enrolling them in prestigious institutes and programs for domain programs. The directors remained alert to discover new, growing talents from their past work experiences, learning, successes, and failures. The directors could sense the candidates' passion and excitement for work. Gradually, they were gaining more confidence and their new competencies like taking lead, listening and doing, increased interest in supervision, technology were becoming evident, said Punani. They also learnt from the open and transparent work environment at BPA. Today, all of them hold independent leadership positions at BPA. Infusing passion in the organization was a difficult process. But the change led to higher commitment to work, devotion, involvement, strong desire to work, and friendliness.

4. Sports infrastructure

Realizing that sports infrastructure and activities were more important for people with disability than other children, Patel emphasized infrastructure for sports. A sports teacher was appointed by Patel on permanent rolls. A table tennis coach was appointed. A number of students were sent to London for TT training. Students also participated in 2010 Olympic Games. They went to Jordan, Thailand, and Bangkok also. Many of them won medals.

5. Inclusive schooling

BPA set up three inclusive schools. Children or adults with disability and other people studied together, played together, danced and sang together, and worked together in the same school under the guidance of specially trained teachers and other teachers. The two groups of students and also teacher plurality increased the feeling of togetherness and both sides looked to the other as normal. Students with disability started believing that others were their partners and so did the other group.

In these three inclusive schools, 90 special students and 465 other students studied. The Savinay Sammilit Vidhalya (inclusive school) was experimenting with groups below poverty line and much above the line students. BPA-MSM experimented with retired homeless disabled people and other group to work together and called the experiment Karam-Yog Centre. KCRC successfully experimented with skills training for self-employment for people with disability and other groups in the farm sector.

6. Centers of special needs

BPA raised special needs centers to cater to adults and children with severe mental and multiple disabilities such as deafdumbness, autism, cerebral palsy, multi-sensory impairment, and mental retardation. These eight centers required special skills. BPA created them and inducted new blood into the organization. Now they offer help available from toddlers to old senior citizens.

7. Financial developments

BPA raised money by selling products that disability trainees produced. During the seven-years period of assessment, performance improved across most of the financial parameters. Balance sheet expanded by 86.84 percent, immovable property rose by 232.50 percent, income by 133.68 percent, receipts by 58.97 percent, and movable property by 50 percent.

Institution development and HRD

BPA trained its staff adequately in their respective domains, but not in management. It did well to train and develop the beneficiaries. It had humane orientation toward staff. All this provided no doubt a healthy background to HRD, but it was not enough.

BPA was yet to take up HRD system professionally and review it from time to time. Looked at this way, BPA would not rank high. For example, BPA needed to develop key behavioral skills such as sensitivity and empathy, service orientation, collaboration and team-building, improved recruitment, performance appraisal, and manpower planning. It retained 60 percent of the retiring employees beyond superannuation. BPA would run into crisis if it has to identify future leaders. It would have to professionalize the staff and management. Multitasking could improve effectiveness and coordination.

HRD department was needed with a professionally trained head. Sponsoring of managers for select management development courses was needed. In fact, there was need for a National Institute of Institutional and Management Development and Research in Disability Management for the voluntary sector. BPA's reputation and respect being what it was, BPA could initiate the challenge and mobilize sector's support and of National Association of Blind and approach the government. NAB was the most competent to take forward the initiative once BPA took it.

Information technology

Information Technology initiatives were no doubt taken by BPA. Online donation option was available and it improved transparency in dealings. The Technology Center was headed by an able head, though he was not professionally trained. The center trained only the people with blindness in courses creating necessary skills for employment in software used by the blind people. There were 20 window-based computers with standard software such as MS Office. There was one internet line to 35 people, making the computers slow. Key people were connected on intranet. Annual accounts and balance sheet were processed on the computers.

Staff of BPA, however, needed to sharpen their IT skills. There was scope for improvement in many areas such as training in software, hardware and networking, human resource system computerization, integrated pay roll, donation management, and disability management.

Conclusion

Lending Wings relates to enabling the disabled. It provides a sketch of Blind People's Association's biography. BPA is an emerging voluntary organization of Gujarat. Though a very old organization, it is modern in technology use and application of concepts such as decentralization and hiring qualified people. In fact, it is a lesson for similar voluntary organizations that organizational obsolesce is not a factor of ageing. It can always be avoided. Further BPA founder could see the ignored area of extending help to the mentally affected in the early 1990s. His successors keep scanning the environment to identify the different needs and new segments of people with disabilities to serve. This is another area if not for research but for proactive search of the needs of the people in respective areas.

Patel was himself passionate to serve the blind and the crippled. Punani devised a special scheme to create passion in half a dozen officers and all of them are in leadership positions today. This too needs to be noted by voluntary organizations. It is worth experimenting.

It is often said that tough situations require tough handling. But in other times, participative approach is the leadership style which produces encouraging results. Punani's effort to orient the people with his participatory style found quick acceptance. I recommend his style for voluntary organizations.

Patel worked with missionary zeal, and many others, especially the seniors and Patel's confidant friends, showed the same zeal. Some of the six passionates created by Punani have strong missionary zeal. Missionary spirit and attitude to sacrifice time, labor, and compensation are surely the back-bone of continuous successes in the voluntary organization. Voluntary organizations should find their own ways to promote missionary spirit. The same would be true of professionals' recruitment and developing professionalism in the organization.

KCRC, NHEH, and Special Needs Centers give BPA immense magnitude and capacity to serve various disabilities, not just blindness alone. BPA can find pride that it covers all 10 disabilities that exist in India. BPA delivers 49 services and activities.

It is disturbing to use words like "blind," "handicapped," or even "disability." These words convey a negative message. BPA needs a respectful substitute. "Under-privileged" is better but yet unpleasant. Probably, BPA may prefer "especially abled" word as KFC does and may decide and sell the selected word inside and outside. The entire sector needs to work on the issue.

Last but not the least, technology and HRD users are going to be the winners in the future. BPA and other voluntary organizations need to adopt them.

References

Punani, B., & Rawal, N. (1997). *Jagdish Patel: The visionary* (p. 248). Ahmedabad: Blind People's Association.

Verma, K. K. (2015). *Saga of an emerging voluntary organization in Gujarat.* Ahmedabad: Blind People's Association.

24 Consulting and institution building as a journey of self-discovery

Ganesh Chella

Udai personified the true spirit of process consultation, thought leadership, and institution building. Interacting with him during my Indian Society for Applied Behavioural Sciences (ISABS) professional development journey had an immense inspirational effect on me. It is therefore a huge honor and privilege to be able to share my experiences as a consultant, a coach, and the founder of an institution in the Udai Pareek memorial volume.

Caveat

Before I begin to share my journey, it is important for me to qualify that these are merely my personal experiences and may or may not be true for others. Also, not all of what I did can be termed as successful. In fact, I learnt a lot more from my failures.

Having been "on my own" for over 15 years, many of my friends in the profession who are considering exiting corporate life to become consultants have been keen to listen to my experiences before taking a decision. Thanks to these conversations, I have begun to develop in my mind a conceptual framework of what it takes to establish a consulting practice and build an institution. This chapter is my first attempt to share this framework with others.

> The seeds must be hardy but the sun and the soil do made the difference.

As a young management trainee at Cadburys back in 1984, I had the opportunity to participate in the monthly meetings of ISABS which took place at Cadbury House under the leadership of Aroon Joshi, one of the founding members of ISABS and the then Western Regional representative. Thanks to these meetings, I had the opportunity to meet and interact with quite a few senior organizational development (OD) practitioners who were all ISABS professional members. Their knowledge, their reputation, and the body of work they were doing were truly inspiring. As I attended these ISABS events and pursued my professional development, not only was my interest in OD further fueled but also my desire to be like one of them.

These inspirational interactions coupled with my first few years of professional life helped me to recognize that my career anchors of autonomy, creativity, and professional expertise could be realized only as a consultant.

As I met many of the legendary practitioners in the fields both in executive roles and as independent consultants, it was clear to me that building my professional competence was a prerequisite to become an independent practitioner. The first building block of this professional competence was human process skills and OD competencies that I gained from ISABS.

As I continued in my corporate career I began to see transformational changes in the world of business. Businesses were struggling and cost pressures and competitive forces were fierce. Organizations were beginning to embrace approaches such as total quality management and business process reengineering and were slowly losing faith in classical OD as a means of adding value to their businesses.

This led me to believe that strong business and leadership capabilities would need to be my second building block to become an effective consultant and OD practitioner. Therefore continued corporate experience was important.

What I learnt: My career anchors must be in congruence with the demands of a consulting career. Also, career anchors need to be nurtured and strengthened over a sustained period through developmental investments.

Pull and not push factors

Given my need for autonomy and creativity, there was a strong temptation to consider becoming an independent consultant whenever I was uncomfortable in organizational roles. There was a time in my professional career when I was very unhappy and had significant value conflicts. At that point, I was hugely tempted to call it a day and start my consulting practice. However, good sense prevailed and I decided to wait for the right time.

I finally decided to quit my corporate career and go on my own when I was enjoying my job the most and was bubbling with confidence. This I realized significantly contributed to the confidence with which I was able to meet clients and generate business and grow my practice.

What I learnt: It is important to start my entrepreneurial journey when my self-worth and confidence is high and not when it was bruised. I should not succumb to push factors but must wait until the pull factors are strong.

Determining my core offering

The concept of the hedgehog (proposed by Jim Collins) appealed to me enormously and helped me in clarifying my own thinking about my core offering as a consultant. I wanted to do something that I was passionate about, I could be best in the world at and would made economic sense.

It was very clear to me that I did not want to be in the business of executive search or other transactional services or anything that would call for mass production and standardization. I also felt that training alone was not comprehensive enough and strategic enough for a sustainable OD consulting practice.

I decided to offer comprehensive HR and OD solutions that would help client organizations to become effective. This would call for work at the system and process level. I felt that high growth entrepreneurial businesses would be most open to accept such an offering to support their growth ambitions. The tag line "*building organizations around great ideas*" was born out of this passion.

I was also keen to work with large organizations on more sharply defined OD engagements.

In terms of my style, my training with ISABS and my exposure to Edgar H Schein's work convinced me that a process consultation approach was what would work best—something I would enjoy.

What I learnt: Being clear about my core offering and how it can solve compelling problems that have not been solved before is very important.

Commitment to thought leadership

My early research indicated that thought leadership was the best investment I could make to remain relevant and respected. The rigor of conceptualization and documentation that I learnt during my professional development journey in ISABS fuelled my interest in research. Also, for six years, I was on the Board of ISABS as Dean Publications and was responsible for bringing out *Here & Now*. All this helped solidify my interest and competence in thought leadership. I launched *Under the Bonnet* a quarterly research-based journal and managed to publish 16 printed editions of it over the years. These were hugely appreciated by my clients and the larger professional community.

I was also a columnist for *The Hindu Business Line* and wrote a hundred articles for them.

Having written a lot I was excited about becoming an author and publishing books. I spent a full year researching about the idea of a helping organization. This culminated into the publication of my first book, *Creating a Helping Organisation: 5 Engaging Ways to Promote Employee Performance, Growth & Well-Being.* More recently, I co-authored my second book titled, *Are You Ready for the Corner Office: Insights from 25 Executive Coaching Experiences.*

What I learnt: Leading thought through research and publication are invaluable investments to stay relevant, have one's voice heard and be respected.

Consultant as an entrepreneur—My blind spot

As I began to grow my consulting practice I realized that I had to hire more people and as I hired more people, I realized that my cost base was going up and I needed to acquire more business. I was beginning to fear that the autonomy I was enjoying in terms of the kind of work I chose to do and the kind of work I declined would be under threat given revenue pressures.

I reached a point where I was running faster to stay in the same place. The one big reason for wanting to start my own consulting practice was autonomy and I did not want to give that away.

Around this time, I was beginning to get offers from a few other consulting firms to consider a merger or acquisition. For a moment, the financial possibilities were enticing and the fact that someone considered my firm valuable made me feel good.

I soon realized that I had to take some big decisions. I realized that I was going through what Noam Wasserman calls *the Founder's dilemma* in his brilliant *Harvard Business Review* article.

According to Noam Wasserman every founder goes through the dilemma of whether he wants to be rich or be king. Rich would mean give away equity and autonomy in return for financial gain and king would mean staying small but staying in control. My mind was very clear—I wanted to be king. However, I still needed to think like a businessman and not just as an OD practitioner.

What I learnt: Professional competence and functional expertise are not enough. The ability to be entrepreneurial and spot challenges, take risks, and think beyond my competence are equally important and the good news is that this could be learnt.

Adapting to change

A big part of our work as OD consultants is around helping clients manage change. I soon realized that I needed to manage change too. The environment around me was changing and I had to quickly accept this reality and adapt myself and my consulting practice.

For one, the HR functions in organizations were changing rapidly. With scale and size, HR functions in organizations were looking for specialist service providers—specialists in employee engagement, specialists in job evaluation, specialists in compensation and benefits, specialists in assessments, and so on. At least for medium and large organizations, their preference was to work with a range of specialist service providers rather than with deep generalists or trusted advisors.

The entrepreneurial landscape was also changing. With venture/private equity funding, many entrepreneurs were becoming a lot more short-term in their orientation and wanted quick solutions. I saw the need to redefine my consulting offerings and also look at new opportunities for specialist services.

I quickly redefined my consulting model keeping in mind these emerging realities.

What I learnt: As an OD consultant, I need a taste of my own medicines quite often. Being adaptive, giving up my pet ideas, and embracing change is painful but essential. Otherwise, the winds of change might wipe out what I have so painstakingly created.

Learning to build an institution

Even as I reinvented my consulting model, I was keen to start a new venture by taking advantage of emerging opportunities, of course keeping in mind my hedgehog.

In the course of my consulting work with many entrepreneurs and high-growth startups, I began to notice a pattern. Many of them would approach me with the plea that they had human resources (HR) problems. After some discussions, it would become evident that their problem was not HR but leadership. Their inability to adapt their leadership style to the changing situation was often the real problem. I was convinced that there was a huge need and opportunity for Executive Coaching.

As a first step I worked with Dr Skiffington in Australia to get certified as an Executive Coach back in 2001. A few years into my journey as an executive coach, I realized that there was a huge need and opportunity to create an India-centric coaching institution, an institution that is sensitive to the helping needs and help-seeking behaviors of Indian executives. I socialized the idea with some of my professional colleagues and they responded positively. This time around I was keen to do it differently. I wanted to build an institution that could grow in scale and stature, unlike the boutique consulting firm I had created. With three other cofounders, we created what is now called Coaching Foundation India Limited.

In cocreating this institution, I was confronted with several challenges.

First, we faced a lot of resistance in promoting coaching as a helping relationship in a country where a stigma was attached to seeking help from formal sources.

Second, it took a lot of effort to create an Indian institution to offer such a service in an environment where everyone believed that all good things were born only in the United States.

Third, it took a lot of commitment to stay our course with a venture that was born more out of passion and less out of commercial considerations.

Finally, at a personal level, my struggle was to learn to respect and value differences in styles and approaches among the founding team. Having worked solo as a consultant, it was a new and different experience working with peers, cofounders, and other members on the Board.

I learnt to play to my strengths and also leverage the strengths of others. I learnt to build and leverage networks. I learnt to sell our services. Of course, I learnt a lot about helping relationships. As I look back, I realize that this phase of my professional life contributed very richly to my growth as a person and professional. After nearly 10 years of effort in building CFI, I can also say that I am a far better entrepreneur today than I was 10 years ago.

What I learnt: The key to building an institution is surrounding myself with people who share my vision and values, whose goals are aligned with mine but are superior to me in many areas and different from me in their preferences and learning to respect and value all this.

Entrepreneurship is infectious

Having created a boutique consulting firm and a national coaching institution, I should have taken a pause. But that was not to be. The experience of creating something was too addictive.

One of the areas where I had done a fair amount of work was around HR capability building. A lot of my consulting work required me to mentor and train HR teams. I was also witness to a lot of "HR bashing" in the print media and public debates, which hurt me a lot as an HR professional. I was moved to do something about it.

I spoke to some of my other professional colleagues in HR and mooted the idea of creating an institution that would focus on HR capability building. The idea excited them. Very soon, we had a network of six very senior HR professionals each one of them doing their own things but coming together for the common cause of HR capability building. This is how totus HR School was born. In the last three years totus HR School has been able to work with over 300 HR professionals, supporting them to enhance their capabilities.

With a sharp focus of only doing in-company programs, we have been able to make a good beginning in creating new knowledge and enhancing HR capabilities on the ground. These are still early days, but I think we are on to something truly exciting.

What I learnt: Some ideas are so compelling that it draws the right people together and I end up becoming a mere instrument in the hands of a larger power that is leading me to act on them.

Some closing thoughts

As I look back over the last 16 years of consulting and institution building, I am able to distil certain lessons in terms of what helped me in this journey.

a. A deep sense of passion and conviction and belief in what I was doing was important to stay the course.
b. The ability to spot a burning need and convert it into a tangible offering was certainly very important and has helped me in whatever success I have achieved.

c. Constantly investing in my own personal and professional development has been so important to stay current and innovative. Starting with my professional development journey with ISABS, constantly investing in acquiring new capabilities has been extremely valuable.
d. It is so important to stay actively connected and engaged with a wide network of like-minded people. There is so much joy in working with others.
e. Given all the ups and downs I have gone through, the greatest gift that has helped me to bounce back is the gift of knowing how to manage myself—the gift I received from ISABS.
f. In the field of compensation and benefits, the term total rewards is often used to emphasize the fact that the rewards that a person gets include not just money but a whole host of other factors including the pride, the affiliation, the autonomy and control, and the job itself. The concept of total reward has huge meaning for me. The ability to live a congruent life, the ability to do what I like, most of the time, the ability to feel choiceful, the ability to do work which puts me in flow-state have all been more important than the money I have earned.
g. Luck and timing are also critical. As I look back I believe I have been very lucky to have come in contact with doyens and visionaries in the field of organization development very early in life. People like Udai Pareek, Rolf Linton, T. V. Rao, and many others have been a great source of inspiration and help in shaping my formative thoughts and ideas about what kind of a consultant I really want to be. For this I will remain ever thankful and grateful!

PART III

Social Development and Nation Building

25 A ninth metaphor: Social catalysis

Tejinder Singh Bhogal and Rosemary Viswanath

Introduction

Gareth Morgan in his bestseller, *Images of Organization* (1998), uses multiple metaphors to understand organizations and management. He uses metaphors because, as he tells us, "Metaphors give us the opportunity to stretch our thinking and deepen our understanding, thereby allowing us to see things in new ways and acts in new ways." In this way, as Morgan would say, metaphors become a powerful tool for creating an understanding about (what we now recognize as) organization and management.

Morgan points out at least three advantages of using metaphors to see and shape organizational life:

- In approaching the same situation in different ways, metaphors extend insight and suggest actions that may not have been possible before.
- The insights generated by different metaphors are not just theoretical. They are incredibly practical.
- Metaphors lead to new metaphors, creating a mosaic of competing and complementary insights.

In his book, Morgan offers eight distinct metaphors to understand organizations. These eight metaphors compare organizations to machines, organisms, brains, cultures, political systems, psychic prisons, flux and transformation, instruments of domination.

We suggest that each of these metaphors is powerful in its own way, and can be used to understand and design action in organizations. Given that both authors work extensively in the social development and social change sectors, we also considered what other metaphors could be introduced that would capture some aspects of the realities of this sector that the given metaphors do not capture as well. Therefore, we suggest the additional metaphor of a *social catalyst* to help understand and shape some of the challenges and realities facing social change organizations in India today.

Defining social change

Social change is a process through which a community is significantly better off than before. There are two broad aspects of being better off:

- Provision of fundamental needs: This includes the provision of adequate drinking water, food security, education for children, health care, shelter, and clothing
- Provision of fundamental conditions for wholesome existence: This includes the availability of adequate and timely credit, adequate and appropriate livelihood, a protected physical and biological environment, provision of adequate and timely justice, absence of discrimination and conflict, violence and exploitation on the basis of caste, class, religion, gender, etc.

While all of the above are directly or indirectly enshrined in our constitution, and the government is expected to ensure them, the fact of the matter is that our society falls short on both the counts. The government is supposed to provide appropriate education, but teachers do not reach schools, or they behave with students from marginalized communities in a way that puts them off education. The government is supposed to ensure that no discriminatory behavior or oppression occurs with the marginalized, but the police looks the other way when oppression actually happens; and does not take to task the oppressors or violators of law.

This is where the social change organizations come in with their attempts to change either the affected community or the government system or society at large. As they are usually neither the affected community, nor a part of the government, their role is mostly that of a catalyst.

Understanding catalysis

Defining catalysis

As per Wikipedia, catalysis is the increase in the rate of a chemical reaction due to the participation of an additional substance called a catalyst. With a catalyst, reactions are faster and require less energy. Because catalysts are not consumed, they are recycled. Often only tiny amounts are required.

Catalysts work by providing an (alternative) mechanism involving different transition states and lower activation energy. One of the characteristics of chemical catalysts is that the catalyst remains unchanged. However, in a social change catalysis process, the catalyst does not remain unaffected. While the role of being the catalyst remains unchanged, most change catalysts are also informed and changed by the process of being a catalyst. They receive insights and learning in the process that help them to sharpen their own skills.

In some cases, catalysts may be affected in ways that are not functional to the catalysis process, particularly if they introject[1] some of the stressful emotions that are created in the change process due to different forces and resistances in the system. An example of this is a feminist organization working with women. Staff often had to counsel and support women who are the victims of violence and those in distress. Repeated such efforts left the staff emotionally depleted and burnt out as there was no institutional system to replenish their own selves. This led to cynicism and a sense of the world outside being bad and not willing to change. This in turn affected more creative approaches to catalyzing change and their world got divided between victims and oppressors.

Inhibitors, poisons, and promoters

Substances that reduce the action of chemical catalysts are called catalyst inhibitors if reversible, and catalyst poisons, if irreversible. Promoters are substances that increase the catalytic activity, even though they are not catalysts by themselves.

Inhibitors are sometimes referred to as "negative catalysts" since they decrease the reaction rate by deactivating catalysts.

The example given earlier of the feminist organization shows that a similar process exists for social change catalysts. This is often seen in many international aid organizations—when the inhibitors in their role of social catalysis are too many they suffer from what is termed compassion fatigue! When the organization recognizes the usefulness of the social catalyst as a metaphor to understand itself, it pays attention to systemic inhibitors and poisons and devises institutional processes and systems to recognize them and counter them! Inhibitors of the social catalysis process include an unwarranted focus on projects, and mismatch between stated and practiced values.

Theories and forms of social catalysis

Theories of social catalysis

Social catalysis can take both violent and nonviolent forms. The violent French Revolution fundamentally changed French and European society,[2] but for the

[1] Introjection is a psychoanalytical term usually indicating a process wherein the subject replicates in himself or herself behaviors, attributes, or other fragments of the surrounding world, especially of other subjects. It is usually a form of unconscious defense mechanisms.

[2] See, for instance, Eric Hobsbawm's *The Age of Revolution.*

purpose of this chapter, we are not concerned with this form. Among the non-violent forms, we can broadly talk about the following methods[3]:

- *Expert or technology led (the modernist approach):* In this approach, external experts tell the people what is missing, and set up systems that will benefit them. It is assumed that when the outside expert provides information, and provides a system, the poor and marginalized will understand the information, and use the system to improve their condition. Most of efforts in India, through the government, fall into this category. The limitation of this approach is that it does not make the process of personal transformation clear, nor is it a process of cocreation, which takes the capacities, and potential of the "users" of the system into account.
- *Gandhian:* The Gandhian approach relates external freedom to internal freedom, social and political revolution to the inner revolution where man conquers himself. The focus of the change is on the poor—to help the individual build up self-respect and self-confidence, and to strengthen their unity. However, Gandhian thinking also envisages a willingness among the rich to change;[4] it also exhibits an inadequate appreciation of societal structures.[5]
- *Frierian:* This approach is inspired by Brazilian educator Paulo Friere's "Pedagogy of the Oppressed," which was influenced in turn by Marx and Fanon. This is similar to the Gandhian approach in that it considers that the key task of the individual is to change from within, but envisages this through process of conscientization and critical thinking. Given the status-quoist interest in maintaining power imbalances, this was necessary as oppressors were unlikely to take the lead in bringing about fundamental social change. It is important to recognize that in situations of oppression both parties lose humanity and dignity. The process of social catalysis can, therefore, also be seen as one of the humanization processes, in which the humanity and dignity of both the exploited and exploiter are enhanced.

Most organizations with which this organizational development (OD) chapter is concerned adopt either the expert/technology, Gandhian or Frierian approaches.

---

[3] Two more methods could be added: political and Mindellian. We have not covered the political approach as it is mainly followed by political parties—most of them seem to be aiming for political power, rather than fundamental societal change. The Mindellian approach to resolving conflict between the discriminator and discriminator seems very interesting as it posits a fundamental internal change in both. Unlike, Friere, this approach does not center around the exploited taking the lead. However, we exclude it here as there is little evidence of this approach having been used by social change organizations.

[4] An example of Gandhian thought in action was the celebrated Bhoodan Movement by Vinoba Bhave, in which he marched around the country to get rich landlords to donate excess land to the landless. Although over 1 million acres was donated, most of the land turned out to be well-nigh uncultivable!

[5] See, for instance, the debate between Gandhiji and Ambedkar, in which Gandhi downplayed the need for changing the caste system, a major oppressive societal structure, of our country.

Forms of social catalysis

It is important to repeat the point that social change organizations are those that are involved in changing the conditions not through providing services by themselves, but by changing the community or the government or wider society or both.[6]

There are four possible forms of catalysis that social change organizations can engage in:

- In building the technical capabilities of community members: This technical capability can be in the area of devising, designing, and appropriately using technology, including information—e.g., capability building through new information.[7] For instance, community members, living in drought-hit areas, may learn how to maximize harvesting of rainwater, and using the same for enhancing their agricultural production.
- In getting the community to change its behavior vis-à-vis each other, as well as its norms.[8] This might include communities deciding that they need to distribute the limited water available in the village watershed not as per the size of land holdings (wherein the rich would become richer), but as per the size of their families! (As was done by Pani Panchayat, Maharashtra). Here, it might be worthwhile adding that the focus of change, as per Friere, needs to be the poorer, discriminated person, rather than the one who is better off. Friere says that even though the "exploiter" is as divested of his humanity as the "exploited."
- In getting the State (State or Central Government) and other institutions such as corporations[9] (Harvey, 2007; Korten, 1995) to change their policies and ways of working.[10]

---

[6] This is not to say that there are no organizations in the social sector that are not providing services (e.g., *Akshaya Patra* which provides mid-day meals to 1.3 million students every year), but that such organizations are not to be considered social change organizations. At best, we may call them charity organizations, or even contractor organizations with a good heart.

[7] Examples of this kind of catalysis are practiced by the following organizations: PRADAN, AKRSPI SRIJAN, MYRADA, DHAN in the area of livelihood/NRM/savings and credit; CINI, Jan Swasthaya, and Sahayog in the field of health; Mobile Creche and SOS in the area of child welfare.

[8] Examples of this type of catalysis are organizations, such as SWRC, Prayas Centre for Labour Research and Action, Timbuktu Cooperative, and Sri Bhuvaneshwari Mahila Ashram.

[9] The role of corporations in consolidating economic power in the form of global capitalism pursuing profit has been well documented by David Korten in his books. He argues for a more human form of markets—healthy markets that are the key to more humane just and compassionate societies. David Harvey also gives brilliant accounts of the implications of neo-liberalism and so-called free markets. In neoliberal ideology governments also support the expansion and power of corporations at the cost of many other sections of the society such as marginal farmers, informal workers, other weaker sections who then resort to migration out of their homes to urban and peri-urban areas or even suicide, with little social security, food security, or access to minimal health and education.

[10] Examples of this kind of catalysis are practiced by organizations, such as CSE and Greenpeace on issues of sustainable environment, EQUATIONS on tourism impacts; Jagori and Naaz for women's rights, Prayas on improving criminal justice and CHSJ and CEHAT on public health, and NCDHR and HRLN on Human Rights issues.

- Working to change entire systems (government plus society): There are two parts of this change. In the first part, the poor and/or marginalized may get together and understand how a current system is harming their interests. They may start a movement to introduce a new methodology of working that is fairer. In the second part of the change, as the government bureaucracy and powerful corporations resist introduction of this methodology, other sections of the society may come out publicly in favor of what the poor and marginalized are.[11]

There are two broad approaches to changing individuals, communities, and the entire system[12]: a teaching,[13] and a facilitative or cocreative approach. While the former approach is broadly synonymous with the "Banking Approach," the latter is synonymous with the "Critical Thinking" Approach.[14] Unlike the former, where outsiders impart knowledge and skills to "ignorant" local people, in the case of the latter, the community members are respected, determine their own goals, and ultimately control the process.

The fundamental difference between the two approaches can be summarized in the following table.[15]

The key dimension to focus on is the one of expected outcomes. As numerous real-life examples have shown, programs that run through teaching rarely make a difference in the life of communities. For instance, a technically very simple program of operating check dams did not work when implemented by the Government Irrigation departments that operated on the teaching/expert paradigm. A technically more complex program of Lift Irrigation succeeded when it worked on a Facilitation/Co-creation paradigm.[16]

Influencing the government to change its policies often requires an approach of identifying key bureaucrats and politicians who are sensitive to particular issues (and are able to take a more long-term view) and through them institute changes in policy or its implementation, which in turn will have a positive impact on social change processes.

---

[11] The most striking example of this is the work on Right to Information. This started off as a movement initiated by a very small organization in Rajasthan—Mazdoor Kisan Shakti Sangathan, in 1995, and culminated in GOI's RTI Act in 2005, which covered the entire nation.

[12] Facilitator or Advocate: What is the difference, by Ndunge Kiiti and Erik Nielsen, from The Art of Facilitating Participation, Sage Publications, New Delhi, 1999.

[13] The terms used in the chapter are "Facilitative" and "Advocacy." However, as advocacy means something fairly different in other contexts, we have decided to use the term "teaching/expert" rather than "advocacy."

[14] Friere (1972) spoke of these two approaches.

[15] Adapted from Kiiti and Nielsen. pp 62–63.

[16] Based on the author's (TSB) own experiences in Chhattisgarh. The check dam required that the community members should lower a wooden shutter, as the monsoon season was about to end, so as to maximize the storage of water. They never did so! In contrast, a far more complex, lift irrigation project with over 140 farmers, and 10 pumps of 20 hp each, and six pipes more than 6 km long, is being operated for the past 18 years, by a similar tribal community in the same area. The difference was that in the case of the check dam the community was not consulted or involved in any way. They had not even asked for the check dam. In the latter case, the lift irrigation project came up because the community wanted it. It took two years for the project to get designed, approved by the bank, and cleared by the Revenue Department; the community was involved in this project at all stages.

Dimension of approach	Facilitation/cocreation	Teaching/expert
Context	Process is the focal point; agenda defined by the process	Predetermined product is the focal point; agenda is defined by the desired product outcome
Problem	Defined by the community; facilitator assists in exploration, understanding, and definition	Predefined by outsiders; promotes dissemination and transfer of knowledge from experts
Approach	Process begins with the community; fundamental belief in community members; solution to emerge from local context	Process begins with external organization; fundamental belief that solution is necessary from external sources; solution not tailored to local context
Strategy	Facilitator stimulates critical reflection and dialogue; encourages people to find and use own voice.	Unilateral information transfer and communication that is also filtered and controlled by outsiders
Expected outcomes	Locally appropriate action leading to increase in local capacity; improved local decision-making; strengthened commitment in locals	Outcomes not necessarily appropriate since they were pre conceived and based on predetermined agenda; weakens local decision-making; sense of detachment among locals
Attitudes and values	Respect community members' ideas; do not claim to have answer to problems	Institutional ideas overshadow that of community members; do not see need for collaboration; make knowledge claims and attempts to impose ideas.

Organizational readiness for social catalysis

The focus of catalysis can be the individual, or the group, or the larger system. This will depend on the form of social catalysis the organization is mostly focused on. In turn, this will require an understanding of "how change happens" at these various levels. Defining the system in which change needs to take place, understanding the points of leverage, and recognizing that change catalysts essentially shape processes and not agents or content, would be important competencies for social catalysts. This is a key issue as often the social catalyst focuses on the victim to such an extent that it forgets that its key role is to work on the *processes of change*—and do whatever it takes for change to happen at the level of the individual, the community, or the system. The focus on the victim or the oppressed to the exclusion of the system in which this oppression takes place may generate solidarity but may not do very much to make change happen. This form of thinking is often seen in some activist organizations wherein asserting the political ideology of, say social justice or human rights, is established quite clearly. However, less attention is paid to what may be the interventions that may make the system change.

In the case of influencing large and powerful systems such as the government or corporate interests, the interventions and forms of engagement may need to be carefully crafted as social catalysts begin with the reality that they work with massive power and resource imbalances to begin with. One of the authors (Rosemary Viswanath) was involved in a long-term change process with an international environmental network. An eternal dilemma within this large and powerful network is whether they should adopt an inside or outside strategy. Should they participate in large global consultations and negotiations on issues such as climate change or should they be among the protestors outside the venue of these large talk shops! Those in favor of protest believed they would be betraying the victims and diluting their politics by engaging with the enemy. Those in favor of influencing the negotiations—however skewed the power imbalances were—believed that it was necessary if they wanted the system to change.

Another example is the organization EQUATIONS[17] which works on influencing tourism policies and models of tourism to make them more people centered and minimize the negative impacts that tourism has on local communities. EQUATIONS is a social catalyst as it is not an organization actually located in a tourism destination and its staff are not members of affected communities. Given its strong empathy for those disadvantaged by a one-sided tourism development, it tends to not engage at times with those who cause the damage—which is usually the powerful section of the tourism industry. However on using the metaphor of the social catalyst, it began to relook at its definition of the system and realized that some form of engagement with the "enemy" was required as the enemy was part of the system and could not be wished away! The key then was to find points of leverage to dialogue or confront the enemy and help them to see why change was necessary. This required strategies on research, dissemination, participation in select industry events, and a strong media strategy. It also implied leveraging sensitive bureaucrats at the state and national levels and at times "educating" them on the realities and perspectives from the position of those impacted.

So how are organizations such as these helped when they use the metaphor of a social catalyst to understand their role and its implications in the context in which they exist? The following sections help look at this issue in more detail.

[17] www.equitabletourism.org. One of the authors (R.V.) worked as its director from 2005 to 2012.

Learning to catalyze

To catalyze is to know what reaction is taking place at present, what the usual reactions are, and what kind of catalyst is to be added that would make the reaction (i.e., the desired response or change) happen faster, better, more successfully, or more sustainably. Social catalysis is much the same.

Let us take the example of an organization that wants to change attitudes with respect to domestic violence. Feminist organizations working in the 1980s and 1990s tried to deal with this problem head on by getting women to organize, hold protests against this violence, and try to force action against men who engaged in such acts. While these interventions had some success, these also led to resistance to what the women were trying to do: they were labeled as not in line with traditional Indian values. In short, the change reaction that the feminist groups wanted, kept on requiring an increased level of energy and engagement in order to succeed. And, as Kurt Lewin points out, the more you try to force change, the stronger is the force of reaction.

Consequently, feminist organizations of the 21st century do not just work with women, but with men and mixed groups of men and women exploring gender relationships, patriarchy, and gender stereotypes. Working with men both young and old, they encourage them to reflect on the kind of lives they would like to lead, the kind of relationships they would like to have, the role of violence in their lives, and its impact on them as well as on women. In the above case, the catalyst is the feminist group, and the system (location of reaction) is the men and women of a specific community.

In another example, individuals of a community are not able to take up a new livelihood because they do not feel confident of handling the marketing. Some organizations try to solve this problem by taking up the marketing on their own. Even if this approach turns out to be successful, it only leads to the community becoming dependent on the organization. This is not a very useful intervention in the long term as the role of the catalyst got converted into the role of an actor or prime stakeholder.

Thus, social catalysts need to pay careful attention to understanding their role and not getting "seduced" by the probable short-term gains of becoming direct actors. Given the increasing pressure of delivering "results" in the short term to continue receiving development aid or grant funding, many social change catalysts succumb to this pressure and may not even be fully aware of the change in role or "corruption" in role.

Competencies or characteristics required for social catalysis

When one places the metaphor of social catalysis at the core, there are characteristics and competencies required at both the individual and the organizational level that will enable such a role.

- A deep empathy for the affected persons in order to connect with the affected persons and understand their situation forms their perspective.

- An empathy and understanding of those causing the problem. This is often overlooked but is as important. By labeling "those causing the problem" as the enemy and not understanding them or engaging with them, social catalysts limit the possibility of evolving effective change interventions.
- What follows from the above two is a willingness to work with all parts of the problem—that is, a systems approach![18] What may get missed is the reality that in dynamic, complex, and interdependent systems, as most social systems are, it is well-nigh impossible to isolate the processes of victimhood and exploiter so neatly as there are collusions and unconscious patterns that perpetuate these dynamics on all sides.[19]
- An understanding of the dynamics of the community, of the system, as well as the specifics of interpersonal, subgroup, and intergroup relationships. This in turn will help build hypothesis that can be tested on understanding of the dynamics of change in this particular context.
- An understanding of, and skill in making different interventions in the community or system. Interventions may include organizing any or some of the following: training or reflective workshops, retreats, seminars, community level meetings, dialogue and negotiation spaces with different stakeholders, protests, campaigns, the use of media (e.g., in making documentary films or other means of publicly disseminating perspectives and information), advocacy, making presentations, counseling, engaging in one-to-one discussions, mentoring, etc.

Building the social catalysts strengths

The next issue is, having acknowledged the metaphor of social catalysis as relevant to their identity, so do organizations build organizational and individual capacity to play this role. In the case of the feminist organization having recognized that compassion fatigue was setting in, they developed systems and processes that would help regenerate individuals and help them not to introject these disempowering and negative social processes that they had to

[18] A systems approach to change is evident from the work of the Tavistock Institute in the 1940s and 1950s. Socio-technical systems design helped grapple with emerging changes in the organization's context or technology. Seminal work was done in English coal mines (Trist & Bamforth, 1951), in Calico Mills Ahmedabad (Rice & Miller, 1953), and in W. R. Bion (1960s and 1970s) among many. Paulo Friere's work in the 1960s and the 1970s on structural and systemic change, unconscious processes work in organizational life, and the role of social defences against anxiety (Jacques, 1950; Lyth, 1970). The organization in relation to its environment, highlighting the role of the turbulent environment (Emery & Trist, 1963) and implications for shaping that environment through collaborative activities were the pioneering influences to systems approaches to change. Several people now work actively on systems psychodynamics approaches influenced by the Tavistock Institute. Other key, but more recent, influences are process-oriented psychology and Worldwork by Arnold Mindell, and systems thinking and organizational learning work by Chris Argyris, Peter Senge, and Otto Scharmer.

[19] A chapter by one of the authors (RV) exemplifies this process in the case of Dalit empowerment and the systemic psychodynamic processes underlying them (Viswanath, 2009).

deal with on a daily basis. Working with the metaphor helps the organization to set up and check the relevance of its systems and processes and their alignment to the metaphor. For instance, some of the organizations the authors have engaged with have done the following:

- Explore the alignment of their stated vision and values to what they practice. EQUATIONS, for example, has the stated value of dialogue for many years. Recently, the current staff team explored to what extent dialoging "with the enemy" was essential if they were to play a catalyst role of policy change in favor of a more people-centric approach. Other organizations have made it a practice to revisit their vision mission and values periodically to check the extent to which they are aligned to the present context and role.
- Build understanding of social processes. Organizations do this through formal structured reviews of their work, and analysis of their strategies and impact.
- Build the strength of the individual to take the pressure of social reality—often in the social sector this is done through retreats, informal mentoring, and capacity-building programs that focus on both ideology and personal development. Formal certifications such as the Organizational Change Facilitation Program,[20] Community Process Facilitators Program,[21] Group Relations Conferences, and Training of Trainers on Participatory methodology are also interventions in this direction.
- Build capacity for empathy and to work in support and solidarity. This is often done through the formation of platforms and networks, and working together on campaigns. Unlike the corporate sector (which focuses much more on competitive relationships) the nonprofit and social development sector has evolved many complex forms of organization toward the goal of working in collaborative and solidarity formations. These require nuanced understanding of distribution of power, forms of communication, trust building, and developing appropriate structures.[22]
- Another way of build understanding and empathy is to spend extensive time in the community. Many organizations place their inductees for significant time of their first year in the organization, staying in villages to understand various socio-political processes. Harnath Jagawat, who founded Sadguru Water Development Foundation, spent the first two years of his work, walking up to 30 km a day in order to meet up and dialogue with villagers in order to understand their problems (Hailey & James, 2002).

[20] The OCFP program organized by the HIDF, Bengaluru.

[21] The CPFP program organized by ISABS.

[22] A chapter by Nunez and Wilson-Grau (2003) offers an excellent conceptual framework for understanding these processes.

A re-look at projects

The mission of nonprofits is expected to be social change and ideally social transformation. Donors provide support to nonprofits in this social transformation, by providing them with money to implement specific time-bound projects. However, over time, donors, looking to ensure accountability, have tended to emphasize linear approaches to performance management, damaging the ability to be effective catalysts of social change (Edwards & Hulme, 1995). In short, an over-emphasis on projects actually leads to inhibiting the process of catalysis. The reasons for this are:

- Projects have a fixed and short-time duration; community change is likely to continue much after the project gets over.
- Projects tend to over-emphasize measurement, and therefore simplify their contexts in order to do so. As a result, the complexity of the context in which social change is to happen is not taken into account.
- As part of the need to simplify and measure, projects also tend to focus on a limited part of reality, which is mostly focused on the indicators they agree on to prove that short-term change has happened. As a result, they sometimes miss out what may be more important for the community at a particular point of time (e.g., an agriculture program may be taking place while the community has been hit by a debilitating attack of cerebral malaria, or there are local political changes, or outmigration trends have increased).
- Driven by donor requirement, organizations stop looking at actual change or catalysis taking place. There is a tendency to evaluate functional accountability (use of resources, return on investment, and immediate outcomes) rather than strategic accountability and the longer term impact of social change intervention. In short, no one seems to track actual social change, or the process of social catalysis.

The more organizations think in "project" terms, the less they are able to look at the process of change, and understanding its nuances. This is where metaphors come in handy. A metaphor or metaphors for social catalysis would help organizations work with the nuances of these processes. Projectization of NGOs results in a more straitjacketed approach shrinking and even debilitating the capacities of these organizations to be real social catalysts.

There is another consequence of rushing to finish project deadlines. Activity orientation, which is seen as more productive, takes the center stage and discussions and reflection comes to a halt. There is hardly any deep dialogue on the specific impacts or the ways of working. There is little discussion on what changes are coming about in communities or in their contexts.

Synchronizing external and internal principles of working

There is a long tradition in the development field of the transformative power of learning. In this tradition, diverse sources of knowledge, inclusion of many people, working and acting together are seen as critical for learning and change.

Understanding the complexity of the social situation requires extensive and deep discussions between those who are working directly with the community and those who have ostensibly a "managerial" role. This tradition is exemplified in the Frierian approach to "dialogue": "To enter into dialogue presupposes equality among participants. Each must trust the others; there must be mutual respect and love (care and commitment). Each one must question what he or she knows and realize that through dialogue existing thoughts will change and new knowledge will be created."[23]

A large number of people join social change organizations because they resonate with most or all of the above principles. When these principles are not adopted inside organizations, it leads to subtle demotivation. There are many examples of social sector organizations breaking apart because its members felt that they were not being adequately heard and respected.

In order to ensure that the internal is synchronized with external principles, the organization needs to ensure the following:

- Ensure a practice of frequent and relevant dialogue-making in organizational life. This may be done through workshops, retreats, regular meetings that go beyond project monitoring, etc. A discussion that goes beyond project-based results.
- Members cannot just relate to others as professional colleagues. They need to relate to them as complete human beings, people who are dealing with challenges at work, at home, and in the larger society. If a person talks about gender equality, then it is relevant to ask whether that person is following gender equality at home, and in the broader society. If the person talks passionately about not discriminating, then it is important that the others can see that this person is truly not discriminating in the broader society and at work.
- It is seen that while many social change organizations are committed to learning of community members, there is no such emphasis internally: there is dissonance between what the organization promotes externally and what it does internally. Among the many factors that militate against internal learning is that such organizations have limited resources and do not like to invest scare resources in intangible concepts such as learning. A study identified that the primary means of learning for most successful organizations was the conscious reflection and analysis of own implementation experiences (particularly when things went wrong). Most of the organizations used a mix of regular meetings, retreats, workshops, and seminars to promote shared learning and to disseminate new ideas (Hailey & James, 2002).
- Learn to live organizational values. There are three aspects of living organizational values.

[23] http://www.freire.org/paulo-freire/concepts-used-by-paulo-freire

a. Dominant societal values may be at odds with humanistic values; these societal values may represent distortions in the social structure. By adopting humanistic values, individuals and organizations make a political statement.
b. Individuals need to live values that they espouse within the organizational space.
c. Organizations need to walk the talk as much as individuals. To take two examples, Jagori, a feminist organization, hired a hearing and speech impaired person in their accounts department, and then to ensure communication was possible that the person fitted in, everyone else learned to speak in sign language! In HRLN, a visually impaired person heads the group that works on Disability Rights, which works with immense effectiveness. The other colleagues in the group have also made the effort to change some ways of working, but that notwithstanding the head is far from being a token figure.

In the 1960s, a powerful slogan emerged among the so-called second wave of feminists: The Personal is Political. What this said was that the personal problems faced by women were not because of either personal inadequacy or the specific situation faced by women, but because the entire social, political, economic, and cultural edifice of society was such that the women inevitably faced problems. To deal with this problem, the action had to also be taken at a political level, as no amount of tinkering at the individual level could solve the problem societally. This had another level of interpretation also—the political is also personal—that it was not enough for individuals to work for macro change but that this had to reflect in their personal stances as well—in the values they embodied and in their personal lives).

Understanding the process of leadership

A more nuanced process of looking at the process of leadership is called for in understanding social catalysis. In fact, we argue that perhaps all acts of leadership are an act of social catalysis. Leaders who are catalysts situate themselves at the boundary of the system—helping manage the interface of the system with its external context. They help members of the internal system get a better hold of the reality—for instance, the reality of whether the system or the organization is in reality walking its talk on its values. Or the reality of whether the organization's strategies and interventions are indeed catalyzing social change or are merely self-serving. Or the reality of whether the organization has the adequate skills, competencies, and resources to meet its desired goals and play its desired role. A more charismatic form of leadership would, in fact, often come in the way of a social catalysis role—as such leadership while effective in certain moments or crisis, can come in the way of people taking the authority to acquire skills and mobilize themselves. Effective leaders help the systems to focus on their primary task, and appropriate metaphors are powerful ways for systems to articulate what this primary task may be.

One catalytic approach to leadership is the one espoused by the Harvard academician Ronald Hiefetz (1998). The Adaptive Leadership approach is to influence the community, or organization (wherever the leader is working) to face—and deal with—its problems. This adaptive work consists of the learning required to address conflicts in the values people hold, and to diminish the gap between the values people stand for and the reality they face.

Adaptive work takes place through the exposure and orchestration of conflicts or internal contradictions within individuals and constituencies. In short, the Adaptive Leader follows a strategy of promoting dis-equilibrium; of highlighting, rather than pushing under the carpet, the underlying stresses.

Adaptive Leadership is a difficult path to follow, and requires a great degree of conviction and moral strength in the individual. Examples of Adaptive Leadership in History include that of Mahatma Gandhi at the national level (e.g., his willingness to abandon the non-cooperation movement when the Chauri Chaura incident went against his principle of nonviolence).

Conclusion

In this chapter, we have focused on the power of the metaphor as an almost primal force or way of thinking that creates meaning and helps understand one's experience often in subterranean ways. Its advantages are that it opens up the possibility of new perspectives influencing ways of approaching or tackling the current situation, leading at times to transformative change.

However, in our enthusiasm for using and recommending this powerful tool, we have not lost sight of the fact that by its very nature metaphors are also limited. In the use of a metaphor admittedly one aspect or one part of the situation is emphasized. Often this is valuable even with its limitation as this is the part that may have been buried or the part that the organization unconsciously "forgot." This can happen, for instance, when an organization or leader creates dependencies by playing a messiah instead of a catalyst.

Further, a limitation of using one metaphor rather than multiple would result in limited understanding of some of the following common situations:

- When leaders start pursuing an approach that makes them personally famous, but does little to strengthen organizational capacities.
- When cliques form in the organization, cliques that make the organization decide to give unwarranted salary raise to all, without considering long-term financial sustainability.
- When striving for increasing dialogue in the community, the members become wooly headed, and forget to follow established protocols of working.

While the metaphor helps understand by throwing light, sometimes in startling ways, it also tends to "shape" one's views of the organization—that is, the metaphor becomes the sole framework. Einstein wisely referred to this process when he said, "Our observations are shaped by the theory through which we see."

It is critical that the organization does not get fixated by and to a metaphor but has a more playful and curious relationship to it, to yield insights. In that sense, the best work with metaphors is almost paradoxical—they have the ability to throw up both complementary and competing insights, and the organization needs to have an open mind to acknowledge that both probably co-exist. In our endeavor to have simple systems, we often subtly "discard" the contradictions! This then leads to the ever present danger of the metaphor becoming an ideology—an oxymoron in itself!!

Metaphors are best used for a diagnostic reading, to help look at the familiar with freshness, to open the doors and windows of the mind. The convergence to a critical evaluation, to the arriving at what needs to be done, the design and implementation of new processes and systems can only be a next step.

References

Edwards, M., & Hulme, D. (Ed.). (1995). *Non-governmental organizations—Performance and accountability: Beyond the magic bullet.* London: Earthscan Publications.

Friere, P. (1972). *Cultural action for freedom.* Middlesex: Penguin Books.

______. (1972). *Pedagogy of the oppressed.* Middlesex: Penguin Books.

Hailey, J., & James, R. (2002). Learning leaders: The key to learning organisations. *Development in Practice, 12*(3/4), 398–408.

Harvey, D. (2007). *A brief history of neoliberalism.* New Delhi: Oxford University Press.

Heifetz, R. A. (1998). *Leadership without easy answers.* New Delhi: Universal Book Traders.

Korten, D. C. (1995). *When corporations rule the world.* San Francisco: Kumarion Press and Berret Kohler Publications.

Morgan, G. (1998). *Images of organization* (Executive Edition). Thousand Oaks: SAGE Publications.

Nuñez, M., & Wilson-Grau, R. (2003). Towards a conceptual framework for evaluating international social change networks. Retrieved from http://www.mande.co.uk/docs/Towards%20a%20Conceptual%20Framework%20for%20Evaluating%20Networks.pdf

Viswanath, R. (2009). Identity leadership and authority: Experiences in application of group relations concepts for Dalit empowerment in India. In E. Aram, R. Baxter & A. Nutkevitch (Eds), *Adaptation and innovation theory design and role-taking in group relations conferences and their applications* (Vol. II, pp 179–196). London: Karnac Books.

26 Process competencies for social development interventions

Zeb O. Waturuocha

Introduction

A common friend told Udai about me but when we met, it was as though we have known each other for a good number of years. He autographed my copy of his book *Training Instruments in HRD and OD*, 2nd edition. He wrote:

> To Zeb, with nostalgia of association and friendship, and high expectations in future.
> Udai

His hug was full of warmth and his voice was irresistibly inviting. The simplicity of this unassuming man with series of feathers on his cap mesmerized me. One thing I was convinced about is that he never expected me to match him in any aspect of life but what was the "high expectation?" It was later that I realized that it was an invitation to live up to my goals and to be the best of what I am, a message I got from a similar significant person in my life.

This chapter is not about the achievements of Udai because I am no body to recount all that he acquired and stored in his portfolio, yet presented himself with great demeanor, dignity, and discreet. This chapter is not about what Udai advocated and stood for because that would take volumes to complete. This chapter is about those qualities of Udai that I admired without telling him, that I observed without his notice, and that I copied without his knowledge, but practiced with great difficulty.

In the paragraphs that follow, I share why this opportunity to contribute to a book that would pay tribute to Udai is so exciting and so self-fulfilling for me. The focus of this chapter will be on the Process Competencies that are required for Social Development Intervention (SDI). The chapter is based on my personal experience in social development as well as a process facilitator.

Brief background

I have mentioned how Udai and I met each other. I was the Dean for Social Development of Indian Society for Applied Behavioural Sciences (ISABS) when Udai died in 2010. He was the Founder President of this institute and contributed a lot to its growth and effectiveness. He was a brand and was among the "Pride Names" of ISABS (at least for me) at that time. Fascinated by his persona and what he had done for ISABS, I initiated a book in his memory. T. V. Rao, Rolf, Yawar Baig, Abhad Ahmed, Madhukar, and Paul Siromani were the first people to submit their manuscripts for the book. For reasons that still remain mystery, the idea was buried and I considered it dead before burial. It is when I read T. V. Rao's announcement regarding a book in memory of Udai that I realized that the idea was just incubating somewhere and the day I decided to contribute a chapter was the day I considered it "A Dream Come Through." I am indebted to T. V. Rao and all who sat together to moot this idea and for the opportunity to write a chapter in this book.

Why process intervention?

I have chosen the title of this page not only because I wanted but also because Udai considered himself a "Process Facilitator." In paying tribute to Udai, the HRinIndia group posted an article in its blog with a rider "please find given below an article from ISABS."[1] I have taken the following statement attributed to Udai from this article:

> I see process work as enriching my professional role as a trainer and a consultant as much as helping me as a person and in my personal world (friends, family). I have found several challenges which beckon me to new voyages: moving beyond intra-personal processes to group processes and societal processes, searching the Indian heritage to learn the dynamics of process work in different settings, extensive use of process work in various aspects of the society, addressing urgent social issues (differences, marginalization, harmony, collaboration, equity, empowering) through process work and so on. It is exciting to work with younger colleagues who are the torch bearers to usher us into the next century, which we hope will be brighter and more humane. http://businessmanager.in/cover_story.php?Cover_story_id=248

I have had the privilege of being a process facilitator for individuals, corporate establishments, as well as the social development sector. In this chapter, I will share my understanding of process facilitation competencies as it applies to the social development sector. I am hopeful that while reading this chapter, those who knew and worked with Udai, those who read his books or had cause to associate with him, and/or those who are reading about him for the first

[1] http://hrinindia.blogspot.in/2010/03/tribute-to-dr-udai-pareek.html

time in this book will be able to visualize within themselves as to how he fits to these competencies in his introduction, interactions, and interventions.

What is process intervention?

An intervention is a deliberate process by which change is introduced into peoples' thoughts, feelings, and behaviors. The overall objective of an intervention is to enable self-confrontation in a nonthreatening way. The objective is to support the individual or group of individuals to become aware of behavior patterns that they follow, explore the impact of these patterns, and so make an informed decision so as to continue the pattern or change it.

What is social development?

Social development is about putting people at the center of development. It is about enabling people take part in those decisions that affect their lives and environment. In other words, it is a commitment that development processes need to benefit people in the belief that the way people interact in groups and the society, and the norms that facilitate such interaction and shape development processes.

The following definition was approved by the IFSW General Meeting and the IASSW General Assembly in July 2014,[2] as the global definition of the social work profession:

> Social work is a practice-based profession and an academic discipline that promotes social change and development, social cohesion, and the empowerment and liberation of people. Principles of social justice, human rights, collective responsibility and respect for diversities are central to social work. Underpinned by theories of social work, social sciences, humanities and indigenous knowledge, social work engages people and structures to address life challenges and enhance wellbeing.

The social work profession's core mandates include promoting social change, social development, social cohesion, and the empowerment and liberation of people.

What is competency?

A competency is a set of defined behaviors that provide a structured guide enabling the identification, evaluation, and development of the behaviors in individual employees. The term "competence" first appeared in an article authored by R. W. White in 1959 as a concept for performance motivation

[2] http://ifsw.org/get-involved/global-definition-of-social-work/

(White, 1959). Competencies are measurable practice behaviors that comprise attitude, skill, and knowledge (ASK).

Social development competencies comprise those attitudes, skills, and knowledge that help people develop critical consciousness to reflect on sources of oppression and/or privileges based on race, class, language, religion, gender, disability, culture, and sexual orientation. It implies enabling people develop action strategies toward addressing such structural and personal barriers that are central to emancipatory practice wherein the goals are the empowerment and liberation of people. This implies the culture of recognizing the inherent worth and dignity of human beings, doing no harm, respect for diversity, and upholding human rights and social justice.

Social work is an attack on the denial of rights of the people. These rights can be seen under three headings:

1. Civil and Political—freedom of speech, conscience, and freedom from torture and arbitrary detention;
2. Socio-economic and Cultural—rights to reasonable standard of living, levels of education, healthcare, housing, and minority language;
3. Natural world and the right to species biodiversity and intergenerational equity.

These rights are mutually reinforcing and interdependent, and accommodate both individual and collective rights.[3]

Process competencies for social development intervention

Human beings are inherently social. Developing competencies in this domain enhances a person's ability to succeed as well as to positively influence mental health, success in work, and the ability to be a citizen in a democracy. The term competencies represents the individual's potential capabilities and at the same time an integrated learning (Ellström, 1997). Ellström's concept of competence is broad and covers cognitive (typical subject area qualifications and skills), affective (motivation and emotion), psycho-motor skills, personal factors (self-perception and self-worth), as well as social factors (cooperation, communication, and management). All these elements are part of the competencies concept. The focus with the concept of process competencies is generally on the affective and social factors, and the process competencies are therefore part of the subject area competencies. They have to be regarded as an integrated process. They represent both technical skills and potential for personal development. These competencies are essentially affective and social capabilities—in other words, they cannot be acquired solely through performance, but they are performed and should be evaluated in their practical context.

[3] ifsw.org/get-involved/global-definition-of-social-work/

Having worked in the field of social development especially with organizations working with tribals, physically challenged people, slum dwellers, elected gram *Panchayathi* women, self-help women's group, youth groups, etc.; worked as a consultant for both national and international funding agencies in studying and evaluating their funded projects, I have considerable experience and knowledge that helps me to shape what I consider the competencies required for process intervention in the social development sector.

1. Ownership and identification
2. Understanding and practice values and ethics
3. Authenticity and transparency
4. Diversity and inclusion
5. Empathy and humanness
6. Learning and research orientation
7. Walking the talk
8. Enhancing, enriching, and engaging
9. Evaluating and reforming

I will illustrate these competencies through expanded writing and where possibly how I perceive Udai in vis-à-vis these competencies in his, what I would call Human Development Interventions through ISABS and through his lectures and writings.

Ownership and identification

I am convinced that one cannot identify with what one does not own; hence, the first competency would be to own up the choice you have made so as to be able to identify with the people. Without this identity, your impact will not reach the people as they would be seeing you from the angle of your own gain and not their own benefit. As you own your choice and identify with the people, your motivation to succeed is high, people see no selfish motive in your efforts to work with them, they do not relent to identify and work with you. If you do not identify with the people, the impression is that you work "for" and not "with" the people. This would involve being aware of and practicing social and ethical boundaries, intense self-reflection, seeking feedback and working on them, demonstrating professional demeanor despite identifying with the people, and engaging in long-term learning of how things change and steps to follow.

Understanding and practice values and ethics

It is essential that the social worker recognizes and manages his/her personal values in such a way that it does not underrate professional values. Ethical decisions are the ones made after due consideration to standards of practice. Social development in many cases involves the downtrodden, the marginalized, and the challenged. Values and ethical practices ensure that the vulnerability of this group of people is not exploited. Even in times of conflict that might be based on caste or creed, applying ethical reasoning in conflict resolution is a necessary behavior. Values are important and lasting beliefs that have major influence on a person's behavior and attitude that serve as guidelines in interpersonal relationships. Ethics on the other hand are the basic concepts and fundamental principles of decent human conduct. I am not sure there is any other evidence of the practice of value and ethics by Udai than the personal experiences of those who knew and worked with him and his own documented stances.

Authenticity and transparency

Essentially, this competency requires that the practitioner leave no stone unturned in his/her dealings with the people. Once an eye of suspicion is cast on the individual or group of social workers, the trust that people have on him/them begins to erode. Authenticity implies doing everything with heart and mind, with no ulterior motive and letting those concerned know each and every step and statement of purpose. Here the practice of self-disclosure, openness to feedback, and perceptiveness (PES Scale; Udai, 2002), and a concept developed and practiced by Udai Pareek himself, becomes very valuable.

Diversity and inclusion

This competency calls for recognizing the extent to which a culture's structures and values may oppress, marginalize, alienate, or create or enhance privilege and power. It requires that the social worker gains sufficient self-awareness to eliminate the influence of personal biases and values in working with diverse groups especially in this era of fanaticism, recognize and communicate their understanding of the importance of diversity in shaping life experiences and seeing themselves as learners while engaging those working with them as knowledge source.

Empathy and humanness

Empathy is the ability to "feel with" another person, to identify with them, and sense what they are experiencing. This competency requires the cognitive ability to understand other people's emotions, or the ability to imagine what they are feeling, by "putting yourself in their shoes." It is the ability to make a psychic and emotional connection with another person, to actually enter into their mind-space. When we experience real empathy or compassion, in a sense our identity actually merges with another person's. Your "self-boundary" melts away; the separateness between you and the other person fades.

Learning and research oriented

There is no end to learning and a social development worker needs to be a keen learner and research-oriented individual capable of application of learning into work. As the society responds to social development interventions, new challenges and new opportunities emerge and the development work is required to understand these changes and act accordingly. Documenting the process and using this experience to predict future intervention is an essential part of this competency.

Walking the talk

This is the competency of authenticity, congruence, and integrity. It means that you live by example. Do what you say you will do. When people find you incongruent, they will not follow you as the mimicry brain is focused on what is done than on what is said. Hence, this competency is of vital importance in order to establish integrity, trustworthiness, respect, self-worth, and personal power. People respect and adore those who live up to their words.

Enhancing, enriching, and engaging

Social development implies enhancing knowledge of the people to become aware of what they already know, helping them to gain clarity of their own ability and power. It also implies enriching the people with new knowledge, making them curious to learn more, and enabling them use this knowledge through engaging them to do it for themselves. It means that over a period of time, the people are able to do their work for themselves; hence, creation of dependency is not part of the social development intervention work.

Evaluating and reforming

It is not proper to see the above skills as all-inclusive because social development is a dynamic process; hence, the skills and competencies are quite flexible to be able to respond to environmental changes. This competency requires alertness on the part of the development worker in response to changes, constant measures, and reformation of strategies as the situation requires.

Process competencies cannot be acquired purely through mental activities as they are not analytical, technical, or scientific abilities. They are rather expressions of the individual's personal approach to learning and managing along with a variety of other abilities, such as working with others, independence, effective communication, behavioral changes, planning and directing, and self-evaluation. They represent a meta-cognitive level of both action and knowledge. They represent a form of knowledge in action, which may be tacit, as it may be difficult to put into words.

> A man literally is what he thinks, his character being the complete sum of
> All his thoughts. As the plant springs from, and could not be without, the seed,
> So every act of a man springs from the hidden seeds of thought and could
> Not have appeared without them. (Allen, 2014)

Udai's practice and behavior seemed as though the competencies mentioned above have been designed while drawing from his own life. If this is your thinking, you are right. I mentioned in the introduction to this chapter that he mesmerized me whenever I come in touch with me and I am left "WoW," what a man? He demonstrated professional demeanor in his behavior, appearance, and communication and engaged in long-term learning through research and writing.

In my experience of Udai in relation to ownership of and identification with the Indian Society for Applied Behavioural Sciences (ISABS), what comes to my mind is this quotation by Lauren Graham:

> I feel real ownership in this show. I feel very invested in it. I care very much about it. I don't feel any more like a hired hand, you know? It's a strange feeling – I feel personally responsible for how the story goes. What happens. What the weaknesses are. And so in a way some of the changes gave me opportunity to have a voice in a different way.

When Udai attended ISABS events, even as many participants might have heard about him because of his presence in the larger human relationship society, yet some of them doubt when it is hinted to them that he is Udai Pareek. The doubt is actually not a doubt: a question of how can he be so simple, so mingling, jovial, and cordial. He demonstrated love for the institute and for the people and did give people equal opportunity to get closer to him. In his membership and association with ISABS, I experienced Udai as one who is well informed through his own reading, research, and practice, resourceful, and proactive in responding to the evolving societal contexts requiring ISABS to always look forward. As would be expected from a person who is constantly searching for and spreading knowledge, Udai advocated continuous discovery, and providing leadership in promoting sustainable changes in service delivery and practice to improve quality of offerings.

In addition to meeting Udai in ISABS programs and meetings, there two specific places that I met and worked closer with him and these places have shaped my image of him.

Appreciative inquiry workshop

I remember when he, Sankar Subramanian, and Wasundra Joshi facilitated an Appreciative Inquiry (AI) workshop which I was a part of in Ahmedabad. Sankar had demonstrated the AI interview process by interviewing Udai. As he was sharing his experiences in responses to questions, I kept wondering how he can gain such self-awareness that could eliminate the influence of personal biases and values in working with diverse groups, how he could be so meticulous to understand and recognize the importance of difference in shaping life experiences, and how he could be broad-minded enough to accommodate all that encountered him.

Among the specific take away I experienced after listening to Udai in this interview were

i. It is important to recognize and manage personal values in a way that it allows professional values to guide practice.
ii. Tolerate ambiguity in handling ethical issues and apply ethical reasoning to arrive at principled decisions.
iii. Recognize the extent to which a cultural structure and values may oppress, marginalize, alienate, or create or enhance privilege and power. This was more meaningful and applicable to me because I was being of African origin, and I was still finding my way to fit into the Indian society in general and into ISABS in particular. This experience as shared by Udai was of very help to me in looking at the difference in culture and how it can empower or dies-empower individuals.

It was through the story of his experiences that I understood several of the definitions and meanings associated with AI. AI is a methodology for discovering, understanding, and fostering innovations in organizations through the gathering of positive stories and images and through the construction of positive interactions. After listening to his story I realized that AI is also a methodology for discovering, understanding, and fostering innovations in self through the same process. According to Cooperrider and Whitney D, "Appreciative Inquiry is a tool for connecting to the transformational power of the positive change core by opening every strength, innovation, achievement, imaginative story, hope, positive tradition, passion and dream to systematic inquiry." (Cooperrider & Whitney, 1999). Throughout his life and in the moments that I met Udai, I experienced him talking and working from strength and passion.

Global OD summit—Mysore, 2006

Udai, as though what he has done for the practicing managers, organizations, and individuals is not enough, he pioneered the idea and formed the Asia OD Network which had its Global OD Summit in Mysore in September 2006. Udai was the mentor to the Summit and instrumental in bringing big names to the summit. "Udai at the right time introduced Roland Sullivan (Roland) from USA to the GODS team. Roland assumed the task of getting the big names in the field of OD as keynote speakers. He succeeded in enrolling Professor Peter Kostanbaum, Professor Michael Beer of Harvard Business School, and Dr David Bradford of Stanford University as keynote speakers."[4]

[4] http://www.aodn.org/about.htm

As I was closely associated with the organizing this summit, I had the opportunity to observe Udai at his professional attire. I could see the critical skills of intervention mentioned above put in practice by Udai. These include

- Collaboration with the Summit Committee for effective policy decision regarding the nature and shape of the summit
- Substantive preparatory sessions that required him to travel frequently to ISABS meetings of visit to Mysore
- Developing a mutually agreed-on focus of work and desired outcomes

Respect to diversity is one aspect of human behavior that I sense conflicting from time to time; for example, as Udai presented himself as an "atheist," and had to work and mingle with people of different faiths and religions. This stances call from different values and beliefs but in nowhere have I heard or read that Udai had any argument relative to beliefs and values. I rather experienced him facilitating constructive confrontation and change in cases in which cultural beliefs, values, and traditions tend to violate basic rights of individuals. His Theory of Extension Motivation is a classic demonstration of his all-inclusive approach to life. This concept has been well captured in a speech delivered by T. V. Rao, a close associate of Udai and who succinctly summarized it as follows:

> [T]he concept extension motivation simply means a need or a desire to extend oneself or the ego to others and relate to a larger group and its goals. It means a motivation for helping others, working for larger goals that benefit larger groups or society. It also means an ability to sacrifice one's own comforts and desires for the sake of others. It is this powerful motivation that has led to many great people to make sacrifices for the good of the larger community. All great preachers and saints have leaded a simple life and taught people to lead simple life.[5]

Tribute to Udai

I count it as a blessing knowing you as throughout the years of entry to the field of Applied Behavioural Sciences, when I was at my most vulnerable state, when I was still struggling to admit myself to different people with different cultures, you were among those who came forward to assure me of your ever availability as and when I required. You were an educator, mentor, philosopher, writer to many, but to me you were "a friend" and it is your friendship that has had greater impact on me because I tried to emulate your Being and not much your Doing. From you I learnt this great lesson."

[5] http://www.iimidr.ac.in/iimi/images/IMJ/Volume2_Issue2/. Address delivered on March 23, 2011, in the Udai Pareek Memorial Lecture for the Jaipur HRD Network Foundation, Jaipur.

"I learned that courage was not the absence of fear, but the triumph over it."
—Nelson Mandela

References

Allen, J. (2014). *As a man thinketh.* Chennai: Divine Bliss Publications.

Cooperrider, D. L., & Whitney, D. (1999). Appreciative inquiry: A positive revolution in change. In P. Holman & T. Devane (Eds). *The change handook.* San Fransisco: Barren-Kochler Publishers, Inc

Ellström, P. E. (1997). *Kompetens, utbildning och la¨rende I arbetslivet* (Competencies, education and learner in working life). Stockholm: PUBLICA

Pareek, U. (2002). *Training instruments in HRD and OD* (2nd ed.). New Delhi: Tata McGraw-Hill

White, R. W. (1959). Motivation reconsidered: The concept of competence. *Psychological Review, 66*(5), 297–333.

27 Extension motivation and its applications for laboratory education

Paul Siromani

My privilege in working with Dr Udai Pareek

I feel greatly privileged to have worked with Dr Udai Pareek on many occasions, more particularly in spreading the "extension motivation" training that he had developed. I will give more details about this later.

Before doing this let me mention that another area of common interest, based on common values that we held, was a concern for working toward greater social justice. This led Udai to join me and the others in a workshop I was helping to organize in Jamkhed in 1975 on "Community Health and Social Justice" for organizations and groups affiliated to the Voluntary Health Association of India.

This concern also led Udai to support and join in working with nongovernmental organizations (NGOs) serving in the rural areas. I remember the training on extension motivation we jointly conducted in Konarak organized by DEEDs Trust for members from NGOs.

As far as I know, Udai had developed the extension motivation training module, and was anxious to try it out with those who would welcome it and practise it. In 1991, I was able to get two organizations in Bengaluru, Indian Society for Applied Behavioural Sciences (ISABS) Bangalore Region and the Training Task Group, to sponsor the first workshop on extension motivation.

Subsequently, Udai Pareek and I facilitated extension motivation labs in 1993 in Calcutta, in 1995, in the National Summer Event of ISABS and December Event in Jaipur. We also facilitated for a mixed group organized by different organizations—in 2006, in ISABS Calcutta region, in 2008 organized by an NGO, Deeds Trust in Konarak, Orissa, and also in 2008 for teachers organized by Young Horizon. The only extension motivation we facilitated exclusively for the corporate sector was in May 2008 for eRevMax, a software company in Calcutta.

Let me explain what extension motivation is all about.

What is extension motivation?

Extension motivation can be expressed as the concern and urge to extend oneself beyond one's limited or narrow interests and desire to forego or postpone gratification of one's wants, for the benefit of others. This is reflected in the concern for other individuals, for groups or organizations to which a person belongs, and also concern for the society at large.

All organizations in the corporate or voluntary sectors are concerned about motivating their personnel to work for organizational or social goals, which go beyond an individual's own perceived interests, and to promote collaboration, and commitment to super-ordinate goals. Extension motivation addresses itself to this.

Extension motivation is also reflected in cooperation with others for the achievement of a common goal, faith, and trust in members of a group, and involvement in goals, which concern not only oneself, but also large groups, community, or society. Various dimensions of extension motivation may be: helping, collaborating, empathy, risking one's comfort or safety for others, sacrifice, patriotism, hospitality, etc.

Relevance of extension motivation

McClelland, in a lecture at Indian Institute of Management Ahmedabad in 1974, acknowledged the importance of this motive that he called the "concern for the common welfare of all" for economic growth: the theme or concern for the countries that subsequently developed more rapidly. That is, these stories frequently described people being influenced by the wishes and the needs of others. Furthermore, it is probably in this way that one may most easily explain the correlations that have been found between investments in health and education and subsequent rates of economic growth.

All organizations, in the corporate or voluntary sectors, are concerned about motivating their personnel to work for organizational or social goals that go beyond an individual's own perceived interests. Extension motivation promotes collaboration, greater participation and motivation, and commitment to super ordinate goals.

Components of extension motivation

The following are some of the aspects that training in extension motivation would deal with: super ordinate goal, empathy and sensitivity, tolerance for ambiguity, understanding of "who am I" or self-concept, humility, compassion, team work, values relating to concern and care for others, leadership with attention not only on the task but also giving importance to the person, and also in using power in leadership not to coerce but to persuade. Another important aspect is the locus of control, whether it is internal or external and whether it is stable or unstable.

Training

The main purpose of an extension motivation lab is to help participants to increase their own Extension motivation behavior. The schedule is meant to facilitate this process. Extension motivation is being treated here with its components, such as empathy, compassion, helping, forgiveness, and humility. Learning takes place in a cohesive and supportive group. Several activities are planned to help participants to know each other, and move toward group building. To make the group more cohesive, a few sessions on agenda-less process work can be designed. The program is based on the principle of Self-Assessment through Feedback on Instruments (SAFI). The participants are likely to change if they get feedback by looking at data generated by themselves, with the help of these instruments. So several instruments are administered to the participants, and through interpretation of the scores and explanation they are left to reflect on the implications of their own scores. They can then discuss if they want to improve some of their behavior.

The laboratory provides a variety of ways to help the participants to "experience" some aspects of extension behavior. One way is to use simulation exercises or games. In order to change, exercises have been designed for them to experience and assess their tendency to collaborate individually and collectively with others. Some exercises are designed to help them to learn action planning by analyzing different forces. This is done by Force Field Analysis. Similar exercises are planned to develop competence in giving feedback with empathy, and collaborate by practising extension orientation (helping other people to be effective). People also learn when they share with each other significant experiences, which are generated by critical incidences by the participants on successful and less successful extension behavior.

Finally, all learning from the program should end in realistic and satisfying plans to advance their extension behavior, and apply it in their real back home situations. This is done by the last exercise of action planning. This part must be given adequate time, focus, and attention to enable participants to make meaningful, realistic, and high impact action plans.

Experiencing, reflecting, hypothesizing, conceptualizing, and experimenting are the chief ingredients of the process of learning in a laboratory setting. The EM Lab will utilize experience-based learning as the main training method.

A model training schedule would be as follows:

1. Preparing the environment/facilitating learning and exploration

    a. Thematic Apperception Test (TAT)—meet in groups of three or six. Stories to be written individually and shared in the group.
    b. Answer the MAO-B questionnaire—individually.
    c. Micro lab—the total learning community coming together in small groups and introducing oneself.

d. Scope of the workshop—the facilitator sharing the scope of the workshop and sharing what is extension motivation.

2. Raising awareness through exercises and questionnaire

a. Regarding knowing self deeply and taking responsibility.
Exercises on TAT, MAO-B, and Who Am I:

i. TAT
ii. MAO-B
iii. CPP
iv. Who Am I
v. Personal Effectiveness (Johari Window)

b. Sensitivity toward others (extending oneself)

i. Unstructured group process session
ii. Locus of Control/ ASUFA (Attribution of Success and Failure) Inventory
iii. Humility
iv. Empathy
v. Compassion
vi. Altruism
vii. Tolerance for ambiguity
viii. Exercise on broken squares
ix. Exercise on "Win as much as you can"
x. Values
xi. Tower building
xii. Scoring TAT and MAO-B

c. Diagnosing and analyzing blocks for extension motivation, using

i. Force field analysis
ii. Critical incidents

3. Acting on the awareness / learning and preparing for change

a. Rewriting the story with highest possible scores (where all the 10 factors are present)
b. Writing "Who Am I," "what has made me what I am" (how one has projected the self as a creator or a puppet)
c. From force field analysis identify the forces in personal and organizational life which are negative and how they can be changed to positive extension motivation.

Elaboration of the Training Schedule

I. Preparing the environment/facilitating learning and exploration

1. Thematic Apperception Test (TAT): Writing one's story using the Thematic Apperception Test pictures individually helps one to interpret one's own perception of pictures and draw from one's life experience.
2. Motivational Analysis of Organizational Behavior (MAO-B): Answering this question and interpretation will help the participant to understand one's own dominant motivation, behavior, and performance at work.
3. The Coercive and Persuasive Power (CPP): The CPP scale measures the value given to and need for coercive and persuasive power bases.
4. Personal Effectiveness: The questionnaire and the model of Johari Window brings to awareness our interpersonal processes in relation to what is known and unknown to self and others.
5. Write on who am I: "What has made me what I am," this exercise could indicate, a) has one projected oneself as creator/actor or puppet/pawn? b) How much extension motivation imagery is there?
6. Micro Lab: The total community come together and get into small groups in which each one introduces himself or herself in respect of their name and role.
7. Scope of the Workshop: The facilitator explains what extension motivation is—gives an overall idea of the contents of the workshop and the schedule.
8. Scoring, explanation, and learning on TAT, MAO-B, CPP, and WHO AM I TAT will be scored taking into account the following dimensions.

   a). Does the story contain imagery of extension motivation? For example, super ordinate goal, empathy, helping relationship, sacrifice, etc.
   b). Extension imagery (EI)—score 1
   c). Need (To do something) (N) —score 1
   d). Activity (A) —score 1
   e). Internal blocks (IB) (Lack of ability)—score 1
   f). External blocks (EB)—score 1
   g). Overcoming blocks (OB)—score 1
   h). Linkages (L) (Relating to others or working with others)—score 1
   i). Feeling (F+) (Feeling for involvement – elation)—score 1
   j). Feeling (F−) (depression)—score 1
   k). The theme only extension motivation (T)—score 1

MAO-B: Below 50 means avoidance is more. Discuss how one can reduce the avoidance behavior.

II. Raising awareness through exercises and questionnaire:

1. Scoring of TAT, MAO-B, CPP, etc.
2. Unstructured group process session: This will help to bring to awareness the human processes at the intra-personal, inter-personal, and the group levels.
3. Locus of control/ASUFA (Attribution of Success and Failure) Inventory: The Loco Inventory helps individuals to know how far their thoughts and actions are controlled by them or others/external factors. The locus of control orientations is reflected in the way people feel about what happens in the organization; how much control they, other significant or neither (being a matter of luck), have in important organizational matters. These matters relate to success or effectiveness, influence, acceptability, career advancement, and rewards.
4. Humility questionnaire: This questionnaire helps one to understand how far the ego plays in one's life to be anxious to take credit to oneself or recognizing self only as against giving credit to others where it is due.
5. Empathy quotient test: Few skills in life are more important for building strong relationships. Seeing things from other's perspective.
6. Compassion index test: Helping others puts one in direct contact in other people's lives. Compassion for those one helps has both negative and positive aspects. The self-test helps to estimate one's compassion status. It also helps one to understand how much at risk one is of burnout and compassion fatigue and also the degree of satisfaction in helping others.
7. Self-report altruism scale: Altruism can be defined as an individual performing an action that is at a cost to themselves, but benefits either directly or indirectly, another third-party individual, without expectation of reciprocity or compensation for that action.

   Having super ordinate goals is important for extension motivation. Exercises that bring to awareness and emphasize the need for empowerment of the other, the larger picture, or taking into account the whole are "Tower Building," "Broken Squares," and "Win As Much As You Can."
8. In "Tower Building," the participants are grouped into pairs or triads. One person is blindfolded and is required to build a tower of wooden square blocks, one over the other. The other one or two persons act as helpers or leaders who are to assist in the building of the tower. The learning is in how they help or exercise their leadership. Does this helping activity result in making the person who is helped dependent or empowered and confident?

9. In broken squares, the learning is that while one is focusing and taking responsibility for one's own task, one has to also be aware of and have concern for the others in the team, and help where necessary, for them to complete their task.
10. In the exercise on "win as much as you can," the emphasis is on cooperation—I win—you win attitude rather than a competitive, I win—You lose attitude, as present in most games where a team wins only when the other team loses. In a family, organization, community, or society such a competitive win–lose motivation leads to weakening the overall objective of the whole.
11. In dealing with values, in the training, the object is to bring to awareness the underlying causes for many of one's motivation and behavior which one has imbibed and practices since one's childhood. This can be done in three different exercises like "examining"

    a. How far are our values "Full Values,"

        i). *Choosing*: How far have we chosen our value a) freely, b) from different alternatives, c) after thoughtful consideration of the consequences of each alternative?
        ii). *Pricing*: How far is the value cherished and how far are we willing to affirm this choice publicly.
        iii). *Acting*: This means doing something, not just once but repeatedly and whether this has helped to enhance once personal growth, or

    b. Auction: Using the "action" questionnaire to get each person to decide how much they would be willing to spend for securing the different value benefits.
    c. Discovery Exercises: A third exercise can be taking up different values relating to extension motivation such as trust, egotism, and concern for others. And with the aid of stories looking deeply into oneself to know how far one is giving value to it and practicing it in one's life, and if not, why not?

III. Diagnosing and Analyzing blocks:

1. This may be done using the "Force Field Analysis" or evaluating using "Critical Incidents": In the Force Filed analysis, each individual takes up one or two important situations in one's life and identifies what factors have helped and what factors have blocked or restrained the self from practicing extension motivation.
2. Similarly, each individual may write down critical incidents in one's life that have helped or prevented the acting out of extension motivation.

IV. Acting on the awareness/learning and preparing for change

1. Rewriting the TAT story with highest possible scores (where all the 10 factors are present)
2. Writing "Who Am I," "what has made me what I am" (How one has projected the self as creator/puppet)
3. From Force field analysis identify the forces in personal and organizational life which are negative and how they can be changed to positive extension motivation.
4. Identify five main goals for your future which includes extension motivation and where you develop collaboration

Program evaluation by participants

Given below are some of the responses of participants who participated in one extension motivation lab facilitated by Dr Pareek and me. This was a mixed group from the corporate, NGO, and service sectors.

My learnings from the lab are as follows:

- A deeper look into myself
- Exploration of own inclination, values, and beliefs
- Knowledge of my own extension motivation level and its implications
- Feeling good about my own extension motivation level, and humbled by lack of it
- Need for extension in my work life as well as personal life
- My strength area
- Ways of tackling the weakness
- How specific one can be to extend extension motivation
- Beyond extension Motivation, re-inforced my understanding of other motivation types too
- How I can utilize my extension motivational needs
- How I can implement extension motivation in my work life and personal life
- Various facets of myself in relation to my own extension motivation
- Awareness of my values and how I prioritize my values
- Importance of thinking about larger causes in my life
- How to develop linkages with myself and with larger issues
- Taking risks
- How to be more effective in my work and as a person in the society
- Changing myself the best way
- Sacrifices and collaboration are the prime for team success
- Behavioral change is possible
- Trying to be focused on my life's goals

- Opening-up of more choices as to what I can do, now that I have the time
- Change is possible, take on change
- Have high expectations and see large picture
- Remain self-motivated
- How to understand the behavior of different types of people in the society
- Appreciate others and be collaborative
- Behavior with less power imagery is more effective
- Understanding the needs of other needy people as altruism
- A journey into the world of social service
- Power of collaboration
- Analyzing situations and overcoming blocks
- Team work
- How power can be utilized as a base for extension motivation
- How to measure different components of extension motivation
- Better insights into the social development sector
- Value clarification
- Using my inherent values to move other people and organization
- How to work when there is conflict of values

In conclusion, let me quote from the evaluation forms what some participants wrote about Dr Pareek and his facilitation:

> My journey with my current profession (HR) started with your text book. It is a pleasure to learn from you.
> Great enthusiasm and dedication
> Very focused approach.
> Very humble.
> Very candid.
> Admire your facilitation skills and contribution for our learning.
> Excellent facilitator: A great experience of knowing a lot about your research work and also sharing of your views. A humble and respectful individual.
> Your simplicity and erudition, moderated with care and empathy were exemplary.
> You are extremely encouraging
> Very alert and focused, knows what he wants to say and is not dissuaded from his purpose.
> Learned a lot from the wide range of topics covered and how effectively it was linked to the common topic.

28 Passion for our nation: A blueprint for India of our dreams

Varun Arya

Let me begin by paying my profound homage and tributes to my great teacher Dr Udai Pareek who was truly my friend, philosopher, and guide for three decades from the time I met him as a student at Indian Institute of Management Ahmedabad (IIMA) in June 1981 till his passing away in March 2010.

When I was conceptualizing Aravali Institute of Management starting in April 1999 and wrote to him about it to get his inputs, proactively he replied to me that he would be glad to be associated with it and like to contribute to it. He became a member of our Advisory Council and did more than full justice to it by visiting us at Jodhpur many times, speaking in the various events organized by us, and delivering guest lectures to our students.

I was touched with his humility, agility, flexibility, encouraging, motivating, and supporting nature—apart from his academic excellence.

Background

For the large part of the last two millennia, Indian subcontinent was a leading global economic and spiritual power—*Sone Kī Chidiya*, respected and visited by people from all over the world for education, trade, and quality of life. However, during the last over 67 years since India became independent, its global stature has been continuously undermined by the successive governments, because of lack of good governance.

In 1947 what we accomplished after about a century's (1857–1947) struggle was freedom only from external slavery of foreign rule—*Swaraj* but not *Suraj*. The total transformation of India leading to its development into a role model nation of the world has not unfortunately happened so far.

Echoing thoughts for India of his dreams, the Father of our Nation, Mahatma Gandhi, had said:

> I would work for an India, in which the poorest shall feel that it is their country; an India in which there shall be no high class and no low class of people; an India where all communities will live in a perfect harmony. Women will enjoy the same rights as men. We shall be at peace with all the rest of the world.

Reflecting on sentiments for India of his dreams, Gurudev Rabindranath Tagore had said:

> Where the mind is without fear and the head is held high... Where knowledge is free
> Where the world has not been broken up into fragments... By narrow domestic walls
> Where words come out from the depth of truth
> Where tireless striving stretches its arms towards perfection
> Where the clear stream of reason has not lost its way
> Into the dreary desert sand of dead habit... Where the mind is led forward by thee
> Into ever-widening thought and action
> Into that heaven of freedom, my Father, let my country... Awake.

India's global rankings: Some latest benchmarks

Being a part of global community (Vasudhev Kutumbakam), it is important that we know, respect, and understand India's position internationally, which is as under on some of the critical parameters as per the latest analysis available:

Corruption Index (Transparency International)	: 85/175
Ease of Doing Business (The World Bank)	: 142/189
Global Competitive Index (World Economic Forum)	: 71/144
Global Gender Equality Index (World Economic Forum)	: 114/142
Global Peace Index (Institute of Economics & Peace)	: 143/162
Human Development Index (UNDP)	: 135/145

Building-blocks for India of our dreams

While there can be several building-blocks for India of our dreams, the most important and critical ones are as under:

- Corruption free
- Equality
- Education
- Water and Food
- Health and Hygiene
- Justice
- Peace and Harmony (Law and Order)

- Government to Get Out of Business
- Infrastructure Development
- Merit-based Leadership and Governance

Corruption free

It is really most unfortunate that because of the unholy nexus between some of the vested interests among *netas*, *babus*, and others, the corruption is eating into the vitals of our society and the nation.

To get the work done in the government there have been precisely only four ways:

1. Pay bribe whether	: Around 80 percent people do this, irrespective of they admit it or not.
2. Exercise influence	: Around 10 percent get their work done by exercising influence over someone who has a say with the decision-maker.
3. Court orders through court	: Around 9.9 percent people get their work done orders and if further required, contempt of court cases.
4. Become nuisance value	: Only a maximum of 0.1 percent resort to sitting on indefinite fast, extensively using RTI, going to media, taking out protest march, etc., to get their work done.

The worst sufferers of corruption are the poorest and the disadvantaged sections of our society since they have to often borrow or sell off their scarce assets to pay the bribes or lawyers' fees.

All regulatory agencies in India, without exception, are actually rent-seeking bodies. Instead of ensuring the qualitative growth of institutions and our nation and serving the relevant sections of the society, they have only been ensuring the quantitative growth of corruption in our country. Also the persons appointed to the leadership positions in these regulatory agencies are usually arrogant, highhanded, incompetent, unprofessional with rent-seeking as the only actual qualification. With the blind protection of powers that be and no punitive action generally taken, these vested interests continue to work against people and our nation.

One of the important side effects of corruption has been that the public has been turned into beggars (surrender of self-respect) and so-called public servants have actually become the masters. Mr N. Vittal had so rightly said in the late 1990s when he was Central Vigilance Commissioner of India that corruption was low risk, high-profit business.

As per a survey done a few years back, within the government, the Central Bureau of Investigation (CBI)—supposedly the highest corruption prevention body and outside the government, the media—supposedly the largest means to spread the awareness, also join the corrupt in India.

Presently while there are a number of mechanisms and methods to shield the corrupt but there is in reality no mechanism to punish the corrupt. There is an urgent need to have the exemplary punishments for the corrupt. Corruption-related matters need to be taken to the logical conclusion in the shortest possible time-frame. The assets of corrupt should be confiscated and become property of the nation.

Equality

The Almighty or the Creator or the Nature made all of us human beings.

- With one community: Humanity and
- One common code of conduct: Human Values

Caste and religion are issues of the faith and family of birth. The government has no role in these.

There should not be mention of caste and religion required in any documents or for any purposes whatsoever. Also the caste and religion should not be basis for any benefits or entitlements.

One position, one pension in all organizations all over India; always needs to be ensured.

The Pygmalion Effect as a process for removal of poverty should be adopted.

Among the Constitutional Right of Equality Vs Reservation, we should understand that Right of Equality is more important and sacrosanct, which must be respected. Equality of opportunity must exist, while suitably financially supporting those without means, irrespective of caste, community, culture, gender, region, and religion.

VIP culture existing in the country is the legacy of British Raj. It is illegal and unconstitutional and therefore, it must be abolished.

States should be formed based on the economic capability and administrative viability. There is no reason for any provision to be there for special status to any state. Political parties advocating regionalism should be banned.

Education

Purpose of education is competency-building; to realize the dreams for career and life and not merely to give away a piece of paper as some qualification. Therefore, education must ensure that the children are shaped as good human beings and good citizens.

Presently in a large majority of the government schools, colleges, and universities, there is massive corruption and favoritism. There are many things happening there except the real education.

The situation of education in most of private sectors is no better. A large majority of educational institutions in the private sector are promoted by *netas*, *babus*, and building contractors. With profit as the bottom-line in these institutions, shortcuts are made to somehow complete and fulfill the formality of having the requisite teachers, learning tools, facilities, infrastructure, and organizational networking. Consequently, education here too suffers.

Because of the above sorry state of affairs, there is either Brain Drain or Brain in the Drain. Either way the country suffers.

Educational institutions need to have the employability, empowerment, and entrepreneurship as the defining premises. These should be supplemented with skill development by nurturing and supporting suitable institutions on merit so that the target of 15 crores skilled manpower is achieved by the year 2020, against achievement of only 35 lakhs so far. In addition, we also need to have the development of scientific temper among our people, especially the rural population.

Water and food

Water and food are among the basic necessities for all people irrespective of the status. We need to ensure proper production, storage, and distribution of food for all. This requires extensive revamping of present Public Distribution System which thrives on black-marketing and hoarding.

For the above to be achieved, it is important to have the appropriate farmers' education, which should also help in prevention of farmers' suicides. The needs of large versus small farmers are different and both should be attended and redressed appropriately.

Our country has over 60 percent of land as the wasteland, which alone should be given for the nonfarming purposes such as setting up of factories, institutions, and offices. Arable land should be strictly earmarked for farming alone. Increasing the productivity and usage of organic farming needs to be encouraged and supported.

The key to have availability of water on sustainable basis is to ensure rain water harvesting by every household and organization.

We also inculcate the sensitivity among our people to stop the wastage of water and food.

Health and hygiene

The government has launched *Swachchh Bharat* movement, which is a good step. However, what is more important is to ensure the effective implementation of this plan.

While the rural India has Primary Health Centers but the doctors assigned to these are hardly there and the medicines meant to be given free to the patients are often sold off in the market. This needs to be corrected.

There is generally a practice among the doctors to prescribe the branded medicines only, which are costly and there is a massive commission/margin for all those involved. To prevent this, the doctors must be mandated to prescribe the generic medicines that are economical and hence more affordable.

Medical education has become a thriving business with donations of ₹50 lakhs to ₹1 crore generally demanded per admission, irrespective of the merit of the student. Consequently, many students passing out are not really good doctors and they are more interested in recovering their huge "investment." Because of all these reasons, the healthcare system suffers. Therefore, it is required to make the medical education system strictly merit-driven as it was earlier.

There are now many ways and means available for turning waste into wealth. These should be encouraged to be adopted and supported since these would solve many of the problems while creating another honest income stream for many.

Justice

India's justice delivery system has the courts and judges but where is the justice? Many lawyers are liars and often get sold. Justice is mostly delayed which is tantamount to justice denied. The cost of process of seeking justice is so high that at times, one feels that justice is primarily for the rich and famous.

We have the British hangover of courts' functioning, which has no place in an independent India.

Government is the largest litigant, most court cases exist because of the malfunctioning and nonfunctioning of the government officials, and generally there is pro-government approach of the courts. There is at times violation of laws by the courts themselves (e.g., not handicapped friendly, no suitable provision for public amenities on the court premises, lawyers' strikes, etc.). Contempt of court is a deadly weapon, which at times obstructs the delivery of justice.

To take care of the above, we need to have the accountability, responsiveness, and transparency in-built into the justice delivery system.

Peace and harmony (law and order)

Peace is the most essential prerequisite for growth and development. In view of the police in India still being governed by the colonial legacy: Police Act, 1861, which was meant to exploit and suppress the Indian population. Therefore, the police of independent India continue to behave the way British Police used to. Villagers, disadvantaged, downtrodden, and underprivileged people are the most adversely affected by this.

There are so many legal loopholes in almost all the laws that there are adequately available ways and means to escape the punishment. Despite the Supreme Court judgement, genuine FIRs are not often registered and instead false FIRs and counter FIRs are registered. The investigation system by the police is at times based on appeasement of *neta*s, *babu*s, and *dada*s, rich and famous by the police through collusion, connivance, conspiracies, and corruption.

There is urgent need to thoroughly overhaul the above to make the police as the public-sensitive and law and order-friendly without fear and favor.

Get out of business

The government has no business to be in the business. There has been adhocism: first it was transition from privatization to nationalization, then disinvestment of nationalized units, and thereafter again promotion of privatization.

Industry is meant to create self-sufficiency for the nation for maximum of its needs, generation of wealth for the nation and provide employment. These need to be achieved by prevention of monopolies and disabling the restrictive business practices.

License Raj system is still prevailing which needs to totally go. Honest entrepreneurship should be encouraged and supported.

Infrastructure development

Earlier the emphasis was on *roti, kapada aur makan* but now it is on *sadak, bijali aur pani*. On the one hand, the cost of living is increasing but at the same time, the quality of life is decreasing. This needs to be corrected.

The new government's plans of Digital India and eGovernance can solve many of India's problems, if effectively implemented.

There is a lot of merit in the concept of PURA (Providing Urban-Amenities in Rural Areas), by late Professor P. V. Indiresan and Dr A. P. J. Abdul Kalam, which should be implemented.

In the name of Public–Private Partnerships (PPPs), Built-Operate-Lease-Transfer (BOLT) and Tolls—there is a lot of harassment and inconvenience to the public and also rampant corruption. When the government is collecting such a high level of taxes then why there should be need for the public to pay additionally for the services that are entirely the responsibility of the government.

Large funds are given to MPs and MLAs in the name of local area development, but at times these are used for their personal purposes and benefits. There is need for effective monitoring of these schemes.

The government's plans of adoption of one village by each MP and development of 100 Smart Cities are good ideas, which should be effectively implemented.

Merit-based leadership and governance

The following exist for all government positions but not for the larger governance roles:

- Background check
- Educational qualification and past experience
- Selection process
- Key Result Areas and Performance Appraisal
- Removal, reprimand, and reward system
- Retirement age

It is of paramount importance to have the right persons in the leadership positions for our states and the nation. Then only the transition from *Swaraj* to *Suraj* can take place by having the right system of governance.

The essence

In the ultimate analysis, the people make all the difference. We need to have people with a set of Core Values, who can be the role models.

Everybody talks of Fundamental Rights but nobody talks of Fundamental Duties, which are equally important and integral part of our Constitution. Good citizens have to be necessarily first the good human beings. As the saying goes—attitude leads to altitude.

Spirit of enterprise and self-actualization need to be developed amongst the people. Each citizen must constantly be adding value and also involved in value creation for himself, family, society and the nation.

We need to again develop the passion, patriotism and pride for our nation in all our citizens, while simultaneously inculcating the feeling of Vasudhev Kutumbakam.

Role models

Historically, India was privileged to have provided to its population the role models at various stages of its journey who truly believed in and actively worked for national interest as well as nation-building—such as Maharana Pratap, Rani Lakshmi Bai, Swami Vivekananda, Mahatma Gandhi, Sardar Vallabhbhai Patel, Sardar Bhagat Singh, Rabindra Nath Tagore, Netaji Subhash Chandra Bose, Sir C. V. Raman, Dhyan Chand, Dr Meghnad Saha, Dr Homi J. Bhabha, Dr Vikram Sarabhai, Lal Bahadur Shastri, Mother Teresa, Milkha Singh—to name just a few.

It has been truly a matter of great satisfaction for me that I have had opportunities to closely know and work with a large number of persons from India and abroad like Dr Udai Pareek, from whom I learnt a lot and also who greatly supported my humble endeavors for nation-building. They are indeed the role models and I would like to briefly profile six of them:

1. His Holiness Dalai Lama

   He had visited us at Jodhpur on February 9, 2011, and I had opportunity to meet him, listen to him, and spend considerable time with him. While he is a role model for many characteristics and qualities relevant for nation-building, I would like to highlight the few.

   Our main event with him was at 09.30 am on February 9, 2011, and to keep the time discipline, I requested him to reach at 09.25 am. He reached at 09.10 am and then patiently waited for 20 minutes for the event to begin at 09.30 am.

During his talk and interaction, there were some occasions when he turned toward his assistant to ask the English equivalent of some Tibetan word. To ensure that there was no wrong impression created, somewhere in the middle of his talk he said that his assistant was more knowledgeable in many areas than him but it so happened that he was Dalai Lama.

He mentioned about the continuing problems he was facing to liberate Tibet from Chinese rule but spoke about the constant efforts being made toward the cause in a nonviolent manner.

Despite his being around 80 years, he demonstrated remarkable agility and attention. He maintains extremely busy schedules and travels extensively.

2. Dr A. P. J. Abdul Kalam

It is indeed most unfortunate that Dr A. P. J. Abdul Kalam is no more with us. A truly great human being, he really dreamt of a developed India where every citizen had the career and life of one's choice, by realizing their true potential. May his noble soul rest in peace!

I had privilege and honor to meet Dr Kalam four times. My first meeting with him was in the afternoon of September 2, 2005, in his Study Room at Rashtrapati Bhawan when he was the President of India. The meeting was meant for just 15 minutes but lasted for almost one hour.

Dr Kalam listened attentively to my story of unending hardships, sacrifices, and struggles just because I wanted to establish a top quality educational institution without any compromise. He said that India needed many more persons like me and advised me not to change myself. He also asked me about my family. When I told him that my elder daughter Aakansha was studying in Gargi College of Delhi University at New Delhi but had not come with me for meeting him since as per her, she had classes that were more important than anything else. Impressed with this, Dr Kalam took out a card from his study table drawer, asked me the exact spelling of her name, and gave me the card addressed to her and signed by him to give to her.

I have met almost everybody who has been somebody in India during the last around two decades but he has indeed been one who was concerned about our country and its welfare. I sincerely wish we had a person of his wavelength as the Prime Minister of India!

3. Maharaja Gaj Singh

Maharaja Gaj Singh, whom we respectfully address as Bapji—meaning father figure, has been immense source of advice, cooperation, encouragement, guidance, inspiration, and support to me personally and to our Institute right from its inception. He has always been available to us whenever we needed him. He has stood by the Institute through thick and thin. With passing of years, his role has transcended from being merely a family member of the Institute to the father figure of the Institute.

For me personally, as I conveyed him sometimes, he has been truly like a local guardian. It would not be an exaggeration to say that Aravali

Institute of Management came into existence at Jodhpur and continues to exist here today largely due to the support of Bapji. During my large number of meetings and interactions with him over the last 15 years, it has been gratifying to observe that he is one person who has the progress of Jodhpur and the well-being of its people closest to his heart. The words can never suffice to express our indebtedness to him.

4. Dr Kiran Bedi
   I have known Kiran Bedi for around three decades, had opportunity to spend a day with her and family before relocating from Delhi to Jodhpur in March 2000 and she has visited us at Jodhpur in December 2000. She represents exceptional commitment, courage, confidence, and conviction in so-called weaker sex that, in reality, embodies *Shakti*. During her student days, apart from being good in studies, she was a national-level badminton player. She became the first woman to enter Indian Police Service and demonstrated her suitability for the various positions she held by delivering exceptional performance, even at times going beyond the normal call of duty.

   In view of her refusal to compromise, as a punishment posting she was transferred as the head of country's dreaded jail Tihar, which is home to over 10,000 criminals. There she carried out a number of steps and reforms proactively which lead to historical transformation, whereby she converted a threat to an opportunity, in recognition of which she was honored with Magsaysay Award—called Asia's Noble Prize, so far the first and only police officer to get this prestigious award.
5. Professor P. V. Indiresan
   Professor Indiresan, who had been the Director of IIT Madras, is no more with us—having passed away on February 25, 2013. He was Dean (Undergraduate Studies) of IIT Delhi when I was a student there during 1976–1981.

   The real interaction with and realization of Professor Indiresan as truly a friend, philosopher, and guide came after April 2000 when I decided to establish a top quality management institute at Jodhpur in Rajasthan as a model of no compromise. Right from the conceptualization of the Institute, we had set ethics, equality, integrity, quality, merit, and societal service as the core values of the Institute. We have managed to strictly adhere to and uphold these core values without any compromise despite many severe adversities, difficulties, and problems faced by us. It had been possible entirely because we had the constant advice, cooperation, guidance, mentorship, and support of persons like Professor Indiresan.

   Professor Indiresan had pioneered the concept of PURA (Providing Urban Amenities in Rural Areas) which he firmly believed would solve significantly India's myriad problems. During the various discussions with Professor Indiresan and also in his writings, he had been voicing concerns regarding the rapid degeneration of governance and politics in our country. As per him it was hindering the real development of our nation to such an extent that unless some urgent measures were devised and put in place, the political and administrative deterioration would soon reach a point of no return. According to him such a situation may even lead to chaos of unthinkable proportion.

6. Dr I. G. Patel
   Born on November 11, 1924, Dr Patel held many prestigious positions nationally and internationally and passed away on July 17, 2005. When he was just 28 years, he became the youngest full professor with the prestigious M. S. University in Baroda. He was Chief Economic Advisor to the Government of India for many decades, where he worked closely with all the Prime Ministers starting with Pandit Jawaharlal Nehru and till Mr Atal Behari Vajpayee. He had been the Governor of Reserve Bank of India, Director of IIMA (where later he became the Chairman, Board of Governors—the only IIMA Director to do so) and Chairman of the Prime Minister's Economic Advisory Council. He was the first non-European and first non-White to become the Director of globally respected London School of Economics & Political Science, UK—a post which he held for over six years.

   Dr Patel had a lot of affection for me personally for over two decades till his passing away. When he took over as the Director of IIMA in 1982 (where I was then a student), the outgoing Director Professor V. S. Vyas (who also always had a lot of affection for me) took me personally to Dr Patel and introduced me to him. Since then, Dr Patel took a liking for me and had been my supporter and well-wisher. His letters to me, even when he was Director of London School of Economics and Political Science, were all hand written.

   Despite his holding such top positions, he was an exemplary symbol of humility and simplicity. In 1984, when from IIMA Alumni Association's Bombay Chapter (of which I was then the Secretary), we decided to organize a Felicitation-cum-Farewell Function for Dr Patel on his appointment as the Director of London School of Economics & Political Science, he readily agreed to come from Ahmedabad to Bombay. When I told him that we would come to receive him at the Bombay Airport, he politely declined saying that he knew Bombay very well and would reach on his own. I was at the portico of Taj Mahal Hotel in Bombay to receive him. He arrived by a normal non-AC black and yellow public taxi. When he was handing over the fare to the taxi driver, suddenly a thought crossed my mind at that moment of time as to whether taxi driver knew that the person who was handing over the fare was the same person whose signature existed on the currency notes being handed over. The noted jurist and Senior Director of the House of Tata, late Mr Nani Palkhiwala, had presided over that function.

   In the year 1999, when I was conceptualizing establishing Aravali Institute of Management (latest Information Brochure enclosed) at

Jodhpur in my home state of Rajasthan, I requested Dr Patel (who was then the Chairman of Board of Governors of IIMA) to be a member of our Board of Governors, and he readily agreed. In our very first year, in November 2000 he visited us at Jodhpur. He placed a condition that he would just speak to and interact with our students only and there should not be any other event nor any public function for him. Dr Patel interacted with our students and coolly, calmly, and confidently answered in Hindi (because that is the only language our students, mostly from rural background then, understood) questions on a wide range of topics including currency notes, national economy, international economic scenario, management education, Indian industry, Indian government, etc.

It is highly important that the stories of such persons who truly have passion for our nation are profiled in a book form and made textbooks in the schools so that the future citizens of our country can imbibe and inculcate the values of national interest and nation-building in themselves, so essential for the future of India as a nation.

29 Importance of values in civil service

Inderjit Khanna

The importance of values in the civil service is an often debated topic within the civil service and outside. At the time of independence, the civil service was termed the "steel frame" presumably on the grounds that it would stand firm on matters submitted and also give fair and impartial advice to the political leadership. Has this expectation taken a nose dive? But first, let me say a little about Dr Udai Pareek and my relationship with him.

Udai Pareek and values

It was on July 18, 1975, that I met Udai, Professor Ravi Matthai, and Dr T. V. Rao at Ajmer. I was then working as the Director, Primary and Secondary Education, Rajasthan, and these three professors from Indian Institute of Management Ahmedabad (IIMA), had come as a Working Group set up by the Indian Council of Social Science Research (ICSSR). Considering that the common perception was that the IIMA was an institution whose objective was to train young, bright girls and boys for taking up jobs in the private sector, it had seemed then strange to me to find these three professors seeking collaboration with the State School Education Department for addressing issues related to school teachers. It was only by and by that I realized that Udai and his other two colleagues, all three renowned in their respective fields, were simple, down to earth human beings who were truly interested in observing what was happening in the rural areas and keen to address problems of the delivery system. What started off with teachers, enlarged in scope to look at the problems of the local people in four villages of Jawaja Block in Beawar tehsil of Ajmer district. This then led specifically to groups of tanners and weavers. Here began the story of Skill Development among the tanners and weavers of the four villages of BeawarKhas, Delwara, Sargaon, and Kabra. That has now become another story by which the Jawaja Experiment became "The Rural University."

From the beginning, it was clear that Udai was not only a man with great intellectual abilities, and a great thinker but also a simple human being. A few years later, from 1981 to 1983 when I was at IIMA, the association with Udai was strengthened. As a primary member of the Public Systems Group, I had innumerable opportunities of formal and informal interaction with Udai at IIMA. Despite his own professional commitments, he always found time to discuss matters and offer advice without ever imposing his views on anyone.

When I returned to Rajasthan in 2000 after my second spell of deputation with Government of India, I was pleasantly surprised to find Udai settled in Jaipur. Despite formal retirement Udai was now working harder than before. He had set up office at the Indian Institute of Health Management and Research (IIHMR), Sanganer, Jaipur. Though he was 75 by then he would himself drive every day to IIHMR a distance of 15 km one way, around 7:30 am and return by late afternoon. His zeal for work had not diminished even one mite. I continued to meet him socially and at academic gatherings where he was always in great demand but, as ever, patient and friendly.

In 2008, I again fell back on my association with Udai. The Gita Mittal Foundation (GMF) desired to set up centers of excellence for skill development of youth particularly from weaker sections of society so as to improve their employability. I, therefore, approached Udai who, without batting an eyelid, not only agreed to chair the curriculum advisory committee but also agreed to prepare the syllabus for the HR and soft skills module. As if this was not enough, he also made available his flat for our frequent meetings. All this was at virtually no cost to the Foundation. Udai worked relentlessly over about 6 months to prepare the syllabus which comprised four books, two on psychological processes and two on entrepreneurship. Each topic had one book for facilitators and another for participants. Only Udai could have accomplished this task in this time in such detail and so meticulously. As I watched him during this period, I marveled at his dedication and passion for the task at hand, even at the age of 83 years.

These days there is much talk regarding "The ease of doing business"; let me refer to this in the context of working with Udai. Udai was 18 years my senior in age. Yet, he insisted on being called by his first name and also calling me by my first name. I still have a mail of his dated February 9, 2008, the first para of which reads as follows:

> Dear Inder,
> I hope I can revive first-name relationship as we need to work for a long period of time. I am taking the liberty and I hope you reciprocate.

Can there be an easier or better way of making a person so much younger than you feel totally at ease? Quite obviously, the relationship can only flourish and be strengthened.

Values are learnt in early years of IAS

I now revert to the topic on which I have chosen to write. Indeed, some of the inspiration is derived from the life and example of Udai. In the 1960s, why did we join the civil service? Partly because there were few opportunities then in the private sector, the salary in the civil service then, and even now, was not too bad, but above all, there was a desire to work for the people. Since Independence was a fact not too far removed in the past, a feeling of idealism also prevailed. Compared to current standards of higher age at entry, most of us were around 23 years of age and were freshly out of college with no work experience. I suppose, therefore, it was easier to knowingly, and unknowingly, mold our minds into thinking that was value based.

Values are imbibed from mentors and examples that you see around you. The process begins at childhood with the family and environment playing an important role. This then extends into the school and college with teachers and peers entering the fray as role models. By the time one is ready to join the world of work, most values, good or bad, have already been instilled into the young mind, but for the civil service the two-year training and thereafter the first few years in service considerably add to, or subtract from, the values that one has imbibed in the formative years. For us, the wars of 1962 and 1965 must have had their impact in terms of creating a greater feeling of nationhood and belonging. Of course, this does not mean that we have to have frequent wars only to instill values and a feeling of nationhood in our youth and in the civil service!

What did our two-year training as civil servants teach us? It certainly taught us to lead by example, to work for the people, particularly the poor and distressed, and to base our decisions on law and equity. I can say confidently that neither my family, nor education, nor my training as a probationer in service made me consider issues or base decisions on caste, color, or creed. Unfortunately, today, I am not so sure that the civil servant will acknowledge this. During our training at the Academy at Mussoorie it was frequently emphasized to us that we must not only be honest but also appear to be honest. The latter takes us to little things like misuse of government vehicles, accepting gifts at Diwali, etc. Are these taken as serious issues today?

Let us take the issue of leadership by example. The Director at the Academy was strict, punctual, and fully involved in the training of probationers. Our first jolt was at 9 am on the first day. That was the start of the first lecture. Doors were closed and those who were not inside the class room by 9 am were given half a day's casual leave. Next time it was a full day's casual leave and the third time it was leave without pay. Even in our time some probationers did feel that this was unnecessarily harsh and even silly, but today as a common man who goes to a government office with a grievance do we not feel annoyed when we invariably find employees coming late to work or not in their offices during office hours. The situation today is that we are installing biometric machines to monitor attendance and even then a majority of employees are reportedly not reaching office in time or not available when required.

The Director was also taking lectures himself and was fully involved in all other extra-curricular activities such as horse riding, PT, cultural events, and other functions. Today, as I see civil service officer's training institutes in various parts of the country, some are without Directors, others with Directors who do not stay on the campus and still others who take little interest in the training of officers. In fact, the Hindustan Times of March 16, 2015, carries a news item regarding transfers in J&K. Apparently, the Home Secretary has been transferred as Director General, J&K Institute of Management. The news item states that this is a much less important post. This is then the public perception of what ought rather to be considered as one of the most coveted and prized postings. What type of leadership and example setting can we get in such circumstances?

Another case is that of the Director of the civil services training institute in my home state. This post has been vacant for over eight months now. Can a headless institution provide any leadership or instill any values into the members of the civil service? Here, it is clearly the fault of the state government which obviously does not attach any importance to the institution, its work, its leader, and his role and probably could not care much about the quality of the civil service—an unfortunately sad state of affairs.

Corruption has emerged as one of the most pernicious threats to governance in India. There is a perception that it encompasses all spheres of governance. The bureaucracy, being under constant glare of the public, has, therefore, necessarily to rise much above this perception by not only being honest but, as stated earlier, appearing to be honest. Matters such as delay in investigation of criminal cases, delay in court cases, and political interference result in most corrupt officials being let of the hook. These matters have to be addressed on priority so that the guilty are taken to task, while the innocent are not harassed. This will boost the morale of the civil service and encourage its members to serve in accordance with the good values that they have hopefully imbibed during their formative years.

It is easy to slide down the slippery path of unethical behavior simply because it may seem advantageous at a particular point of time. However, the measure of a person's character is what he would do if he knew that he would never be found out. The malaise of corruption can only be attacked by examples being set at the highest level and by firm and swift action against the corrupt officials. I think the recent results of the assembly elections in Delhi reveal the exasperation of the people with the way in which the government was being run and on the conduct of public servants. These results necessitate action at two levels: first, swift action to punish corrupt and guilty public servants and second for the newly elected political leadership and all elected representatives to set the correct example to the civil service which, in turn, will percolate down the line. Obviously, there is a great sense of hope among the people and we should not lose this opportunity.

The XI Plan document correctly states that some of the values that the civil servant should be mandated to reflect and follow are objectivity, integrity, neutrality, dedication to public service, transparency, exemplary conduct, accessibility, and efficiency. No doubt, each value is necessary, but it is only the sum total of these values which will make a difference. I recall a conversation with the Chief Secretary of a state about 20 years ago. Reflecting on the quality of officers working with him, he said that he had three types of officers. Some were intelligent but with doubtful integrity, some others had integrity but no initiative and only a few had the quality of integrity and that of doing hard work. Some may question this somewhat simplified analysis, but in the overall it is not far from the truth today.

The basic objective of governance is to provide well-being, comfort, and happiness to the people. This can be done when those who are entrusted with the responsibility to govern do so with a sense of responsibility, in an environment of transparency, exhibiting a culture of accountability and as people with integrity.

For this to become somewhat achievable, we must start with a fair and honest recruitment process. Recruitment has to be based on merit and young persons who can be groomed through a rigorous training process and who should then be placed according to their academic qualifications, aptitude, and experience. A square peg never fits a round hole. We hear and read about corruption in recruiting institutions like Public Service Commissions, Staff Selection Commissions, and other such bodies. To start with, therefore, such bodies must only be manned by persons with impeccable personal integrity and, of course, outstanding professional expertise. There cannot, and should not, be any politics in the process of appointment of persons to such recruiting bodies.

The "boss" has not only to provide leadership, but has to encourage and set an example to his subordinates. Much too often we find that the "boss" is either too engrossed in his own work, or overwhelmed by pressures of work such that he finds no time to guide the juniors and subordinates. They therefore tend to drift along themselves. This has to be corrected. "The boss" must mentor his juniors.

Here is some free advice to the young officers. For their personal work they should themselves regularly visit government offices like the municipality, urban authority, electricity office, water works, transport office, the hospital/dispensary, bank, etc., and experience real-life situations. This should be incognito. Only then will they realize what difficulties a common man has to face. Hopefully then, they will respond differently to people.

Three periods of values

I have often looked at the period of the last 68 years, since Independence, as being divisible into three subperiods. The first is that of the 1950s and the 1960s. In this period, notwithstanding the holocaust of partition, there was euphoria due to independence, of idealism and of willingness to sacrifice. This pervaded the political class, as also the civil service. The country went through three wars—the Chinese aggression of 1962, and the two wars with Pakistan of 1965 and 1971. Each event made people of this country more conscious of their duties and certainly less demanding in terms of their rights. The nationalist spirit was high on the agenda, while the materialistic trend was far from the forefront. As a consequence, the values imbibed by the younger generation of that time were presumably of a higher order than those of today. The parents of that period having been witness to the freedom struggle were naturally made of "sterner stuff" and this must certainly have been passed on by them to the children of that period. The generation that grew up in this period during the 1950s and the 1960s carried along with them values of that relatively higher order into their work life. This helped them to perform their duties in a relatively better manner.

The second period is that of the 1970s and the 1980s. During this period the country witnessed the Emergency. Things changed, certainly not for the better. The civil service was not isolated, rather, it was to an extent affected by this jolt. Another event that comes to mind relates to the public "firing" of the then Foreign Secretary by the then Prime Minister. That Foreign Secretary resigned the same afternoon. Can the civil servant of today take such a decision or have we gone to the other extreme of being weak, spineless, and of lobbying for postretirement jobs? If the latter, what example are we setting for our juniors and are we not then compromising our own position and office in the "twilight" years of our service.

The third period is that of the 1990s and the first decade and a half of the 21st century. It started with the economic crisis, and saw the rise of caste considerations and also the emerging importance of regional parties. The liberalization on the economic front may have resulted in some people moving up the ladder, but the race for materialism has certainly resulted in people choosing to "cut corners" to move ahead in the material rat race. In the process, the attitude has changed from one of nationalism and idealism to one of materialism, individualism, selfishness, and to "hell with the rest." In such a scenario values are often ignored, forgotten, or conveniently lost sight off. Where, in fact, the civil service should have been standing firm on issues, it has rather tended to bend backward.

Let me cite two very simple examples from everyday life to show a deterioration of standards and values over the years. During the 1950s there was a respect for the authority of law. I recall that after dark, riding a bicycle on the streets of Delhi without a lamp would invite the street constable to stop you and deflate the cycle tires. Instant punishment, with no recourse to appeal. Today, we see constables at cross roads and yet people merely drive through red light traffic signals. Recently, I saw a man in Jaipur driving a two wheeler past a red light signal with his child on the pillion seat. There were three constables standing in one corner who ignored the happening. Trivial one might say but what is the lesson being passed on to the child—"I couldn't care a damn for the law." Is this the example parents should be setting to their children? I should think not, but this is the reality of today.

Yet another interesting and fairly recent incident needs to be related. A dear friend was driving his car through an important crossing having right of way due to the green light traffic signal. A young boy driving a two wheeler came from his right, ignoring the red light signal, and hit his car. When my friend got out and tried to reprimand him, the young boy retorted "just because you have the right of way due to the green light signal, does that mean that you will run over me?" What values are these youngsters growing up with? In the ultimate some of them will join the civil service, starting with these very values—a somewhat sad state of affairs.

Yet another recent incident, on April 1, 2015, while I was traveling by a bus from Delhi to Jaipur we stopped at mid-way for a 15-minute break. At the billing point of the food counter I found a young boy aged around 20 years suddenly come from my left, thrust out his hand with the money and order his meal. When I asked him to join the queue he retorted rudely asking, why was I being impatient. I responded that it was he who was being impatient and inconsiderate. During this brief dialogue the middle-aged lady standing behind me saw her opportunity, came forward, and ordered her meal. What upbringing does this reflect in these two persons, one generation apart? The boy obviously could not care a hang for systems and procedures and was only interested in having his own way. The lady, one generation older, was looking for an opportunity to cut corners. This is one of many examples that one comes across today. It indicates deterioration in standards, upbringing, and values. From among this group of people will emerge the members of the civil service. That being so, what values can we imagine will percolate into the civil service? Needless to say, we need people with better values than this in the civil service.

Value-based early years in civil service

Let me now go back in time and share two experiences while I was the Collector in the district, during the 1970s. Communal incidents are always sensitive and difficult to handle. They can prove embarrassing to the Government and create awkward situations for civil servants, particularly those working at the field level in the districts. Invariably, such incidents occur without advance notice and therefore require immediate, mature, and effective response. The first incident relates to the year 1972 when I was posted as the Collector in the Banswara district in Rajasthan. Around March 22, the superintendent of police (SP) informed me that since the previous day, a girl aged 4 years, belonging to the Nagar community, was reported missing. Immediately, there were rumors and rumblings converting the incident into a communal one. Not only was this possibly due to inherent mistrust between the two communities, but such feelings were probably aggravated due to the presence of two very senior leaders, of all India stature, of each religious community at that time at the district headquarters.

Additional police force was sent for and arrived from Udaipur early on 23rd morning. The situation was extremely tense with allegations against the police and rumors that the girl had been sacrificed. Large crowds had collected at various locations, but the police arrangements were adequate, yet we had to ensure that under all circumstances peace prevailed. Fortunately, around mid-day the body of the girl was found in a well that was near her house. Even then taking her body out of the well and having the post-mortem conducted was not an easy task, because thousands of people had collected and were still making allegations that she had been thrown into the well after being killed. The family and public refused the post-mortem being done by the Government doctor. We therefore arranged a doctor of the family's choice, one from the neighboring mission hospital and another well-known doctor. They were clearly and firmly told that the post-mortem should be fair and reflect the correct cause of death. Their report revealed that the death was clearly the result of drowning. Firmness and impartiality were probably responsible for maintaining the peace and resolving the issue. In the evening, I had separately called on each of the two religious leaders to apprise them formally of the outcome of the matter.

The second communal incident relates to the year 1974 when I was posted at Bhilwara, as Collector. Pur is a suburb of Bhilwara situated about 10 km away. The issue related to a particular location at which each of the two communities had their religious building. The dispute had been going on for many years and would create tension whenever a religious festival of either community occurred. In October 1974, Id-Ul-Fittar was on Friday, 18th and Dushera on Friday, 25th. The proximity of these two festivals naturally made things more difficult. While adequate security arrangements had been made for the Id festival, around 10 am on 18th I received a phone call from the Sub Divisional Magistrate (SDM) who was camping at Pur that there had been a clash between the two communities and that the mosque had been desecrated. The SP and I, along with more force, reached Pur by 11:00 am. The situation was very tense with groups of persons standing in the streets around the disputed site. We first distanced them from each other and then slowly got them to disperse. Once tempers calmed down, we imposed orders under section 144 CrPC around mid-day.

Returning to Bhilwara by around 1.45 pm, I informed the Chief Minister of the situation. Even at that time communication was so fast that by then a delegation of the Muslim community had already met him. Around 5 pm we returned to Pur where, by then, the situation was quiet, but tense. Some loud speakers had been started but we got them put off. A deputation of Hindus had met us at Pur. Twelve cases had been registered. The SDM, driver, 7 constables, and about 20 others were injured, of whom 3 were hospitalized. We visited them, having returned to Bhilwara by around 9.30 pm. For the next few days, a number of visits were made to Pur and many rounds of dialogue were held with different persons of various groups.

On the afternoon of October 21, at the request of some persons, the SP and I went to Pur to try and resolve the issue. After discussions with the local people, finally a decision was reached to dismantle both structures and make a vegetable market at the disputed site. Both communities were to be given new sites elsewhere to establish their respective religious buildings.

The next day 22nd at the district headquarters there were protests from students and public at large against the decision taken in the previous evening at Pur. We again went to Pur in the afternoon of 22nd along with some public persons from Bhilwara who had been protesting against the decision taken yesterday. Discussions were again held with the local residents who were firmly of the view that the decision taken on the previous day should stand. Immediately thereafter dismantling work started and went on through the night. Some students of neighboring Gangapur did stop buses, but the SDM was sent to resolve the issue.

On 23rd afternoon we returned to Pur and held a meeting with the local people. By then the work had been completed. We were happy that through repeated dialogue with both communities and different sections of the society, perseverance and firmness, an issue that was recurring repeatedly on year-to-year basis was finally resolved. The media also came forward with reports that a long-pending communal issue had been resolved through the efforts of the Collector, SP, and local administration.

An article by Julio Ribeiro, of the Indian Police Service (IPS), in the Indian Express of October 5, 2013, provides interesting reading. He observed that today's probationers possess better knowledge than we possibly had when we joined service, he in the fifties and we in the sixties. True, but he also states that what really differentiated us from today's entrants into the service were values we cherished and the greater accent on justice and integrity. He further rightly observes that the menace of corruption has affected all government services in our country and that the root cause is political corruption. These views and the thoughts that I have stated earlier indicate a level of decline in standards of the civil service over the years. The contrary is what is necessary and expected through proper values in the civil service today.

Another major reason for the decline in standards of public servants is that of a crisis of character and the following values. This has pervaded society as a whole, the political class, as also the civil service. The latter are very much part of that society and environment and, therefore, they are also prey to this fall in general standards.

How do we reverse the decline?

Is there a possibility of reversal of this trend? Not being a pessimist I would say that this is possible but it is a long and difficult road ahead. Common knowledge is that bringing down a structure takes little time, but re-building the structure always takes much more time. Acknowledging that over the last 68 years there has been a fall in the value system in society in general, and within the civil service in particular, where do we begin? The start has necessarily to be with the leadership. In a democracy the leadership is provided by the political elements. That being said, the other responsibility is that of the family and then from within the education system. Children spend 15 to 20 years in this system and a fair share of their waking time is spent under the care of the teachers. A heavy responsibility, therefore, rests on the shoulders of the teachers. Just like the politicians and the parents, the teachers must not only be good leaders but also set proper examples to their wards.

Let me now look for the silver lining in the context of the future by relying on recent examples, three from officers currently in the civil service and one of a 10-year-old child, reflecting hope for the future. The first example is an experience of an Indian Administrative Service (IAS) officer as Collector of a district in western Rajasthan some 10 years ago. It is common knowledge that in a democracy both the political executive and the permanent bureaucracy should work in tandem for a common cause. This Collector faced a powerful minister who was known for his strong arm politics and patronizing elements in the government to perpetuate crony culture. As Collector it was his duty, however, to ensure that the rule of law prevails and the loyalty of the personnel is wedded to the government in public interest. His value system had taught him not to tolerate indiscipline and wrong doings, come what may.

There was, however, a revenue functionary who was very close to the minister. He would care less and indulge in all kinds of activities undermining every single authority in the district. No one dared to discipline him but accepted humiliation at the hands of this junior official. There were very serious complaints against this official, but they were ignored. The Collector knew that any action taken against the functionary would evoke an immediate reaction from the minister. Quite obviously, he had to be doubly sure that once he took action it would stand the scrutiny of law and that he should not be put into an embarrassing situation or be compelled to withdraw such an action. Therefore, he drafted and issued a suspension order giving details of almost 10 serious charges.

As expected, the minister got furious and brought much pressure to revoke the suspension. He also tried to malign the Collector's reputation through all possible means and is apparently still doing the same whenever he gets a chance. Fortunately, the minister could not muster much support from the public. He tried to get relief from the courts, but there he was rebuked because of the details contained in the suspension order and, in fact, the delinquent official was ordered to cooperate in the enquiry. Indeed, in future years the Collector has lost some opportunities of better postings because the minister ensured that he did not move to such positions. Though this officer does feel that compromising and being pliant is slowly but steadily becoming the order of the day; nevertheless, he is proud that he did not budge and still holds such values close to his heart. That is, indeed, what is required from every civil servant.

The second example is the one taken from a news item in the *Rajasthan Patrika* newspaper of July 7, 2015. This news item states that a young officer of the IAS picked up a young girl, who lay bleeding on the road having been knocked down by a speeding motor-cyclist, and took her to the hospital. Other bystanders there just watched, each one asking the other to send for an ambulance or the police, to take the injured girl to the hospital. At the hospital the officer, not having initially disclosed his identity, was subjected to considerable questioning and humiliation, but finally the girl did get medical attention. This act of a good *Samaritan* does stem from good upbringing, good education and reflects an individual who has compassion for those in distress and certainly values to follow. I have checked the background of this officer. He hails from a rural area, has been educated in Government schools, and his parents were teachers, his father subsequently having moved to the police department and retired as an additional SP.

The third experience is that of a young officer from the Rajasthan Administrative Service. In 1992 when he was SDM in a district, the district was backward in comparison to other districts in the state. The population of scheduled castes was quite high. As is known, they are generally a very low-income group and are by and large landless. In the years from 1968 to 1970, considerable government land was allotted to these landless families, but the possession was not given to them as other powerful persons of the village had encroached on the said land and they were not willing to vacate the land. These allottees raised their demand for possession of land at all levels, but no one was willing to resolve the issue. The young SDM took up this as a challenge, surveyed the land, spoke to representatives of the political parties, and arranged for adequate police force. While efforts were on to give the allottees possession of the land, the powerful persons made many attempts to intimidate the allottees by resorting to firing in the neighboring fields. However, the young SDM persevered. Ultimately, all the land was vacated and the possession handed over to the allottees. It was ensured that the ensuing crop was sown by the allottees. A clear case of, where there is a will there is a way.

Another example is from Delhi. Recently, a mother and her two children (aged 10 and 5 years) were shopping and a particular toy, said to be in great demand, was bought for each child. Another lady, a stranger to this family, who happened to be shopping in the same area came to them and enquired where they had got that particular toy. She said that her son was shortly going to celebrate his fourth birthday and he had been requesting her for this toy also. Immediately the elder of the two children came forward and offered her toy to this lady to give to her son on his birthday. The lady was considerably surprised to find this young child so willing to part with her toy to a total stranger. The mother of this 10 year old was herself surprised that her daughter so willingly parted with this toy that she had been wanting for quite some time. Clearly, this 10-year-old child had already imbibed the value of sharing, of being ready to give rather than to receive. Obviously, such a quality could only have been imbibed through examples at home and at the school. The hope is that as the child grows and moves into the world of work, such values will not be lost on the way, rather more such values would be added to her personality.

These four examples, three of presently serving civil servants and one of a 10-year-old, do give reasons for hope in the future.

To conclude, in the context of values in the civil service there are many areas to address but I would mention only a few. Training, both induction and in service, has come to be severely neglected over the years. The age of entry to the civil service has also been increased from time to time and this has adversely affected the character of the service. On a recent visit to the Lal Bahadur Shastri National Academy of Administration at Mussoorie, I was told that the current average age of probationers was around 29 years, and some were as old as 35 years of age. Such persons generally would have set notions on many, if not most issues. Inculcating fresh ideas and values into their minds is obviously an uphill task.

Tenures are also far too brief in most cases, with the result that officials rarely have time, to make any impact. Posting and transfers today are the result of a high order of political interference. There is also normally no system to reward or punish. Once one enters the civil service one can safely move to the top of the ladder even without working, so why work at all? Punishment is rarely heard of and is far from swift. The investigation in cases of corruption is prolonged and the cumbersome legal procedures do not instill any confidence among the public that delinquent civil servants will be punished.

All these are aspects that may sound trivial, but are imperative to implement if this decline in values has to be arrested and reversed. Coupled with this is the dire need for correction among the political class. They have to lead by example.

The above would indicate that there has been a fall in standards and values in our country over the last 68 years and this has percolated from the society right down to the individual. The civil service has neither been protected nor excluded from this process. However, it cannot be denied that good values are necessary in members of the society and, so also, among members of the civil service. Obviously, this will require perseverance and a lot of effort but that is unavoidable if we wish to have good governance. I can only conclude with two lines of the hymn "Abide with me" which was one of Gandhiji's favorite hymns. These lines are "Change and decay in all around I see; O thou who changest not, abide with me."

30 Dilemma of industry–academia interface: Will the twain ever meet?

Aquil Busrai

The need

Indian economy is facing a paradox. On one hand, there is a rapid industrial growth that calls for skilled and experienced workforce. On the other hand, there is a challenge in quality of talent that is entering the talent pool. The momentum that the economy has gathered will keep it moving for few years but in long term, there is a danger of talent mismatch to fuel further growth or even maintain the current growth pattern.

The root cause of this dilemma lies in a lack of interface between the industry and academia. Neither of them has fully understood and addressed the issues and challenges facing the other stakeholder. Consequently, sizeable number of so-called "qualified" engineers or business managers are entering the employment market each year but in reality a large percentage of this talent pool is not ready for immediate deployment by industry. This leads to additional cost in retraining the talent, hired presumably with the requisite knowledge. Besides financial costs, it also questions the credibility of the academic institutions from where these talents are hired. This dichotomy affects the very growth pattern and confidence of investors and also creates a misconception that the academia world is isolated from realities of business.

Global experience

In my four-decades experience in Human Resources with blue chip organizations like Unilever in Kenya and India, Motorola in Asia Pacific, Shell International in Malaysia, and IBM in India, I have had the privilege of having a ringside view of the talent market. The story of gap between what is "produced" by academic institution and what is "needed" by the industry rings a similar bell in many Asian and African countries. Though in Europe and Northern Americas there is some evidence of connectivity and relevance between curricular designed by academia and its application to industry requirement.

The interest in this subject started more than two decades ago when I was involved in talent acquisition in large numbers, across various geographical locations spread over several countries. In some instances, the frustration of not getting the "right-fit" was detrimental to growth plans of the organization and stunted expansion plans. Honestly admitting, at that stage of my career, this frustration was selfish and reflected my discontentment only from my very own perspective of not getting the right number or quality of talent. I had missed the bigger picture of how this could be changed. Two decades was sufficiently long time for me to introspect and realize that lot could be contributed to increase the collaboration, not merely to improve the hiring intake but—more importantly—increase the quality and caliber of the talent pool in India for sustainable advantage.

Having had personal experience of hiring significant number of fresh talent from various institutions in India as well as academic institutions in Asia Pacific countries and also UK and USA, I have been fascinated by the positive impact a well-crafted curriculum has on the talent that is produced by the academic institutions. At the same time, I have been perplexed by the detrimental impact and the lost opportunity that an ill-designed or irrelevant curricular has on the quality of the talent that enters the employment market.

It serves no purpose to blame any of the two stakeholders—academia or the industry—for the above dilemma.

World of academia has rich knowledge and abounds in high-value concepts. Most of them are strongly committed to share this very strength with the student community they interact with. Industry needs basic talent with high-quality knowledge and is thirsty for conceptual perspective to many business issues. Yet, both are at seemingly cross-purpose.

Reasons for the disconnect

The predominant reasons for disconnect lies in the fact that there is lack of interface between the industry and academia. Most industry practitioners visit the academic campuses at what I call the "harvest time"—time to recruit. For many it is the only time they have had any dealings with that academic body. After going through the recruitment process, many are highly critical of the quality of the students they have interviewed and even the subject matters they are taught. Most of these executives who express these views have never thought of offering their opinions and expertise to the academia during the year, which would help in reshaping the curriculum. Nor have they offered to work with academicians to discuss the change the industry desires and how they will participate to facilitate in that change process. Pontification by these executives has further alienated the academia from industry.

Many in academia, in their own wisdom, have chosen to cloister themselves with the set idea of teaching what is predetermined in the curriculum, irrespective of its relevance to the needs of the industry. Many have very limited exposure to a real-life business scenario. Even those who have had industry exposure have not made efforts at self-renewal and kept abreast with the dynamics of change that is occurring at such a rapid pace in the industry, marketplace, government policy, and global economic balance.

Admittedly, the above observations are somewhat generalized and do not apply to many enlightened academicians and industry practitioners. There is also strong evidence wherein industry leaders have taken initiatives to mold academic institutions to deliver on what is relevant to industry. And there are also academicians who possess first-hand knowledge of the impact that their contribution and knowledge can have on industry. Having said this, it is critical to note that these observations are not merely casual commentaries. They are presented to highlight the purpose for which this research study was undertaken.

Some research results

In order to delve deeply into this fascinating world of contradiction, I undertook to do serious research—first of its kind to examine industry–academia interface—for my doctoral thesis. Four plus years of studying the dilemma from inside the system provided valuable insights and helped me to understand the reality with empirical evidence and data. This research study was an interesting journey, especially for a practicing professional manager like me who had secretly wished in younger days to join the ecstatic world of academia. The genesis of the research was to determine whether there exists sufficient and genuine collaboration between industry and academia. And whether such a collaborative exercise impacts the quality of the end product—the talent pool.

The framework of this research was to study industry of different size and varied product ranges across geographical locations. The academic institutions studied included management and technical stream and covered premier as well as smaller and upcoming institutions. The aim was to clearly determine the current status and gaps, if any, in the collaborative efforts.

Many organizations have claimed participation and involvement with academia. Similarly, many academic institutions have also claimed close working relationship with industry. Care was taken to ensure that the stated intent and practice was verified with empirical data and evidence of such collaboration. Special emphasis was laid on determining whether such collaborative efforts were at all relevant and whether it impacted the output.

Meeting various stakeholders face to face to get the right perspective was important part of the process. To accumulate live examples and demonstration of collaboration in practice was also important. Meeting student community and recruiters in large numbers provided that most essential litmus test whether what was claimed as collaborative efforts was in fact producing the right and desired results; and whether the fit between skill and knowledge required by industry was actually being provided by the academia.

Studying best-in-class practices of academia involvement in many organizations was a refreshing reassurance that the partnership has potential to be further strengthened. High level of awareness of what the industry needs, in minds of many academicians was also indicative of the sustainability of this partnership.

The entire exercise of systematically collecting information, data, and interviews with various levels of practitioners and thinkers and critical analysis of various initiatives has provided deep and real knowledge. This was further strengthened by a verification process. Researching best practices and understanding practical issues in executing desired initiatives was rich learning experience and provided that much needed grounding to make this research study both intellectual and practical.

All data and claims were verified to ensure authenticity of the study. The study was not restricted to merely determining the current state but also to understanding the aspirations and views of industry practitioners and academicians for going forward. Ideal situations that would benefit India Inc were kept in mind during interactions with the stakeholders so that this study transcends from being a mere academic exercise into something more evolving that could shape the future of industry–academia interface and collaboration.

Our research study confirmed that over 42 percent of respondents from academia felt that the sole purpose of industry–academia interface was to procure better placement for students. Nearly 31 percent felt that industry academia collaboration was sufficiently demonstrated with executives visiting campuses and addressing students. Thirty-nine percent of academician felt that summer internship was sufficient to get insights into industry practices. Less than 6 percent were excited about the prospect of taking sabbatical and working in industry.

Even on the question of relevance of curriculum to the current requirement in industry, we observed some serious disconnect. Two of the ten IT-related engineering schools that we studied were still teaching Cobol and Fortran. When questioned about the relevance of these subject matters in today's times, we were offered an explanation that "strong foundation" was a must for students. Those shaping the curriculum had missed a whole generation of IT evolution and thereby rendered their program defunct.

Reaction from industry woefully, was no different. Forty-seven percent of industry respondents felt that visiting campus and meeting students and faculty for an evening of interaction was adequate contribution to industry–academia interface. But 38 percent felt that agreeing to take summer interns was adequate contribution to build that collaboration. Less than 2 percent acknowledged that they had either skills or inclination to sit with academic experts and help redesign the curriculum framework. Just about 1 percent of industry respondents felt it would serve any useful purpose to take sabbatical and spend 3–4 months in an academic institution helping them with curriculum content and also delivery.

It was therefore not a big surprise to note that 41 percent of students interviewed felt that visit by industry executives to address student community or take sporadic classes was sufficient evidence of interface. Over 47 percent students in first-year MBA schools felt that industry sponsorship of events like Marketing Fair or an HR Conference was sufficient to conclude that collaboration was high between Industry and Academia.

Above evidence clearly indicates that Industry and Academia need to realize the importance of mutual interdependence and create platforms where both get an opportunity to interact and appreciate each other's potential to contribute.

Some positive experiences

Many organizations have committed themselves to participate in the process of actively engaging with the academia world as part of its corporate responsibility. They have set aside resources to facilitate growth and improvement of academic institutions. These resources extend beyond financial assistance and include providing research facility, access to corporate data, and opportunity for project work that provides real-time knowledge of the industry. Several organizations, in technical domain, have shared technical breakthrough with academicians and have also invited them to join in cocreating future course of action. Others have provided valuable access to practices and processes, which in turn has given insight to academicians on the efficacy of conceptual work they have researched. Providing assignments to students has been a single most significant contribution to create that essential connects with real-life world of the industry. Such assignments, be it of few months or even longer, have shaped the minds of students in realizing the gap between what is acquired from the academic domain and what is applicable in the industry space. Many academicians have done an outstanding job of guiding the students during this industry exposure phase to examine industry issues and correlate them with theory or concepts. In return, they have also increased their own knowledge of current industry practices and have embedded the same in their repertoire.

Notable academicians have redefined existing practices and the accepted wisdom by challenging its validity and relevance. This has often resulted in path-breaking change of direction in the way we work. Such seminal work has impacted the industry in very significant manner and has changed the course of action for the future. Industry has applauded and embraced such contributions, being fully aware that practical experience is no substitute for strong conceptual framework. Both must coexist if that magical breakthrough is desired.

Way forward

In order to recapitulate with ease, I have a simple acronym to offer—C A R E, where C stands for conceptual and theoretical knowledge; A for application of that conceptual framework; R for relationship of mutual respect and dependency that must be cultivated between Industry and Academia, and E for enduring or long-term collaboration between the Academia and Industry.

Academicians are strong in concept they should then partner with Industry leaders to examine Application angle of that conceptual knowledge. For example, Professor Govindarajan of Tuck Business School served as General Electric's first Chief Innovation Consultant and Professor in Residence from 2008 to 2010. While working at GE, Govindarajan co-authored a paper entitled, "How GE Is Disrupting Itself" with Chris Trimble and GE's CEO Jeffrey Immelt. "How GE Is Disrupting Itself," which introduced the idea of reverse innovation.

Industry leaders are naturally inclined more toward Application but they need to open their mind to Conceptual input that they can obtain from their academic partners. Relationship can be built and strengthened by Industry leaders taking time off from daily humdrum of shop floor and office and imbibing the atmosphere of campus. While in the campus, they should seek to acquire fresh and state-of-the-art knowledge and concept on issues they are facing or likely to face in industry. Duration of such an arrangement will be circumstances specific.

Similarly, academicians should opt for sabbatical and spend quality time at industry and the market place. This will give them valuable insights into realities of business and fluctuating business life cycle. Academicians from institutions like the IIMs took a break from their professorial responsibilities and ventured out to work in industry. Professors from IIMA are known to have worked for short or long periods of time inorganizations like the Bharat Earth Movers Limited, Infosys, Titan Industries, and the like. This is quite a known practice in B-schools abroad where close collaboration exists between academia and business. Professor Nirmalya Kumar, a Professor of Marketing at the London Business School, joining the Tata Group as a member of the Group Executive Council is another indicator of the interface.

Another pragmatic approach is through a process of a Task Force consisting of both Academicians and Industry practitioners working together on a common issue and finding a solution that is both conceptually strong and practical to implement. Power of such partnership is obvious. True relationship can be built if such task Force works jointly on live issues facing the industry. The HR Committee on Public Sector Banks appointed by the Ministry of Finance (popularly known as Khandelwal Committee) consisted of both practitioners and academicians associated with IIMA and IIT. The competency model developed jointly by National HRD Network, XLRI, and CII is a joint effort of Industry and Academia. The only unfortunate part that blocks institutionalization of such work is the tendency on the part of both the academic leaders and industry leaders to make them personalized and their reluctance to share with the rest of the community in their own institutions. Professional bodies however can play a significant role in having MOUs with academic institutions rather than individuals include a dissemination clause.

The government cannot be a mute spectator in addressing this dilemma arising out from gap in industry–academia interface. It would be essential for the Government to initiate forming a tripartite body of Academician, Industry practitioners, and Government officials connected with education to meet together every three or four years and review the relevance of the curriculum offered to validate its relevance to changing needs of the industry and also to keep in tune with the technological evolution. This will ensure that what is offered by the academia is in line with what is required by the industry.

Industry could play a significant role in this journey by encouraging few interested managers to pursue PhDs in subjects relevant to industry. Even a sabbatical to managers interested in exploring conceptual solution to business issues could be a breakthrough. Sponsoring such scholars will bring academia closer to industry and open up new vista of collaboration with the academic world. The work done at the Academy of Human Resources Development (AHRD) is noteworthy in this aspect. AHRD initiated a Doctoral program for practicing managers with an interesting design of their having to spend six weeks at a stretch to times to learn theory followed by writing papers and ending up with a Dissertation for Doctoral level fellowship. The program with the help of XLRI has graduated over 25 candidates in a few years time and most of them progressed well in their careers including taking up academic assignments.

In addition, small actions can go a long way to build the connect. There is increasing need to connect research with practice. Topic selection for PhDs based on identified needs of the industry or a particular corporation can help the process. PhD. on peripheral subjects does not help anyone, except the candidate to qualify for appointment or promotion. For example, the Doctoral work on Industrial Relations in Bank of Baroda by Dr Anil K. Khandelwal was facilitated by the bank itself as a priority area. Dr Khandelwal who later became its Chairman and Managing Director acknowledged that the Bank's commitment helped him to have access to higher echelon of management. Eventually, the research insights helped him to carve out a new paradigm in IR. As CEO he acknowledges to have initiated many reform processes. Mr H. N. Arora's work on leadership was facilitated by Escorts and led to the discovery of value-based HR interventions yielding greater results. The support provided by L&T, BEML, Madura Coats, etc., helped Mr Nagabrahmam to discover how innovations in HRD or MBO are determined more by departmental leadership than organizational leadership. The facilitation given by State Bank of India leads to many Doctoral dissertations in pioneering institutions like the IIMA. Organizations need to sponsor managers for PhD. programs and this itself will be useful in bridging the gap and create a pool of manager scholars which is so necessary for the ambitious India.

On the part of academic institutions, industry experience for faculty should be considered a plus in his career growth. The problem with our academic institutions, at least with some better ones like IIMs, is that most faculty members are interested in consultancy projects than in soiling their hands with the actual problems. Currently, there is no incentive either. It is interesting to note that in recent times some of the IIMs like the one at Ahmedabad have started inviting practitioners from industry to take up term appointments as Professors of Management Practice. This is a step in the right direction.

Will the twain ever meet?

All that has been described about the partnership and collaboration between academia and industry practitioners must have one incontrovertible objective—that of producing high-quality talent that will add value to industry. This will be achieved only when quality and relevance of education imparted produces talent that will be germane to the needs of the industry. To facilitate movement to this new orbit—industry will have to abandon its hallowed perch and identify itself with the issues and challenges faced by academia and proffering a platform where industry participation in shaping the curricula and its delivery becomes a shared responsibility between the industry and academia.

Such collaboration has to be consciously sustained over a long-term period failing which it will result in a mere ad hoc and sporadic activity, which is unlikely to generate any positive impact. Nor is it likely to produce any tangible outcome that will benefit either of the stakeholders. Consequently, such collaborative efforts will degenerate into a public relations type activity wherein pseudo-participation will only result in increasing the chasm between industry and academia.

Ground reality in India today is not all that dismal though. There is a growing consciousness—albeit driven by self-interest—to address the issue of matching the quality of education with what is required for immediate deployment in industry. Both academia and industry are aware that if this issue is left unattended then the talent pools will unquestionably shrink and India Inc will lose its competitive advantage in the international arena. To maintain its competitive edge, India will need to reconfigure the process by which it produces talent, which in turn will give a fillip to dominate the world economy through knowledge supremacy.

There is high potential to take this collaboration to the next level that not only will benefit the two sets of stakeholders but also contribute to nation building in an indirect and subtle manner. It was humbling experience to note the dedication and passion with which many academicians and industry practitioners are involved in this journey of collaboration and the enormous impact they are creating. But there is more to be done. There is need to spread this collaboration to a deeper and wider echelon, and to ensure that such research work is translated into an action plan for enhancing the collaboration between industry and academia in a structured, pragmatic, and sustainable manner.

The journey, therefore, has more begun at this stage rather than ended.

31 Social and organization leadership with whole system transformation in India

P. Vijayakumar, Mary Jane B. Balasi, and Roland L. Sullivan

The eminent, illustrious, intellectual giant and prolific author Udai Pareek, PhD, brought the concepts of organization development (OD), human resources, human resource development, organizational behavior, social psychology, applied behavioral science, and feedback first to Asia from his intensive labs at National Training Laboratories (NTL) and matriculation at major USA Social and Organizational Psychology University programs such as Columbia, Michigan, CASE, and UCLA (watch video of the "History of OD in India" by Udai Pareek at http://is.gd/uKbp2w).

Pareek was the first Asian to have received from the NTL the title of distinguished "Fellow." National Training Laboratories in the linage of Lewin is the birthplace of the human change movement that is now sweeping the globe.

Pareek first met one of the authors of this chapter (Sullivan) in the early 1960s. When the popular book, *Practicing Organization Development* (now in its fourth edition, 2015), was first conceived, Sullivan and Rothwell researched the world to determine the *guru* in each critical significant element of organization development. When it came to the section of individual development, without question, Udai Pareek was their first choice. Pareek's chapter for their book titled *Individual Development in OD: Human-Centric Interventions* became an important contribution to the world of change agents. Therefore, the current chapter is dedicated to the significant contribution Pareek has made to the growth and development of organizations.

Pareek spent his life preparing for the age of quickening that we are just beginning to enter. One cannot even imagine what India will look like in 50 years. Transformation is about to blossom at the speed of imagination as organization transform themselves to seize local and global extraordinary opportunities. Individuals and organizations that can transform will thrive; the rest will be left by the wayside.

To support such transformation we all are challenged to positively change the way change interventions occur. Why? Because it is common knowledge supported by W. Warner Burke, who is known for defining the field of OD and heading the Graduate Programs in Social-Organizational Psychology at Teachers College Columbia University, that change efforts are failing 60 to 70 percent of the time (from the personal email exchanges of R. Sullivan and Warner Burke).

Like many emerging economies, India is facing a complex, ambiguous, and rapidly changing business characterized by technological, economic, social, and political uncertainties. These forces pose unprecedented challenges to the organizations specifically in the context of their growth and development.

A few months ago, The Drucker Institute (Global Drucker Forum, 2014) held a global transformation conference with 2500 organizational leaders in Vienna. They concluded that the change work currently going on was not adequate. It is just too slow for the arising age of quickening. As they concluded the conference, the realization occurred to them that managers of the world at large do not have enough of a clue about how to actually lead transformation in a way that will increase positive business results.

Another source supporting the need for transformation comes from the latest IBM CEO research. Following is the summary of the findings of this research:

- We must create more open and collaborative cultures—encourage all employees to connect, learn from each other in order to thrive in the world of rapid change.
- This is now a continuous feedback kind of world, we need the organizational nimbleness and agility to respond.
- The time available to capture, interpret, and act on information is getting shorter and shorter.
- We need better information and insight, but what we need most is the capability to act on it.
- Leaders need to mobilize their collective brainpower for innovation.

In general, a professional who is specializing in OD or the members in the OD team handles and transformation-related assignments in organizations. In India specifically, we see a lot of diversity in the structure of OD teams either as a separate department or OD as a function integrated with other functions such as learning, talent, or strategy. Many organizations have different OD teams for different businesses, while certain organizations have a centralized OD team. Traditionally, OD has reported to HR. But it is time to realize that OD is systemic and it is about the whole organization.

OD initiatives in organizations have traditionally emphasized on parameters of efficiency and effectiveness. Now, we believe that practitioners also need to focus on the importance of basic humanistic values with a focus on strong leadership in democracy and social justice to create and sustain an engaged workplace.

There is an unprecedented dynamism in the workplace and these situations demand fast, effective, integrated, and sustainable change initiatives. Changing market dynamics, technology, and employee demographics are posing further challenges for organizations. In such contexts, change initiatives can be sustainable and effective only when change agents use a whole systematic and systemic perspective with a business imperative to have a thorough understanding of all aspects of the organization as they design interventions to cause immediate change (Rothwell, Stavros, & Sullivan, 2016, p. 3).

There are many instances in the Indian context to exemplify the fact that Organization Development and change initiatives have substantial focus on sustainment of the change journey and how this has contributed to the realization of strategic goals. These organizations focus on learning and holistic people development and they highlight the symbiotic relationship between employees and organization as the means to realize both individual and organization objectives.

The change management initiatives at the Hindustan Petroleum Corporation Limited (HPCL) (Vijayakumar, 2011) and the Vision Community Initiative at Zensar Technologies (Natarajan & Mathur, 2013) are cases in point. Both are typical cases of co-creation where the core tenet of the process lies in the understanding that every professional in these companies has a role to play in developing the strategic goals and the means to realize them (in HPCL) and innovative strategies (in Zensar Technologies). In both cases, employees who are in the forefront of operations or client-facing roles changed the way the companies address the challenges posed by the external environment. They developed very innovative systems and processes to realize the strategic goals. Thus, we must keep discovering new innovative methodologies as we attempt to help executives to leverage the wisdom from the entire system. It is imperative to engage all in disciplined execution.

Learning and current challenges

A recent study conducted at the Centre for Social and Organizational Leadership (C SOL) at Tata Institute of Social Sciences, Mumbai, among 40 OD professionals in India revealed the following aspects for making OD and change initiatives more effective and sustainable.

- OD professionals need to have an in-depth understanding of overall business, the ground realities, systems and processes, people challenges, as well as customer and stakeholder expectations. This is important to influence and manage employees at various levels.
- Uncertainty and ambiguity in many facets is a major challenge for organizations. OD professionals need to assess, analyze periodically, and make bold conjectures to conceptualize and operationalize challenges faced by them to reduce the volatility.

- OD professionals should innovatively leverage technology for communication with all stakeholders. New software such as that from *Co-vision.com* and *Skillrater.com* are examples of what is to mushroom in the next few years.
- Senior business leaders' commitment and involvement on the OD initiatives is very important. Leaders must walk the talk and model the way. The change must start at the top and even if it starts from the periphery, the top management should be cognizant of that.
- OD professionals should be capable of anticipating future needs of organizations. OD interventions should be designed also to meet future requirements to ensure that business and organizations are ready for the future for seamless on-going growth and development. OD practices, leadership development, and strategic decision-making have to be integrated with business realities and results. This is essential for the future of healthy OD (Cady & Dannemiller, 2005, p. 5).
- It is good to have an understanding of the best practices in other organizations, but it is very important to evolve unique next practices keeping the idiosyncrasy of the context of one's own system to stay ahead of the curve.
- Business people at various levels should be important partners in conceptualizing and implementing OD initiatives. The whole of the organization must be engaged and empowered.
- The change journey or intervention needs to be reviewed and renewed periodically. Sustainment is the new norm for OD.
- OD interventions should focus on business needs so that they can be addressed effectively to ensure measured business results.
- It is important to market the OD initiatives among the stake holders to ensure sustainability.
- Research should form the very basis of the design of OD interventions.
- The feasibility and suitability of interventions must be tested real time by the participants.
- Assess the outcome/impact of interventions whenever possible and conduct research that learnings can be generalized to be adopted in other organizations.
- It is essential to attract people with business and other expertise lie employees from line function to form collaborative teams to drive OD interventions.
- Expediate team-based wise judgement and decision-making when we are faced with information overload.
- Diverse markets have a variety of needs and expectations and these need to be understood periodically for business development and organization effectiveness.
- Individual/group learning and change is very important in the OD context.
- OD needs to be extra sensitive to the local culture and to integrate relevant philosophy and wisdom to develop theories in this area.

Social and organization leadership harmonizing with whole system transformation (WST)

Oh India! Oh India! The land of sages and saints. Wake up. Our time has come. Aspire to be leading country of the world in all aspects. India has the human and nonhuman resources to transform country to set the pace for the world. In order for that to happen, each of us must transform ourselves in the context of transforming the groups and organizations that we encounter and live in.

All of us must find our own radically different solution to be successful in the impending acceleration of world change. Many great solutions are emerging. What will be the solution that you play a meaningful part in? We share with you one solution that Pareek believed in, that is, whole system transformation (WST). He was working with us on his deathbed to purport WST to transform rural development in Rajasthan, India. Since then we have continued to test this evolving model intervention in numerous emerging economies. It works. Best of all, we know how to transfer it to others. It responds to the change process plights that we site above. It has the following focus on how:

- to cause and create positive change and not how to manage, control, and plan change.
- to help an organization become more real, true, and authentic. "We define authenticity as the ability to freely express and communicate experiences or said in another way, authenticity is the ability to say and share with others what we really think and feel" (Martinoff, 2013).
- artistically and economically WST helps us sail with speed through the turbulant "whitewater" that Peter Vail articulated (R. Sullivan's personal conversation with Peter Vail in his home, 2014).
- to align a new plan course created by an expensive outside consulting firm or a top management who cascades irrelevant direction.
- to intervene and catalyze complex reconfigurations, not how to plan and implement from the top (Worley & Mohrman, 2013).
- to transfer into the organization a traditional and key organization development competency of setting up the internal consultant to be the most powerful person of the entire organization. We believe the person who controls the process controls the business results.
- to bring back the true focus of OD to the organization rather than to coaching, teambuilding, fragmented interventions, human resource development, or training. The whole system becomes the focus. What is uppermost in the mind of the CEO in the executive team becomes the priority.

WST is a paradigm shift. It is the transformation of recent OD theory and practices and the newer most recent trust in change management. Transformation is where the caterpillar becomes the butterfly. Out of the essential DNA comes the blossoming of a beautiful and news enterprise that cannot revert to its old ways of being.

We need organizations that can adapt to the changing times. They do so when they shift and take a leap forward. The day of incremental change and managing the solution of problems with glorified quality circles from the 1980s with recent project management was a great step taking us to this point. Now we must become more agile in the, adaptable to meet the current state of reality, and prepare for even surprise change.

With WST, executives in the new agile organization have a process to cause positive organizational change. Our experience tells us that the successful executives want to make explicit system-wide decisions that promote adaptability over stability and flexibility over inertia. This occurs while reducing the inefficient bureaucracy existing in their Indian organizations that have us all in knots.

WST is a proven methodology that helps leaders to become stronger and shine. Followers are more eager to support and to be inspired to be up and doing what is priority for the system. Never before in the history of the world have we needed stronger leaders. To be a great leader today one must be connected to the entire system. A leader must listen to the system. The system must listen to the leader. Together a transformative organization emerges.

WST is based on the latest organizational social psychology. Leadership must utilize extraordinary small group dynamics and eliminate the autocratic, fear-based command and control culture. Transparency leads to self-realized solutions. The system becomes a fluid flow of united networked interdependence (Sullivan, Rothwell, Carter, & Balasi, 2013).

Leadership must utilize the unbelievable magic power of large group dynamics where 300 to 700 people dialogue over a time period of at least three days with their leadership as a focused energy is released that will help all aspire and accomplish business results never seen in the organization up to this point.

Transformed individual executives and a unified executive team together with engaging integration and relationship building with a critical mass create spontaneously the new identity and tackle the priority of the moment.

Such critical mass summits dialoging with their leadership require an organization to master engagement and empowerment with their people as they continue to learn how to learn more quickly.

The heart of WST is transformative summits. A key element is having one executive team member along with a design team that has representation from the entire organization to co-create the big system event. As the design team becomes a team themselves they come up with a script that they believe will be most significant conference in the history of the organization. Typically, design team members will say that working with WST is the most meaningful work in their career (Todd, Parker, & Sullivan, 2010).

We no longer have time to spend forming a consensus. Instead, the whole system must sense the common ground and initial direction and cause positive change immediately. Of course, the system becomes agile and has the ability to self-correct as soon as big data says correction is needed. The immediate clear focus for change in the moment unites all with a new sense of urgency and priority.

Figure 31.1. Whole system transformation model

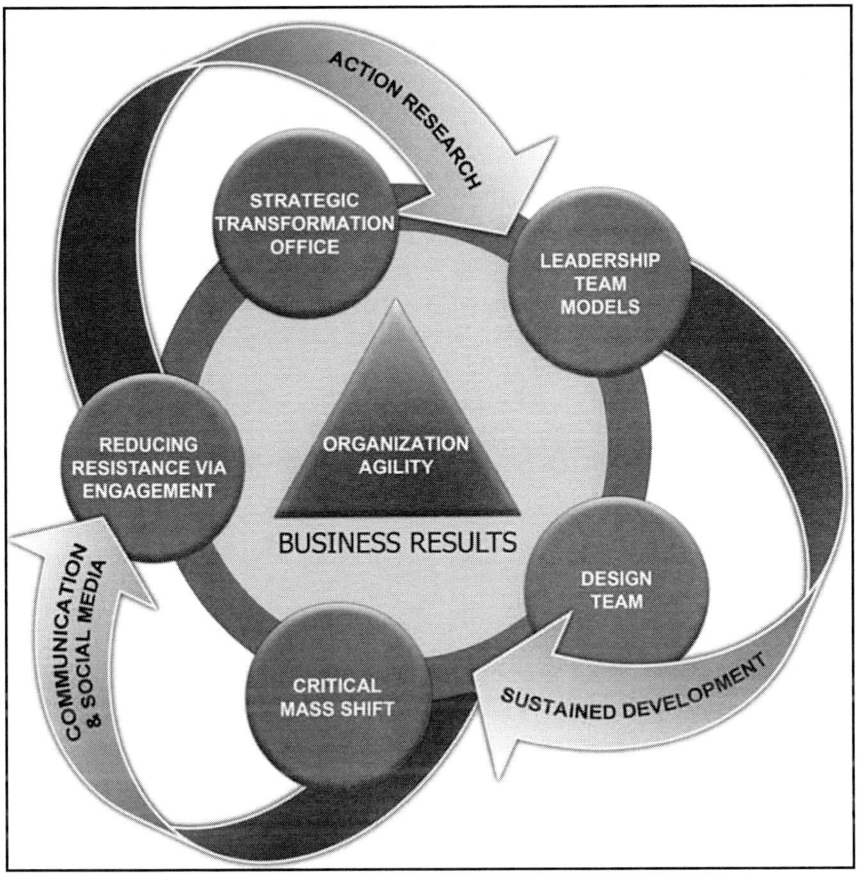

One of the reasons that change efforts fail is there is no disciplined execution. WST with its engaging truthful dialogue in a safe manner locks in people's brains and hearts to a committed and disciplined execution.

The WST process follows contextually relevant norms, which are relatively stable, and established patterns that guide leaders in their interaction. Interaction forms as a mediator among the client system and they react and contribute to collective activity (Weick, 1979). Interaction is aimed at aligning divergent interests of the community around strategic goals through negotiation and communication from all parties and facilitates the becoming nature of the activities related to change. The change agent or the leaders through interaction contribute to the creation of shared meanings among team members. See Figure 31.1.

In sum, WST is a complex journey made simple by best practices and theories developed and integrated into a single approach that enables and empowers human talent in organizations to accomplish faster, cheaper, and sustainable positive change. It marries leadership and the organization through the use of small and large social groups resulting in enterprise-wide transformation. It involves and engages the entire system, allowing ownership of the process (people support what they co-create), suggesting and causing change to happen in the moment, and focusing on the alignment rather than planned or segmented change.

Bibliography

Cady, S., & Dannemiller, K. (2005). Whole system transformation: The five truths of change. In W. Rothwell & J. R. Sullivan (Eds). A *practicing organization development: A guide for consultants.* San Francisco: Wiley.

Global Peter Drucker Forum. (2014). The great transformation: Managing our way to prosperity. 6th Global Drucker Forum, Vienna, November 13 and 14, 2014. Retrieved from http://www.druckerforum.org/abstract-drucker-forum-2014/ (accessed on April 10, 2015).

Martinoff, B. (2013). *Develop exponential power.* Manila, Philippines: F1C International.

Natarajan, G., & Mathur, R. (2013). Vision communities-employee involvement for sustainable innovation. *NHRD Network Journal*, 6(4), 25–29.

Pareek, U., & Rao, T. V. (1975). HRD system in Larsen & Toubro. Unpublished consultancy report .Ahmedabad, India. Indian Institute of Management.

Rothwell, W., Stavros, J., Sullivan, R. (2016). *Practicing organization development: Leading transformation and change* (4th ed.). New York: Wiley.

Sullivan, R., Gopalkrishna, S., & Rothwell, W. J. (2011). Organization

development: Now and beyond. In S. Ramnarayan & T. V. Rao, *Organization Development: Accelerating Learning and Transformation* (2nd ed.), New Delhi: SAGE Publications.

Sullivan, R., Rothwell, W., Carter, L., & Balasi, M. J. B. (2013). Whole systems transformation: an effectiveness paradigm shift for strategic change. In L. Carter & R. Sullivan (Eds), *Change champion's field guide: Strategies and tools for leading change in your organization* (2nd ed.). San Francisco, CA: Wiley.

Todd, J., Parker. J., & Sullivan. A. (2010). Whole system transformation: Becoming dramatically different. In W. Rothwell, J. Stavros, R. Sullivan. & A. Sullivan (Eds), *Practicing organization development: A Guide for consultants.* San Francisco: Wiley.

Vijayakumar, P. (2011). Development of operational capabilities: Exploring the role of micro-level organizational activities (Unpublished PhD Thesis). Tata Institute of social Sciences, Mumbai: India.

Weick, K. E. (1979). *The social psychology of organizing.* New York: Random House.

Worley, C., & Mohrman, S. A. (2013). Change management is obsolete: Learnings from research and practice about what's next. Retrieved from http://ceo.usc.edu/webinar/webinar_change_management.html (accessed on April 9, 2015).

32 Grassroots to global: Sanctuary for incubating innovations at/with/for/from grassroots

Anil K. Gupta

Spawning a large number of small enterprises provides the greatest guarantee for nurturing democracy. In this age of acquisition and mergers, corporations becoming larger, small enterprises will be a viable unit of analysis in future. Knowledge networks of small producers, entrepreneurs, designers, and marketeers could create as effective horizontal supply chains as some of the large corporations try to achieve. There is a possibility of building an ecosystem which permits large and small start-ups and enterprises to not just coexist but support mutually. A large tree will provide sustenance to a vast variety of small organisms, flowers, fungi, bacteria, birds, squirrels, and other beings without losing lightest advantage of its being a large tree. In fact, it can be a large tree only because so many small organisms nourish the soil, pollinate, and keep its bark clean and functional. Polycentric development of entrepreneurial urge requires availability of autonomous spaces for individuals to try, experiment, fail, and still persist. How has the interface between individual enterprise, society, markets, and state been managed historically? Dr Uday Pareek was a great humanist, spotter of talent, and always nurtured "nobodies" to help them become "somebody." He created a huge body of knowledge which was intended to make organizations more humane and compassionate. I am sure, if he saw the emerging tensions between large corporations and small innovative organization, start-ups, he would have advocated a synergistic, mutually convergent, and helpful arrangement for societal evolution.

During the industrial revolution, several trends emerged, not all of which became the legacy for those who caught up late. The general model was that of competition, ruthlessness, and allowing market forces fuller play. But there were exceptions. In Lyon, a town in France, famous for silk looms, there emerged an interesting model of networking. The demand for silk handloom fabrics even at that time was very fashion sensitive. The loom owners in Lyon knew that when demand for one particular design went up, the demand for others dipped. The owners had a very interesting custom when it came to train their children as apprentice. Generally, the owner would not keep his or her own ward as their own apprentice. They would request some other loom owners to train their children.[1] Thus, in the market place, they compete with each other, in knowledge or learning space, they cooperate with each other. Silicon Valley reinvented this logic. In south Italy, small firm networks also discovered a variant of this model. Whenever a small firm bid for a large order, knowing fully well that fulfilling that order would be beyond its individual capacity, it relied on social network. Once the order was obtained, the concerned firm would invite others to join hands. Till the order was completed, other partners almost became like a division of the principal firm. After that, they started competing with each other. The competition and cooperation can alternate but also take place simultaneously in different life spaces. Similar experience is noticed in Women readymade garment market in several parts of India. Different shopkeepers in a street famous for such clothes invite customers by hanging a variety of attractive suits outside or in front of the shop. They compete with each other in the lane. When a customer enters and shows interest in particular design but want a different shade or combination, they go to a neighboring shop which might have it, and bring it to satisfy the customer need. The inventory of all the shopkeepers in the mane now is pooled by each one to ensure that no customer having entered some shop goes back without being satisfied. They cooperate. We teach competition more than cooperation and leave the potential of different combinations under, or unexplored. One can have alternative, simultaneous, cyclical and episodic combinations of competition and cooperation. Future innovation based enterprises will need to evolve new vistas of cooperation in market place to humanize the interactions and make growth and development a positive-sum game rather than zero-sum game.

Today, when there is an extraordinary need to spur the entrepreneurial revolution around the developing world, we have to invent new models. This would link the grassroots ideas, innovations and institutions with global market players and other stakeholders in a way that globalization acquires a new perspective: Reverse Globalization will have to prevail. Poor will not be seen just as consumers of the products and services marketed by the large corporations. The poor will also be seen as providers of goods and services for which global market space will need to be created.

So far the globalization has meant generally squeezing of spaces for small innovators and entrepreneurs. It has been by and large a one-way street. The Honey Bee Network has been trying to reverse this process. It is trying to create a new ethics and institutional culture in which grassroots innovations developed by often uneducated or less educated or valorised to address global demands. Already a few innovative products have reached five continents. But, a great deal remains to be done. The successful entrepreneurs can mentor the start-ups

[1] Professor Paul Richards, personal communication, 1987.

whether in formal or informal sector. However, the mechanism of mentoring small, scattered and disconnected innovators without access to much education, banking or communication systems is not easy. Distributed mentoring is a challenge that we have to meet, if Grassroots to Global (G^2G) has to become an international reality. In other words, if the triangle of linking innovation, investment and enterprise has to be formed across the world, then transaction costs of each actor will have to be reduced considerably using online platforms. Assume that a European entrepreneur selects an innovation from India and wants to set up an enterprise in South Africa with investment from say, US, then a G^2G model would have come about. Likewise, if entrepreneurs in developed countries can find application for innovation by grassroots innovators in third world, then a poverty alleviation model will emerge which would look at *poor as provider* of solutions. The post capitalist society will essentially be a networked across time and space, loosely coupled, and poly-centric society.

Let me amplify this model: Honey Bee Network approach

From catharsis to creativity

In 1986, when I returned from Bangladesh after a yearlong stay, I began to realise a strange but a severe guilt in my heart. It appeared to me that while I was getting recognition and rewards for the work I was doing with the people, much of the benefits had not flown back to the people themselves. I had shared various research findings of my studies with the people, all this while but nothing much had happened beyond that. Much of my work was still in English language. I had finished two more studies on Impoverishment in Drought Prone Regions and Learning to Unlearn: a study on participatory learning in a tribal region from the banker's perspective. It was clear that the people who had shared their knowledge with me and had contributed to my own personal and professional growth had not gained much directly or indirectly. I tried to argue with myself that I had lobbied for policy changes, had shared my findings with the knowledge providers, and had tried to be authentic and faithful to what I had learnt. However, the explanations did not exempt me from the guilt.

It is around this time that I wrote a small Review on Ethical Dilemma and Value Conflicts in Management Research. It became clear to me that the dilemma may not be original, the solution had to be. Around 1988–89, one day the thought of Honey Bee came to my mind while coming from office for lunch at home. As they say, there was a light afterwards. Honey Bee does what we intellectuals seldom do. It connects flower to flower through pollination. And flowers don't complain when their nectar is taken away. Since we do not often write in local language, the people to people connectivity is not possible. Hence sharing whatever we learn from people back with them in simple language became imperative. Another challenge was that we extract people's knowledge, and innovations and often make them anonymous and become the author ourselves. Almost all of the ethnobiological work and much of the ethnographic work is illustrative of this attitude. Thus, the benefit sharing in a just and fair manner became another major goal of the

Honey Bee Network evolved to overcome these asymmetries in knowledge economy. Whatever we learn from people must be sourced to them. They should not become anonymous. We must share our knowledge with any third party only after their Prior Informed Consent (PIC). People should have a right to know what we did with their knowledge. They should also be able to learn from what we learn from other people through local language communication. If we get any consultancy, award or any other income through exchange or dissemination of the knowledge with or without value addition, a reasonable share should go back to the people.

A philosophy, a value, a social movement

Honey Bee Network began to evolve and bring in large number of people in its fold at village level as well as at professional or institutional level. For some peculiar reasons, majority of the members were drawn not from NGOs but from NGIs (Non-governmental Individuals). It is natural that when a network grows, contestation around the ideology, belief systems, values espoused as well as practiced, is bound to arise. Some people disassociate because the constraints of accountability imposed by the philosophy appear too much. Sometimes, the network also loosens its links with people with whom compatibility of values becomes difficult. But the major energy comes through inclusion rather than exclusion. Over the last 25 years, the Honey Bee newsletter has been brought in seven different languages. Currently there are six regional versions of the newsletter apart from the English version, which are managed by the core team of the Honey Bee collaborators in different states, namely – Tamil (*Num Vali Velanmai)*, Kannada (*Hittalagida*), Gujarati (*Loksarvani*), Hindi (*Sujhbujh*), Malayalam (*Inikarshankan Samsarikkatte*), Oriya (*Ama Akha Pakha*) and Telugu Honey Bee. There is also Chinese language version. The Honey Bee collaborators provide major support to the network in maintaining and operationalizing the informal horizontal relationships. People-to-people learning is also facilitated through *Shodh Yatra*s (learning walks). In addition, Tianjin University of Finance and Economics, China, brings out a Chinese language Honey Bee newsletter.

The Honey Bee philosophy operates through blending of the eight Es, Excellence, Equity, Environment, Efficiency, Ethics, Empathy, Entrepreneurship and Education. It is the ethical basis of the eight Es which has enabled the voluntary spirit of the network to build upon people's knowledge systems without impoverishing the knowledge holders for the last two and a half decades. The network has achieved success to a large extent in connecting individual and communities around the world through local language interfaces, protecting their IPRs and ensuring equitable sharing of benefits in as fair and transparent manner as possible. When NIF or SRISTI use public domain knowledge (that is not sourced from a particular community or individual), they ensure that benefits generated from value addition and commercialization of such knowledge based products are used only for strengthening the inclusive innovation ecosystem. The contested domains of knowledge, that is, private, community and public have to be harnessed in a viable, ethical and transparent manner to ensure that cooperation among contending stakeholders or participants in the innovation ecosystem continues.

NIF has mobilised about 200,000 innovations and traditional knowledge practices so far (not all unique) from over 500 districts in the last six years. Nothing at this scale has ever been achieved in the field of grassroots knowledge, innovations and practices anywhere. It has helped file 730 patents besides numerous plant variety protection cases including a few of which were filed by GIAN (www.gian.org) and SRISTI (www.sristi.org) in the United States. The US patents were filed with *pro bono* help of a Boston based law firm, viz., THT and now called as KLNG. Similarly, all the patents in India have been filed with the *pro bono* help of different IPR firms. The professional community dealing with intellectual property right protection has been very generous with its time. Likewise, the research community has been no less generous. Our average cost of filing patent is hardly ₹15,000 as against market rate of ₹2.5 lacs. Similarly, the average cost of validation and value addition trial is ₹2.5 to ₹3 lacs as against conventionally ₹10–15 lacs per trial. The ethical and social capital of voluntary contributions by various actors motivated by the values of Honey Bee Network has enormous implications for financial and institutional capital. Many colleagues are unable to fathom how a new social movement like Honey Bee Network does contribute towards institutional development and organizational activities.

To provide risk capital for talking innovations to market, Small Scale Industrial Development Bank of India (SIDBI) joined hands with NIF after the Finance Minister made the announcement in the parliament to set up a Micro Venture Innovation Fund of about USD 1 million in October 2003. There is a great deal of discussion on micro finance but when have we heard about the micro venture finance. The implication of this policy gap could be that policy planners and heads of major international financial institutions have no idea of the creative potential that exists at grassroots level. Or they believe, naively I suppose, that these innovations could be scaled up without the support of risk capital. The Micro Venture Innovation Fund (MVIF). More than 75 percent of investment made in 180 projects was repaid by the innovators despite signature, no collateral and practically no supervision. Trust and transparency still work.

More than 80 technologies have been licensed to small entrepreneurs with benefits going back to the innovators completely. What is remarkable about the licensing experience is that several technologies have been transferred without having received a patent on any them. It is obvious that diffusion of technologies must take place through social as well as commercial channels. The diffusion through farmers and artisans own networks have been most effective in past. However, the media exposure has also made significant difference.

NIF has organized three award functions so far and the fourth one would be organized this year. Hon'ble President of India has given the awards in the last three functions.[2] When the head of the state honours grassroots innovators and traditional knowledge holders, a statement is made. When Dr A. P. J. Abdul Kalam invited the awardees to the President's house and personally offered snacks to the elderly awardees, everybody was touched to the care. Never before had extremely common people pursuing uncommon distinctions imagined that they would receive such consideration from the President of one of the world's largest democracies. Subsequently, Dr Kalam gave Ignite awards every year to innovative children at a function organized by NIF at IIMA. He is no more but his spirit of selfless service will sustain the Network. Mrs Pratibha Patil started the tradition of organizing grassroots innovation exhibition at Rashtrapati Bhavan. Shri Pranab Mukherjee, The President of India, raised the whole activity to a new level when his office started hosting the Festival of Innovation (March 12–18, 2016). From grassroots, innovators scouted by the Network now have a dream to reach Rashtrapati Bhavan and not just that. Under a new initiative of Innovation scholar –in-residence program, they stay as the guest of The President of India for two weeks at the President of India's house. A message gores home, the country has begun to care for the creative people.

Dr R. A. Mashelkar, Chairperson, NIF has always believed that "I" in India should really stand for "Innovation." Moved by the impact of the movement, he realized that the rising aspirations of the people will convert into frustrations if value was not added. Accordingly, an MOU was signed between NIF and CSIR (Council of Scientific and Industrial Research) in June 2004 and renewed in 2009 to support research in four areas, viz., herbal, mechanical, food processing and nutraceuticals, and energy. Similar MOU has been signed with Indian Council of Medical Research (ICMR) to add value to herbal knowledge and Indian Council of Agricultural Research (ICAR) to add value to agricultural innovations by farmers. Out of more than 200,000 ideas, innovations, and traditional knowledge practices (not all unique), more than 130,000 deal with herbal applications for human, animal, and crop health. Unless value is added to the unique practices as such or after pooling the best ones, no benefits will accrue to the knowledge providers. Just to give an example of the international interest in commercializing grassroots innovations, in the last few years, NIF received hundreds of inquiries from over 60 countries. Many more enquiries have come from within the country. The idea is catching up. In due course, it is expected that a larger social acceptance will emerge so that minimum ethical protocol will be followed while dealing with people's knowledge systems.

Shodh Yatra

Every six months, we walk from village to village honoring local knowledge experts and grassroots innovators, young children who have extraordinary

[2] Professor Paul Richards, personal communication, 1987. findia.nic.in/scripts/sllatest1.jsp?id=969"http://presidentofindia.nic.in/scripts/sllatest1.jsp?id=969 profiles http://www.4award.nif.org.in/profile<http://www.4award.nif.org.in/profile>

sensitivity about the biodiversity-based knowledge systems, and other change agents. Having walked for about 5000 km, in 35 *Shodh Yatras*, every summer and every winter for last 18 years, a great deal of cross pollination has taken place. We have been inspired by the innovations we came across. The multimedia multilanguage database carried on laptops and projected for wider viewing wherever electricity is available generates excitement among the local community members also. During the 16th *Shodh Yatra* in Kerala, lot of people had gathered around in one of the roadside meetings. After half an hour, we stopped the presentation and asked people to share their insights into local creative people. There was a lull. People asked us to continue presentation and we would not. After a while, several people came forward and told us about fascinating innovations involving modification in steering arrangements for car to make it possible for handicapped people to drive, improvement in vanilla-processing technique to raise the level of vaniline extraction, development of cardamom variety, etc. We realized that raising expectations from ourselves and others seems to be one fundamental tenet of social and institutional change. The recent 35th *Shodh Yatra* in Tripura brought out unique insights into intergenerational knowledge transfer through written records with symptom-wise index. We had never come across such meticulous inventory of knowledge among tribal communities.

Having walked in Tamil Nadu, Kerala, Karnataka, Maharashtra, Gujarat, Rajasthan, Uttranchal, Himachal Pradesh, Orissa, Uttar Pradesh, besides of course, different parts of Gujarat, we have learnt a lot. For last 13 years, a course has been introduced in Indian Institute of Management Ahmedabad (IIMA) on the same subject encouraging students to learn from four teachers: teacher within, in nature, among peers, and the common people.

Cycles of creativity

There is an increasing use of cycle in everyday life in some of the European countries though in India it has remained more popular among poor people. There are lot of variations that are available in the market which have made cycling popular.

Kanak Das has designed a cycle (patent applied), which can harness the energy normally dissipated in the shock-absorbing springs for propulsion. Therefore, when we cycle on an uneven road or a mountainous track, the bumps work for the rider. Such a cycle can eventually be modified into E-bicycles as Kanak Das has tried with limited success.

Saidullah and Chaurasia have developed slightly different designs of amphibious cycles which at very low cost provide entertainment, means of transport, exercise, research support in water bodies, aquatic photography without much noise, etc.

Late Vikram Rathore had designed a cycle-based pump that works in shallow waters, Mansukhbhai has developed a cycle-based sprayer, Prem Singh has attached a cell phone charger, Kamrudhin has converted cycle into a small workshop with grinding, cutting, drill, four girls of class six then, namely, Riya Kothari, Nimran Kang, Kaamya Sharma, and Mehr S. Mehta from Delhi thought about attaching a broom to cycle to make cycl-o'-cleaner, etc., to add dignity to the life of sanitation worker and get rid of age-old practice of manual sweeping. All of these and many other innovations can have applications in developing world but in remote locations even in the developed world. The search for sustainable lifestyles will invariably situate cycle at the center of our energy dynamics. A washing machine cum exercising machine designed by Remya Jose can be a very functional substitute of exercising cycles.

The challenge is for designers to engage with such innovators scouted by Honey Bee Network members and supported by NIF, Society for Research and Initiatives for Sustainable Technologies and Institutions (SRISTI), Grassroots Innovation Augmentation Network (GIAN), etc. Same innovation can be adapted according to the user needs under different socio-cultural conditions.

Let me illustrate how users can find new applications of existing technologies in a creative and innovative manner and thus expand G^2G model.

Enhancing efficiency: Expanding applications

A professor from Boston wanted to do research on biodiversity on the top of tree canopies. She and her students had to make costly platforms for making observations in a quiet manner. When she looked at the tree climber designed by Appachan from Kerala at NIF's website (www.nifindia.org), she ordered four of such tree climbers at the cost of about USD 60–70 each. Now she can do research more efficiently and with greater options. Kevin Daivs used the same climber in Florida and found it very efficient for coconut harvesting. Mushtaq, a young innovator from Jammu and Kashmir, has developed a tree climber, which costs hardly USD 5–6. Indian advantage lies in sustainable, low-cost, and efficient technologies. These solutions are finding applications all over the world.

In most countries, cleaning the sea beaches is a problem for which efficient solutions may not exist. Dr Raman, an entrepreneur, saw a groundnut digger–separator at our website and contacted us to explore its licensing. After seeing field demonstration by Yusuf from Rajasthan through GIAN North, he was convinced that it could have another very interesting application. This was to clean the beaches. What this groundnut separator does is to scrape the soil, pick the pods left in the field after the harvest, stir the same on a sieve so that soil is dropped below and the pods are collected in the sieve. A new application for Indian as well as overseas markets was discovered by a user.

Involvement of users in developing technological modifications has been studied extensively in literature (von Hippel, 2005), but its applications in G^2G model are only beginning to appear.

There is a patent filed by SRISTI in the United States with the help of a pro bono IPR firm, KLNG (Kirkpatrick & Lockhart Nicholson Graham, LLP) on behalf of Mansukhbhai. This is for a multipurpose motorcycle attachment so that one can use motorcycle for ploughing, interculture, and other applications in vineyards, vegetable gardens, and flower gardens where heavy machinery may not be needed or even feasible. This patent (No 6854404) makes it possible for G^2G to happen if somebody can apply this technology in an anticipated or unanticipated context. For instance, we often find airport authorities using 25 to 30 horsepower tractors for transporting baggage from airport to the delivery conveyor belts. A four to five horsepower motorcycle would be sufficient for the purpose. Energy saving has to drive the diffusion of such innovations.

Prem Singh has developed a very low-cost cell phone-based switch for electrical household appliances as well as for pump sets in the field. Such switches exist in the Western countries but at much higher costs. The result is that one does not find many applications of cell phone as a switching device. With the switch costing less than USD 50, a farmer can switch on and off his pump sets whenever electrical supplies are received. It can also be useful for a busy executive who would like to have a hot water bath and/or a hot sandwich after reaching home. She can switch on these devices using her cell phone on the way to home. The smarthomes about which we have heard so much can become an affordable reality. If someone has forgotten to switch off a water tap or gas, such devices can help stop.

There are a large number of other innovations, which have applications all around the world. We have not exploited a new model, which looks at the fortune at the tip of the iceberg.

The popular model of Fortune at the Bottom of the Pyramid considers poor as consumers. It tries to find opportunities for the large corporations to sell things, even the ones the poor may not need, for instance, a one-rupee ice cream. But, the model that we talk about clearly recognizes that poor are not at the bottom of all pyramids. They may be at the bottom of economic pyramid but at the same time, they may be at the top of innovation, ethics, and values pyramid.

Unless we transform the models of thinking, we will not be able to develop new possibilities of G^2G.

What next

AASTIK, an Academy for Augmenting Sustainable Technological Inventions, innovations, and Traditional Knowledge, has been recently set up by SRISTI, to promote research by the knowledge holders and grassroots innovators themselves with or without the support of professionals. The idea has been conceptualized with the belief that research into creativity and innovation is not the prerogative of only the institutional scientists; it can also be undertaken by the innovators themselves. The academy aims to provide the virtual platform to the innovators, academicians, and scholars for presenting, harnessing, and upgrading their knowledge pertaining to innovation, creativity, heuristics of traditional knowledge, etc.

AASTIIK may eventually evolve into an international center of excellence for capacity building of scholars, local community leaders, and innovators themselves and their organizations from around the world. Grassroots innovators (with the help of research fellows and other faculty members) will be imparting training to policy makers, NGO leaders, and others interested in building similar networks.

The grassroots innovation movement in India is spreading rapidly. The National Register of Grassroots Innovations and Traditional Knowledge provides no legal protection as such though PIC does cover contractual protection. There is a need to modify the international IPR regime to make it more responsive toward the need of innovators and traditional knowledge holders in developing countries. Unless the transaction costs are reduced for people to disclose their knowledge, innovations, and practices, why should they share their unique insights and experiences? At the same time, many of the knowledge holders do share their innovative experiences because of their inherent generosity. Should those who share remain poor while others who are greedy become rich? How do we sustain the communitarian spirit without penalizing the local innovators and traditional knowledge holders for putting faith in our institutions? Most international attempts in this regard have fallen short of the ethical and institutional requirements of transparency and accountability toward the knowledge holders.

To restore the confidence, we have to consider several policy measures at international and national levels.

a. *Globally distributed network of mentors:* A website for inviting volunteers was designed, viz., *www.indiainnovates.com* with the help of IIMA students some years ago. Idea was to assist the grassroots innovators in developing their products, designing these better, and testing these for taking these not just to domestic markets but also in global markets. Already technologies have been commercialized in the United States, Singapore, and Pakistan. Grassroots to global seems a realizable dream: so far, all due to domestic resources.
b. *Global-GIAN:* Brazil and China have joined hands with SRISTI to scale up Indian experience. Malaysia and South African science and technology departments are also in touch with us to explore how can Honey Bee Network experience is replicated there. The purpose is to reduce transaction costs of entrepreneurs and potential risk capital investors from around the world to join hands with the innovators. Each one of them may not find the other in the current disjointed system. It should eventually evolve into a virtual Incubator for green grassroots innovations.
c. *International register of innovations and traditional knowledge:* Since 1993, SRISTI has campaigned for International Network for Sustainable Technological Applications and Registration (INSTAR), but such a registry is yet to evolve. People in one part of the world may thus learn from creative people in another part. Such a registry could also provide a low

transaction cost system to clear the demand and supply of innovations, investment, and entrepreneurial support. It can also be agreed under a new treaty likely to be negotiated at CBD/WIPO to provide a quick protection from bio-piracy.

d. *Building upon women's knowledge:* One of the major gaps in our activities has been the inadequate number of innovations and traditional knowledge contributions by women. There are several reasons, which may have caused this gap; one important reason could be lack of women field workers. About 1.2 million self-help groups of women are believed to exist in the country. Even if a small section of such women groups can be reached, and be linked with the innovation movement, it would be highly beneficial. In every monthly meeting of SHGs, the traditional knowledge of women about agriculture, food processing, weaning foods, health, and livestock may be documented. By pooling the best practices, new products can be developed which can be taken to market with the help of investment from Micro Venture Innovation Fund. This will help in three ways: (i) horizontal trading among the groups for products developed by them will lead to solidarity and social welfare, (ii) create market for such products in urban areas, additional income can be mobilized for the women, and (iii) new knowledge-based products may provide additional incentives for people to try other institutional approaches for adding value and generating incomes.

e. *Inventors association (innovator led regional incubator):* The horizontal networking among the innovators through their own associations is something very crucial to evolve a culture of cooperation. Later, these informal networks could be formalized in the form of an Inventors and Innovators Association to provide institutional support to grassroots innovators located in villages and urban areas all over the country.

f. *International/National Technological Innovation Acquisition Fund:* Sometimes, an innovator may not have sufficient resources to scale up his/her innovations or inventions (in private, public, or informal sector). Yet, some of these innovations may need to be diffused for a larger social cause. For instance, improvements in design of a kerosene stove which saves energy may be very vital for national interest, but the concerned innovator (as is the case with most of the innovators with NIF who have improved stove design) may have neither the incentive nor the capacity or both, to diffuse the design among a large number of small-scale manufacturers. Creation of a National/International Technological Innovation Acquisition fund may be helpful to acquire the licensing rights of such innovations and inventions for eventual out licensing these at low or no cost to small-scale manufacturers under technological up-gradation program. Ideally, the rationale behind having a technology acquisition fund is to compensate those knowledge providers whose knowledge has potential for economic value addition. Later, this knowledge pool could be governed by open source philosophy so long as people meet their livelihood needs.

g. *Reforming Intellectual Property Information System:* The oral knowledge, which is not documented in public databases or is not reasonably accessible, should be protected and not considered as prior art. Once this knowledge of any society is assumed to be prior art, no compensation may be due from anyone who uses it for commercial purposes. Every patent applicant in developing as well as developed countries should be required to declare that the knowledge and/or resources used in the application have been used *lawfully* as well as *rightfully* (through prior informed consent). Similarly, there should be *First to Invent* system for small innovators instead of First to File. How can small dispersed, often illiterate grassroots innovators be expected to compete with large corporation in reaching the patent office first? There are a large number of other reforms that need to be brought about to blend the advantage of GPL used in open source technologies with the licensing system followed in conventional intellectual property right system.
h. *Collaboration with schools:* A very large effort is needed for supporting curiosity, commitment, and collaboration among children. NIF has made some effort in this direction. Unless the spirit of creativity is unfolded at an early stage in life, it may become difficult for this trait to be developed at the adult stage.
i. *Linking technology students with problems and solutions of grassroots innovators:* There are at least 400,000 technology students in India. Each one of them does one project in a year. There is no means to ensure that these projects are linked with innovations and traditional knowledge or problems that need to be solved. Even if 1 percent of these projects are based on innovations that have already been scouted and documented, and aim at improving and calibrating them, this would be a great achievement. There is a need to have a web-based database that would have information on such projects. The portal should also have a segmented list of problems which needs urgent attention with particular focus on the problems of women and handicapped people. This would require provision of funds for running a portal and extending small support to the students managed by a committee of eminent scientists and technical experts.
j. *Mobile and stationery multimedia–multilanguage databases and exhibitions of innovations:* These will help immensely in overcoming three barriers that we have identified for learning at grassroots, that is literacy, language, and localism. Multilanguage multimedia kiosks set up as *Gyan Manthan Kendra* (knowledge churning center) will help in bridging the knowledge divide within village or nearby places.
k. *Dedicated labs for adding value to traditional knowledge:* Diffusion of technologies with high social impact suffers a great deal as there are hardly any specific laboratory facilities for large-scale demonstration of such technologies. There is a necessity to allocate separate space and fund for large-scale demonstrations, viz., herbal pesticides, herbal veterinary medicine, small sprayers, ploughing devices, cycle-based devices, etc.

A chain of dedicated labs devoted to value addition in biodiversity-based innovations and traditional knowledge are needed to fill an important gap. In the first 16 years, we could not get 16 experiments done on grassroots innovations in public or private sector labs. We were expected to pay the same cost that large corporations are expected to pay. Things have begun to change in the last two years. We may develop decentralized common pool fabrication laboratories and testing centers for faster validation. This will particularly help the local innovators and artisans so that their knowledge can be blended and upgraded with modern tools and techniques to produce competitive technologies.

1. *Innovation Insurgents:* The idea is that those young people who are angry and are willing to destroy the existing icons of power and authority have a reason to feel angry. We have to accept their frustration as genuine and legitimate. It is true that violent means very seldom can achieve, if ever, positive results. Therefore, the power to destroy has to be transformed into power to create. I submit that the examples of grassroots innovations and traditional knowledge that we have collected from over 400 districts in India provide a basis for attempting this transformation. It will be naïve on my part to suggest that these insurgents will overnight take over the function of innovation augmentation or knowledge brokers and managers to build value chain around their indigenous technologies. The only resource in which poor people are rich is their knowledge (and institutions and values). Unless we build upon richness of nature and richness of their knowledge, the people on the margin will not become major actors in transforming the grim situation in these regions. The struggle of local communities mainly in forest and other marginal regions has been around *jal, jungal, and jamin (water, forest, and land)*. The *jankari (knowledge)* has not been generally a basis for struggle. In the emerging knowledge economy, the knowledge of people is the most precious resource. However, the rights of the people in this knowledge are not yet properly defined, or for that matter even recognized. I am only suggesting that there is a need to listen to the angry youth of our society who fortunately is not willing to be patient indefinitely with the continued injustice and exploitation. Their restlessness is the biggest asset that this initiative intends to build upon. The role of financial, R&D, market, and design institutions is obvious in building value chain and providing back-up support to the *innovation insurgents.*

Ultimately, what is needed is a global movement in which providers of funds will consider their support as fees to be paid to learn from creativity and innovations at grassroots. Policy makers will have to be cajoled at international level to explain their continued indifference to the potential people have, to solve (even if suboptimally) long-pending problems; though sometimes we need to challenge the young and the old to solve the problems, which even grassroots innovators have ignored. The modification in the design of pulley for drawing water is one such example. When we posed the problem of the static design of pulley, which had remained frozen for almost 2,000 years, the innovators came out with many suggestions. Ultimately, Amrut *bhai* came out with an arrangement of a simple lever on the top of the pulley. It enabled women to leave rope whenever they were tired so that they could gasp for the breath. Once they had rested, they could start pulling water again. The bucket of water once pulled up would not slide down into the well as is the case with conventional pulleys unless one kept holding on to that. There are a large number of such problems that will have to be posed to the students as well as other innovators.

It has taken us a long time to reach this point. Others will have to take this movement forward. Building collective leadership is the greatest challenge since many individuals, devoted as they are and have been, have their own life goals to pursue. The network of innovators will have to take charge ultimately of this movement. *Amen.*

I am very clear that Honey Bee Network provides a viable framework for transforming the focus and force of globalization. A recent initiative of bringing community/individual grassroots innovations from China, India, and Brazil together is delivering some results. We hope that more and more young people, start-up companies, investors, and entrepreneurs will explore the untapped potential of G^2G. The poor people will no more be the *sink* of our advice, assistance, and help. Instead, they will become the *source* of solutions (Gupta, 2006).

References

Gupta, A. K. (2006). From sink to source the Honey Bee Network documents indigenous knowledge and innovations in India, innovations (pp. 49–66). Cambridge: MIT Press.

von Hippel, Eric A. (2005). *Democratizing innovation.* Cambridge, MA: MIT Press.

33 Action research in HRD

Vikas Rai Bhatnagar and Rajen Gupta

Introduction

The topic of this chapter is as distinctive and relevant for shaping the future of organization and people processes, as has been the relationship of the second author Professor Rajen Gupta with Dr Udai Pareek. In 1982, when Professor Rajen Gupta submitted doctoral thesis at Indian Institute of Management Ahmedabad, Dr Udai Pareek was the Chairperson of his Dissertation Advisory Committee. This was the first doctoral thesis done in India and on human resource development (HRD) processes that adopted action research methodology. Dr Udai Pareek supported Professor Rajen Gupta in adopting an emerging and controversial methodology for carrying out research in organizations, namely action research. It led Professor Rajen Gupta (1990) to propose a "Multilevel Action Research System" (MARS) for doing action research in large organizations. The first author, having led HRD function in multinational as well as Indian promoter-driven corporations, has used action research approach in his practice to bring positive changes in organizations, while also contributing to creation of knowledge. Deeply influenced by the outlook of Professor Rajen Gupta on building organizational and people capabilities, and mentored by him on action research methodology, the first author now has a firm of his own offering Human Resources and Organization Development services, driven by action research.

As action research influences positive outcomes in organization and contributes to knowledge creation, the audience for this chapter is both academicians and practitioners. In the words of Leonardo da Vinci, "Simplicity is the ultimate sophistication." One of the ways to find simplicity and sophistication is at a higher dimension or at a higher level of abstraction. To elucidate this point, a battlefield in two dimensions appears chaotic but the moment one ascends the third dimension of height, meaningful patterns emerge and deeper insights gleaned. Incorporating the fourth dimension of time or taking historical and futuristic perspectives, newer dimensions of phenomenon unfold. Similarly, at a higher level of abstraction of social, cultural, political, and economic fields, a pattern having similar influence over scholars as well as practitioners of HRD explicates. We will try to encapsulate the shifting paradigms that impinge upon both academicians and practitioners, elude to the implications for each while proposing the way forward for both academicians and practitioners of HRD. We will be drawing from the action research projects done by us in distilling some of the shifting patterns in the environment.

Worldviews and paradigms

Worldviews held by people shape their perceptions and guide actions. A prevailing thought or zeitgeist characterizing each era influences the development of paradigms in people, as well as their perceptions and actions. The spatio-temporal impact of an era varies, ranging from a few years to centuries. A CEO of a company may create a culture that influences perception and actions of employees probably over a period of few years, while philosophers and scientists like Aristotle, Descartes, Newton, and Einstein have influence over people that lasts for centuries. In this chapter, we will broach and explore few predominant worldviews, the shifts occurring therein and allude to the persistence of many in clinging to the rather outdated and defunct worldview, despite the wide commonsensical as well as research evidence to the contrary. These worldviews or as Thomas Kuhn (1970) mentions as paradigms, have similar implications for research as well as practitioners of HRD. It is our view that research into social sciences as well as practice of HRD is at a very crucial phase. It is experiencing "pangs of growth," wherein clinging on to the past methods of research, dominated by positivist science and empirical methods, while leading to plethora of publications, is neither contributing to creating improved practices nor providing a transformative experience for the researcher. On the other hand, thrust in HRD practice appears to being aware of "what's going around" and adopting "best practices" of other organizations, without understanding the unique business, cultural nuances, underlying assumptions, casual factors, and systemic reality of one's own organization. Such practice, apart from suboptimized outcomes, adds to creating complexity and stress in the organization. It is our experience that number of researchers as well as practitioners afflicted with meaninglessness and even alienation is on the rise. Something very fundamental seems to be amiss in our orientation to knowledge creation and its application for creating better futures. As we will discuss the changing paradigms and worldviews, we shall also unravel the objective and underlying assumptions in the previous statement pertaining to "creating better futures." While understanding the shifts in paradigm is important, actual shifts happen when someone perceives an alternate reality that has higher scientific validity, has the courage to state it—notwithstanding the initial resistance, opposition, and even ridicule—has acceptance with wider audience, who seeing merit of improved scientific explanation adopts the new worldviews or paradigms.

We will be discussing and questioning the validity of few paradigms currently in vogue that have influenced social sciences as well as HRD practice. We will draw attention to the limiting effects of these paradigms for social sciences as well as practice of HRD and discuss initial trends of shifts in these paradigms. We will then attempt integrating these paradigms into a holistic meta-paradigm and suggest how its application to social sciences and HRD has potential for providing transformative experience for a researcher and contribution in creating a better world for practitioners. We will be discussing following paradigms: Newtonian–Cartesian paradigm and the quantum-psycho-physical paradigm; neo-classical conceptualization of human being as an economic or rational entity and need for a holistic conceptualization of human being; inert nature of four-dimensional space-time field wherein events take place and organic nature of field derived from panpsychic philosophy. Figure 33.1 shows shifts in the paradigm.

Newtonian–Cartesian and quantum-psycho-physical worldviews

As per Kilmann (2001) dualistic separation of consciousness and matter characterize Newtonian–Cartesian worldview, while positing an objective reality that is independent of observers. Descartes claimed that truth can exist outside of humans, thereby justifying the dichotomy between the process of

Figure 33.1. Major paradigm shifts in social sciences impacting HRD

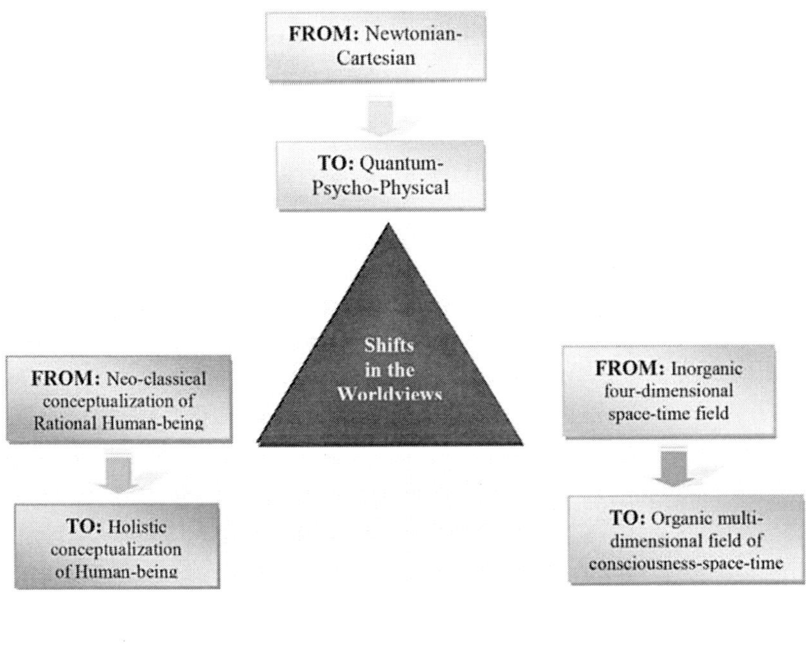

thought and object of thought. This worldview is deterministic, lending accurate predictions of future, based upon the knowledge of previous state of a system. It lends future predictable and hence controllable. If a system is controllable, clear cause–effect relations are established and reality and experience are treated as variables and hence liable for manipulation for soliciting a particular outcome. The predominant implication of this worldview upon science was the researcher being separate from the object of research and entailing a value-free process of discovering facts about the world. This method for knowing about the world was labeled as "scientific method," alluding to other methods of knowing about the world as "non-scientific" by implications. This worldview dominated over many disciplines, including social science and HRD practice. While the empirical method based upon positivist philosophy is quite successful in material sciences and largely in botanical–agriculture science in terms of increasing productivity, the ethical tone when applied to social sciences degenerates from empirical to empiricist. In this tradition, humans are "objects" of study and are researched *on*. In the field of management and organization behavior, the Newton–Cartesian worldview manifests in the scientific management of Frederick Taylor, McGregor's Theory X, and metaphors identified by Gareth Morgan for understanding organizations (Morgan, 1986) such as the metaphor of machine, political system, and instrument of domination. In terms of HRD processes, "prediction" and "control" influence design of majority of its processes. Manpower planning, recruitment training and development, and performance management systems are few examples of HRD processes, wherein the prime design principle is prediction and control.

The key feature of quantum-psycho-physical worldview is unification of consciousness with matter, wherein physical laws intertwine with consciousness and are subject to varied interpretations (Killman, 2001). The ultimate building block of universe is not matter but consciousness and conscious participation causes collapse of quantum waves into material particles (Zohar, 1990). . Quantum-psycho-physical worldview supports panpsychism. Uncertainty is an innate feature of quantum-psycho-physical worldview and probability is the tool to manage it. Nobel laureate Ilya Prigogine (1980). states, "Whatever we call reality, it is revealed to us only through an active construction in which we participate." The linear cause–effect equation of Newtonian–Cartesian paradigm gives way to a systemic interconnection of variables in a quantum-psycho-physical paradigm. Niels Bohr, one of the founding fathers of quantum physics states (Bohr, 1934), "Isolated material particles are abstractions, their properties being definable and observable only through interactions with other systems."

The challenge in designing HRD processes is factoring in elements of uncertainty and utilizing properties of consciousness such as free-will and intentionality of employees in shaping a better future for self and other stakeholders. No wonder companies such as Deloitte, Adobe, Microsoft, and recently Accenture are modifying the Performance Management System, designed on Newtonian–Cartesian paradigm of predictability, control and focused on "rear-view" to quantum-psycho-physical system that is more agile, adaptive, and futuristic. However, these organizations are merely adhering to the postulates of current theories, such as goal-setting theory of Locke and Latham (Locke & Latham, 1990). For instance, the goal-setting theory identifies "giving timely feedback" as one of the important moderators (other being addressing constraints, providing task clarity, and commitment) to effective performance. Certainly, an annual feedback on performance is not coherent with the postulates of goal-setting theory. When companies move from giving employees annual feedback to a feedback on an ongoing basis, based on the task, it corrects a faulty execution. Notwithstanding these corrections, what ails the current Performance Management System (PMS), dubbed as "dishonest annual ritual" by Armstrong (Armstrong & Baron, 1998), calls for questioning and addressing the basic assumptions—double loop learning (Argyris & Schön, 1974)—that grounds current PMS. The limitations of goal-setting theory and approach of Balanced Score Card would not be addressed without transcending from the neo-classical conceptualization of man as an economic or rational entity to taking a holistic conceptualization of human being as spiritual-social-psychological and physical entity and incorporating this in the design of PMS. We will share how in an action research project, we had successfully conceptualized and executed this (Robbins, Judge, & Vohra, 2013).

Interestingly, there are two aspects in the current design of PMS that revolts against the cultural and psychic make-up people from the East. Let us mention them here and you will see how we have gone about addressing these too in our action research. PMS focuses on the outcomes. Terminology such as Key Result Areas (KRAs), goals, objectives all orchestrate toward outcomes. In the oriental world, the virtues of doing one's *karmas* or actions, leaving the outcomes or results to the Supernatural force, ingrains deeply in the psyches of people. The other aspect relates to ego. Anyone who has been through a performance feedback session and has in humility used the term "we" would have been cut short by the reviewer insisting on "forget the 'we,' I am interested to know what 'YOU' have done." In the oriental worlds, taming the ego and attaining an egoless state is one of the cherished and aspired virtues. However, in the current design and execution of PMS, ego is inflamed during the performance review sessions. How does one reconcile with these two dominant design characteristics in PMS—focus on outcomes and bloating of ego—with the deep-rooted ethos of the oriental world?

After we have discussed the shifts in three paradigms and distilled a meta-paradigm feeding the three paradigms, we will share an action research on PMS that holds promise of addressing the current limitations in the system by grounding the design in quantum-psycho-physical paradigm and positive psychology. This action research was presented in the 12th International Conference of HR held at MDI in December 2012 and is also published as a case in the latest edition of book titled *Organizational Behavior* authored by

Stephen Robbins, Timothy Judge, and Niharika Vohra (2013). Let us now explore the shifts in second paradigm pertaining to conceptualization of man.

Conceptualization of human being

The first author, Professor Rajen Gupta, dwells upon the "Nature of Man" in his doctoral thesis. Human being is a complex entity possessing the ability of consciousness and self-reflection. Being a rational-emotional being, there are occasions when he acts non-rationally deluding development of neat sets of logical laws explaining or predicting behavior. Human beings are inclined to seek meaning (Frankl, 1997), purpose, values, and ideals (Ackoff & Emery, 1972). Human beings seek freedom to choose rationally, emotionally according to their purpose, values, and ideals. The very nature of human beings makes epistemology of positivist science and empirical methods insufficient to tackle the realities of organizations and employees. The second author, Vikas Rai Bhatnagar, in an action research related to leadership development (Bhatnagar, 2012) re-conceptualized human being as a Spiritual-Social-Psychological-Physical entity as compared with merely being rational or economic. The Newtonian–Cartesian paradigm manifests into empirical method of attaining and creating knowledge, while acknowledging and leveraging only the cognitive aspect of human nature. Any research based on this truncated conceptualization of human being, discounting the values, meaning, purpose, freedom, and intentionality of researcher, is as insufficient as HRD practice that accounts and engages only the cognitive abilities of employees.

While the limitation of the empirical method is divorce of values from sense perception, action research does not lay any claims to be value-neutral. Rather, being value-laden is one of the cherished aspects of action research methodology. One can easily see how action research, being value-laden and purposeful, is grounded in quantum-psycho-physical paradigm. We have earlier discussed PMS and will continue to give examples from our action research pertaining to it. Balanced Scorecard developed by Kaplan and Norton (Kaplan & Norton, 1996) primarily engages the cognitive aspects of human being, while the purpose remains maximizing financial gains. The causal linkage of developing capabilities to processes to customer satisfaction and finally achieving financial gains assumes fundamental purpose of optimizing financial gains. Morgan's metaphors of organizations as machines, instruments of domination, and political systems are suited to the predictable and control-oriented Balanced Scorecard. When the purpose is financial optimization, where are the checks and balances for the other aspects of scorecard when the Board and shareholders expect profit maximization? Chris Argyris, and Donald Schon were intrigued by the dichotomy between what people at times profess they do and what they actually do (Argyris & Schön,1974). This led to contributing the much-acclaimed theories of action, whereby people operate based upon mental maps they possess about any situation. However, words of explanation we use to convey what we do or make others believe why we do what we do may be at variance to actions actually performed. The former they term *theories-in-use* and the latter *espoused theory*. What is called for is, again to use the contribution of Chris Argyris, double-loop learning, wherein the basic assumptions are questioned. "Purpose" provides a pull to activities. If the purpose of business organizations is profits maximization, then the current design of PMS (continuing with the example we have chosen to develop our arguments) and Balanced Scorecard is valid. However, when we have human beings characterized with consciousness, self-reflection, free will, searching for meaning in life, their individual purpose may be at variance with that of organization's purpose of profit maximization. The challenge is to bring about convergence in the highest purpose of a human being and that of organization by utilization of latest and valid quantum-psycho-physical paradigm and holistic conceptualization of human being. Before we discuss how this was achieved in our action research, let us first also dwell upon briefly on the third paradigm shift pertaining to the field where events take place.

Field where events take place

Kurt Lewin, considered Father of Action Research, utilized and developed the concept of field in studying the phenomenon of changes in organization. Field is an invisible, yet real and powerful, force that exerts influence over a situation and people. What is the nature of this field where events take place? Is it a four-dimensional field of space–time as Einstein had us believe? If so, how do human beings—definitely occupying space–time—but not limited in their characteristics as the first author Professor Rajen Gupta espoused in his doctoral thesis and the second author Vikas Rai Bhatnagar in his action research published by Springer (Robbins, Judge, & Vohra, 2013) and discussed earlier in the "Conceptualization of Human Being" correspond to the lifeless space–time field? In the four-dimensional, inorganic conceptualization of field, Physicalism reigns supreme and no wonder all sciences tend to anchor to physics. Aspects that are unexplained by physics are considered fictitious and not taken seriously. Quantum-psycho-physical worldview lends primacy to consciousness and so does panpsychism philosophy. The second author establishes consciousness being fundamental to cosmology (Robbins, Judge, & Vohra, 2013) and conceptualizing an organic field that appropriately corresponds to human beings possessing entities of consciousness, mind, cognition, and emotions, apart from merely occupying physical space and being present at a particular time. This field is termed consciousness-space-time, making experiences of human being real in the research context as well as in the context of management practice. Such a conceptualization widens the scope of science from mere Physicalism to actually achieving the sociology of sciences by bringing esoteric experiences of consciousness and irrationality of emotions within the purview of sciences. Peter Russell (2003) and other panpsychists such as Deepak Chopra (2008), along with scholars of consciousness studies such as Stuart Hameroff and Roger Penrose (2011),accord primacy to consciousness as the building block of nature. It would be appropriate to have the field wherein events take place as mere consciousness, with space–time a manifestation of consciousness; however, for the time being we may go with terming the field as consciousness-space-time instead of only space–time. As this concept is quite subtle and appeals to paradigmatic shifts in mindsets of people, we will not belabor it further and move on to discussing meta-paradigm emerging from the three shifts in the worldviews enumerated earlier.

Meta-paradigm

A meta-paradigm combines few paradigms, distils their essence, and captures the emergent flavor as a new paradigm. Let us attempt this and then discuss its implications on research and practice, particularly action research and HRD. A quantum-psycho-physical paradigm brings consciousness as an essential ingredient for perception, with intentional choices and free-will choosing out of infinite possibilities. This worldview reconstructs the organic worldview held by Greek philosophers, particularly Pythagoreans, who held the world to be *kosmos* consisting of the physical realm or the cosmos, realm of life or the biosphere, of mind or noosphere and spirit or theosphere (Wilber, 1996). If human consciousness plays an important part in perceiving, selecting, and creating futures, can we continue with the truncated neo-classical model of man being a rational entity. The reconceptualized human being is termed spiritual-social-psychological-physical, with "rational" and "economic" man aspect included in the psychological dimension of human being. How is it possible for a reconceptualized human being to operate and have correspondence with a mere space–time physical field, devoid of consciousness? So the emerging meta-paradigm is conscious human beings, with free-will and intentionality, actively engages with the environment in perceiving, selecting, choosing, and creating future events. With evolution or shifts in consciousness of a human being, perception and the sequential train of activities of selecting and impacting also changes, creating continuous flux and turbulence, justifying the cliché "change is the only constant" or "change is the only permanent thing in the world." Figure 33.2 models the emergent meta-paradigm.

Our tools need to suit the purpose for being relevant. How relevant is value-free pursuit of social science, when consciousness, laden with potentiality and information, is an intricate part of reality. No wonder, plethora of publications churned out by scholar based on empirical methodology and grounded in positivist science have marginal readership and hardly contribute in making positive changes in the organization. Adopting outdated practices by

Figure 33.2. Emergent meta-paradigm shaping social sciences

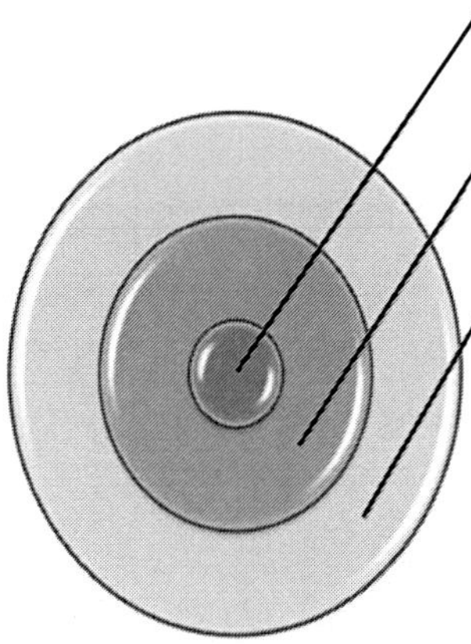

benchmarking, without considering relevance based upon nuances of industry and culture, ails the profession of HRD. Our experience informs us of higher acceptance, impact, and sustainability of an HRD initiative if it is an outcome of an action research project as compared with HRD professionals instituting processes based upon incorporating best practices. Action research is about creating positive changes in the organization while also creating knowledge. It is deeply humanistic in orientation and validates the deep-rooted conviction of Plato that consciousness being a cumulative phenomenon (Plato and Greeks believed in the transmigration of soul) human beings—even those who are illiterate and uneducated—know the answers and when posed series of questions, will eventually respond correctly. An action researcher partners with practitioners, genuinely seeks their views in diagnosing the issues, finding alternatives, digs into the existing theories and draws from them, co-evolves action plan, executes it, course corrects based on the impact, develops more research questions, and the cycle continues. An important aspect of action research is an ongoing meta-analysis of each process step, documenting reflections and finally sharing the knowledge created by having it published.

A useful method of action researching that we developed while carrying out an action research project on PMS in a cement company, is shared later. The technique is drawn from the Father of Action research, Kurt Lewin, who stated, "There is nothing more practical than a good theory." Fundamental assumption of this technique is good theory informs practice and if a practice is not delivering results, theory has to be enlarged, enriched, or rebuilt. Practitioners in multiple situations can apply this technique to improve processes. The steps of the technique are as follows:

1. Identify the gap in practice leading to suboptimized results or an aspect that needs improvement. The action research question and objective emerges from this gap.
2. Ground the existing practice to the theory, clearly identifying theories that have contributed in designing of current practice and that needs improvement.
3. Assess alignment of current practice with existing theories. Often there is a gap between the practice and theory that is corrected post-assessment of the fit between practice and theory.
4. Assess the efficacy of existing theory and enrich it by looking within the discipline and if needed across disciplines.
5. Redesigning the process based upon enriched theory. As an example, in our action research on PMS, this is precisely what we did and the outcome of it all was shift in three major paradigms as discussed in this chapter and coming out with a new practice of PMS that is attuned to the new paradigms identified. Further, once the theory was enriched, we applied it to other HR processes such as Induction and Employee Engagement.

The above technique is shown in Figure 33.3.

As seen in Figure 33.3, there is an ongoing romance between theory and practice in action research, with theory informing practice and practice enriching theory. The thrust in the above technique is enriching theory by carrying out cross-discipline literature review, if the literature in the existing discipline is insufficient to explain phenomenon and based on the enriched theory, redesigning practice that has desired outcomes. In the typical five steps of action research—diagnosis, planning action, taking actions, evaluating actions, and documenting knowledge created—the trust of the technique is in the diagnosis and action planning phase, wherein existing practice is traced back to theories and actions are planned based on enriched theory. This technique truly operationalizes words of Kurt Lewin and is worth reiterating, "There is nothing more practical than a good theory."

Operationalizing the paradigms in HRD

How do these paradigms play out in the context of organizations and HRD in particular? We will cite instances from our previous action research by operationalizing one paradigm shift of Spiritual-Social-Psychological-Physical (S^2P^2) conceptualization of human being. We will see how this paradigm shift is implemented in three processes of HRD, namely Employee Engagement, PMS, and Induction. In fact, employee engagement based on S^2P^2 conceptualization is an approach, while Induction and PMS are examples of processes wherein the new approach of employee engagement, based on S^2P^2 conceptualization of human being, is executed.

Figure 33.3. Process of improving practice by enriching theories

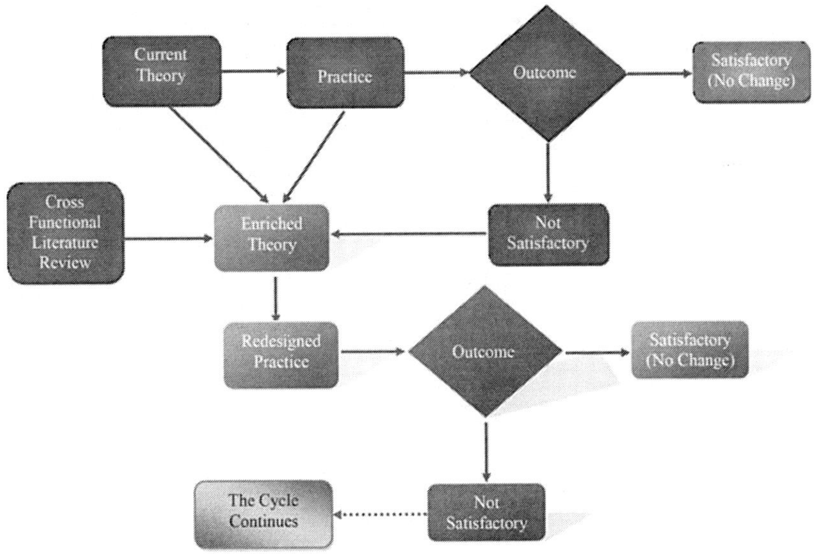

There are two models of employee engagement worth mentioning. One is put forward by Kahn (1990) and was tested empirically by (May, Gilson, & Harter, 2004). . Kahn found out that three psychological conditions are relevant for employees to feel engaged at work. These are meaningfulness, safety, and availability. May and his colleagues found predictors to each of the psychological factors given by Kahn, with job enrichment and role fit being predictors of meaningfulness; rewarding co-worker and supporting supervisor relations being predictors of safety; and resources available being a predictor of availability. The other model worth mentioning is given by Maslach, Schaufelli and Leiter (2001), characterizing engagement by energy, involvement, and efficacy. The approach second author has taken for engaging employees is designing HRD processes aimed at connecting employees at spiritual, social, cognitive, psychological, cultural, and physical levels. Let us discuss how this was done in the process of PMS and Induction.

The broad operational definition of each term is given for having common understanding on them. By spiritual process, we mean what meaning one derives and how intentionality is used for visualizing a future. By social process we mean building of mutual bonding and trust. By cognitive process, we mean the systems, processes, action plans, and resources optimization. By psychological process, we mean the awareness and deployment of individual strengths at work. By physical process, we mean possessing resources such as laptop, seating space, and telephone for performing activities as well as need for physical activity to remain physically fit.

Table 33.1 Deployment of S^2P^2 conceptualization of human being and approach to employee engagement in induction and PMS process

Serial No.	Entity/ Dimension	Induction	PMS
1.	Spiritual	Process for understanding what organization stands for. Making aware of the Vision, Values and deep connect with role.	Process has three parts: Visualization of a desired future (is different from planning), Inspiration and Internalization.
2.	Social	Process whereby relations between immediate supervisor and new joiner as well as other relevant stakeholders are built.	Process of creating goals by interlocking of expectations and based on the Value of helping each other to deliver on the agreed goals.
3.	Cognitive	Process to understand the systems, processes, structure, policies, etc., of the organization.	Process whereby visualized future in the Spiritual process is translated into specific and actionable plans for achieving.
4.	Psychological	Process for developing a deep connect of self with organization and work	Process whereby the strengths of an employee are leveraged while setting of goals and giving feedback
5.	Physical	Resources to work and to remain fit	Process wherein the focus is on making available resources for acting on the action plan

The utilization of S^2P^2 conceptualization of human being for engaging employees and its deployment in the process of Induction and PMS is shown in Table 33.1. Due to constraint of space, only the broad concept is shared and details are omitted in this chapter.

We are happy to mention that deployment of these processes fetched the organization where the second author led the Human Resources function, 10 International Awards, including the prestigious "Dream Company to Work For" award and "HR Research of the Year" award in 2013 by the World HRD Congress.

Superiority of action research projects over consulting

Mark Twain had cautioned, "All generalizations are false, including this one." Yet categorization and generalizations assist in sense making of phenomenon. While we are generalizing and being somewhat critical of consulting of a particular type only, which is devoid of scholarship, we also present an alternative that holds great promise for creating sustainable changes in organizations and people practices. We have earlier discussed Chris Argyris espoused theories and theories in use. Whatever be the espoused theories, the theories in use of consultants (with the exception of *scholar-consultants*), validate what Abraham Maslow alluded to—"I suppose it is tempting, if the only tool you have is a hammer, to treat everything as if it were a nail." Sustaining high costs of offices in posh localities and high employee costs places pressure on consultants for generation of revenue. Making money, which is a play of cash generation, return on assets (margins and velocity), and growth (Charan, 2001) makes a consultant use and reuse existing knowledge in the shortest possible time. Further, the workload does not leave enough time or energies with a consultant to read the latest literature and reflect upon the application of concepts to various situations. A "quick fix" syndrome characterizes gaining knowledge as well as using it by majority in consulting. TED talks and online short videos that generate awareness of concepts and familiarity with jargons become one of the prime means of updating one's knowledge. Since the practitioners too (unless one is a scholar-practitioner) are pressed for time, consultants are able to have the market primarily by playing on the brand and having ready availability of "hammers"—mentioning the analogy used by Abraham Maslow. In quite a few family-driven organizations, hired consultants submit reports endorsing the views of few interest groups, with many pseudo and unethical consultants obsequiously complying for revenue and growth considerations. Scientific truth, validity, and impact are not the governing values for consultants. Seldom capability builds within the system due to a consulting intervention. Statistics on the percentage of implementation on the reports submitted by consultants would be revealing for its non-implementation.

In contrast, let us consider the case of an action researcher. It is value based and purposive in orientation, aimed at making positive changes and improving client situation. Involvement of practitioners as coresearchers is an essential part of action research. The internal capability building commences right at the very beginning of the project. Specific and customized knowledge solves the issues of the client. Efficacy and impact is ascertaining by reviewing actions. Course correction takes place aimed at creating positive changes in organizations. Theory informs actions and actions guide in building theory. There is a method of practitioners themselves carrying action research (Coghlan & Brannick, 2001); however, the time available with practitioners and possibly limited inclination or aptitude to indulge in scholarship makes an external action researcher viable and effective option to consulting.

Conclusion

In this chapter, we have explored the philosophical basis and argued for action research as a superior methodology for approaching organizational science and HRD. Vision, values, teleology, or purpose shapes futures. Action research is value based, is purposive, is steeped in humanistic values, and is humble to intervene only to the extent human vision provides visibility, and once a specific change happens, new research questions that evolve are taken up for further bringing about positive changes in organizations. HRD profession would certainly jump a few orbits in building functional credibility and impact if approach toward designing and instituting HRD initiatives is based on action research as compared with merely adopting good practices of other organizations.

References

Argyris, C., & Schön, D. (1974). *Theory in practice: Increasing professional effectiveness*. San Francisco: Jossey-Bass.

Armstrong, M., & Baron, A. (1998). *Performance management: The New Realities*. London: Institute of Personnel and Development.

Ackoff, R. L., & Emery F. E. (1972). *On purposeful systems*. London: Tavistock.

Bhatnagar, V. R. (2012). Blending Greek philosophy and oriental law of action: Towards a consiousness propelled leadership framework. In G. P. Prastacos, F. Wang, & K. E. Soderquist (Eds), *Leadership through the classics-learning management and leadership from Ancient East and West Philosophy* (pp. 161–180). Berlin: Springer.

Bohr, N. (1934). *Atomic physics and the description of the nature*. London: Cambridge University Press.

Charan, R. (2001). *What the CEO wants you to know*. New York: Crown Business.

Chopra, D. (2008). *The seven spiritual laws of success*. New Delhi : Hay House.

Coghlan, D., & Brannick, T. (2001). *Doing action research in your own organization*. London: SAGE Publications.

Frankl, V. E. (1997). *Man's search for meaning*. Washington: Washington Square Press.

Gupta, Rajen (1990). *Implementing human resources development: Action research into the process*. Jaipur: Rawat Publications.

Kahn, W. A. (1990). Psychological conditions of personal engagement and disengagement at work. *Academy of Management Journal*, *33*(4), 692–724.

Kaplan, R. S., & Norton, D. P. (1996). *Balanced sorecard*. Boston: Harvard Business Press.

Kilmann, R. H. (2001). *Quantum organizations: A new paradigm for achieving organizational success and personal meaning*. Palo Alto: Davies-Black.

Kuhn, T. S. (1970). *The structure of scientific revolution*. Chicago: University of Chicago Press.

Locke, E. A., & Latham, G. P. (1990). *A theory of goal setting and task performance*. Englewood Cliffs: Prentice Hall.

Maslach, C., Schaufelli, W. B., & Leiter, M. P. (2001). Job burnout. *Annual Review of Psychology*, *52*, 397–422.

May, D. R., Gilson, R. L., & Harter, L. M. (2004). The psychological conditions of meaningfulness, safety and availability and the engagement of human spirit at work . *Journal of Occupational and Organizational Psychology*, *77*(1), 11–37.

Morgan, G. (1986). *Images of organization*. Newbury Park: SAGE Publications.

Prigogine, I. (1980). *From being to becoming*. New York: W.H. Freeman & Co.

Penrose, R., & Hameroff, S. (2011). Consciousness in the universe: Neuroscience, quantum space-time geometry and Orch OR Theory. *Journal of Cosmology*, *14, 1–14*.

Robbins, S. P., Judge, T. A., & Vohra, N. (2013). *Organizational behavior*. Delhi: Pearson.

Russell, P. (2003). *From science to God*. Novato, CA: New World Library.

Wilber, K. (1996). *A brief history of everything*. Boston: Shambala.

Zohar, D. (1990). *The quantum self: Human nature and consciousness defined by New Physics*. New York: Harper.

About the Editors and Contributors

Editors

T. V. Rao is currently the Chairman of T. V. Rao Learning Systems Pvt. Ltd, Ahmedabad. He was a Professor at Indian Institute of Management Ahmedabad (IIMA) between 1973 and 1994 and, subsequently, Adjunct/Visiting Professor until 2014. He is the Founder President of the National HRD Network (NHRDN), and the First Honorary Director of the Academy of HRD, India. He was also the President of the Indian Society for Applied Behavioral Science (ISABS). Dr Rao has worked as an HRD Advisor to the Reserve Bank of India, has assisted the Administrative Reforms Commission in reviewing the personnel management practices for civil services, and has also served as a member of the HRM Review Committee of Nationalised Banks set up by the Ministry of Finance. Dr Rao has worked as a short-term consultant to UNESCO, Ministry of Health, Indonesia; National Entrepreneurial Development Association, Malaysia; and the Commonwealth Secretariat, London. Dr Rao has over 50 books to his credit, of which 15 are authored jointly with Dr Udai Pareek.

Anil K. Khandelwal is a former Chairman and Managing Director of Bank of Baroda and Dena Bank. For transforming Bank of Baroda into a valuable brand, he received the prestigious The Asian Banker, Singapore Leadership Achievement award for lifetime achievement and contribution to excellence in the financial services industry, the only Indian banker to receive this so far. He has been rated among the 100 most powerful CEOs for three consecutive years (2005–2008). He is a recipient of Life Time Achievement award from NHRDN. He has been a UNDP Consultant to Banking Reforms Commission, Government of Tanzania, and a Visiting Professor to Asian Institute of Management, Manila. The government appointed him to head a committee to study HR in public sector banks. The report, popularly known as Khandelwal committee, has led HR reforms in public sector banks. His book *Dare to Lead* (SAGE, 2011) has been a best seller. He is currently a Corporate Advisor, Board Member, and Consultant on governance, leadership, and mentors CEOs. He has been an HR advisor to international consulting firms such as Accenture, Booz and Company, and Hewitt.

Contributors

Arvind Agrawal has been the President, Corporate Development and HR at the RPG group until recently and continues to advise the group. He is a PhD from IIT Bombay. His dissertation topic was "Examining the impact of strategic leadership on effectiveness of business organizations." He is an IIT Kharagpur and IIMA alumnus. He has majored in marketing and human resource development. The first 12 years of his working life were spent in the HR function in companies such as Escorts and Modi Xerox. Thereafter, he assumed additional responsibilities for total quality management (TQM) in Modi Xerox, a move which brought him closer to the business environment and issues. The exposure gained in TQM equipped him to take on the role of Corporate Strategy and Marketing at Modi Xerox itself. During the years 1994–1999, he was the Chief Executive at Escorts responsible for the two construction equipment business companies, that is, Escorts JCB and Escorts Construction Equipment Limited. Since 1999, until mid-2015, he was President–Corporate Development and Group HR in RPG Group. He has been active in management and HR forums in the country. In 1992, he was awarded the National HRD Award. He served as the National President of NHRDN during the years 2000–2002.

Abad Ahmad has been involved in initiating and managing change in many organizations from mid-1970s, including BHEL and many others, and founding the South Campus of the University of Delhi. His well-known books are: *Management and Organization Development*, *Developing Effective Organizations: The Indian Experience*, and *Designing and Developing Organizations for Tomorrow*. His latest book, based on an intensive empirical research on outstanding Indian organizations, is entitled *Passion to Win: How Winning Companies Develop and Sustain Competitive Edge?* and has influenced the management thinking for excellence a lot. He has helped a number of leading organizations in public and private sector with top management in vision building, team-building, and senior-level executive development programs with HR and applied behavioural science orientation. Some of these organizations were Hughes Software, Diamond Cements, Eicher, Modi Chemicals, ONGC, NTPC, NFL, Power Finance Corporation, Bharat Pumps & Compressors, IFFCO, TERI, etc. In 2001, when he was still at the University of Delhi, His Highness the Aga Khan invited him to be the honorary Chairman of National Committee of Aga Khan Foundation (India).

After his retirement as Pro-Vice Chancellor and Professor from the University of Delhi, Abad continues to be actively associated with governance and advisory roles with several professional associations such as AIMS, AMDISA, and SAQS; management institutions such as IIML and IMT; and a few consultancy organizations.

Varun Arya is a first-class graduate from IIT Delhi (1976–1981 batch) and a post-graduate from IIMA (1981–1983 batch). Thereafter, he worked for 16 years at senior positions with the leading companies such as Indian Rayon, Reliance, and American multinational DuPont. After having established and successfully run Aravali Institute of Management for 13 years, for the last 3 years he has been associated with his dream project of establishing and shaping up a world-class educational complex, Aravali Gurukul Ashram, spread over around 100 acres of saline wasteland transformed into green campus. It is planned to have a group of institutions in diverse areas of education including a university, envisioned to be a model of no compromise blending the best of traditions with the latest in modernity at Jodhpur in his home state of Rajasthan. Because of his firm belief that education, by definition, is based on ethics, his journey as an educational entrepreneur has been filled with unending hardships, risks, sacrifices, and struggles.

Tejinder Singh Bhogal is the co-founder and Director of Innobridge Consulting Pvt. Co; Dean, Social Development, ISABS; and a Fellow of the Lead-India network. Being a first-batch alumnus of Institute of Rural Management, Anand (IRMA), he worked for 15 years at the grassroots. He organized village communities, while working with National Dairy Development Board and PRADAN around oranges, social forestry, and micro-irrigation. For the last 18 years, he has worked as a consultant and trainer in OD and HR for over 70 NGOs and other social sector organizations. He is a co-author of the book *HR Strategies to Build Effective Institutions for the Poor.*

Visty Banaji is a founder and the CEO of Banner Global Consulting, a boutique strategic HR consulting firm that has provided consultancy services to top-ranking corporates. Until 2010, he was Executive Director and President (Group Corporate Affairs) on the Board of Godrej Industries Limited. Before that, he worked in France & India for Alstom. He started his career in 1973 as a Tata Administrative Service Officer with the Tata Group, where he became Corporate HR Head for Telco (now Tata Motors). He is a recipient of the NHRDN Pathfinders Award 2006 for Seasoned HR Professionals. He has been a member of the Confederation of Indian Industry (CII) National Committees on Human Resources and Industrial Relations for several years. He is also an Executive Committee member of the Employers' Federation of India.

Vikas Rai Bhatnagar is the founder and Chief Action Research Officer of a firm specializing in providing action research services. A visiting faculty at

IIM-Indore and a research scholar with Management Development Institute (MDI) Gurgaon. Vikas—a certified Gallup Strengths Coach—is a sought after executive coach. Leading global consulting organizations have empanelled him for strategic HR interventions. World HRD Congress conferred him "HR Research of the Year Award" in 2013. Before his current phase of consultancy and research, Vikas led HR of MNCs such as GE, Aventis, as well as other Indian organizations. He began his career with the Indian Army, where he was a Captain.

Aquil Busrai graduated in commerce with a distinction and is a gold medalist from Xavier Labour Relations Institute from where he obtained his MBA degree. He holds a post-graduate degree in law and an advanced diploma in training and development. He was awarded a PhD degree in April 2012. He has 42 years of HR experience with blue-chip organizations such as Unilever in Kenya and India, Motorola in Asia Pacific, Shell in Malaysia, and IBM in India. Dr Busrai is a former National President of NHRDN and a Fellow of the All India Management Association.

Somnath Chattopadhyay has done his PhD from Indian Agricultural Research Institute, has worked as Director–Behavioural Sciences at SIET Institute, Hyderabad; National Institute of Health Administration & Education, New Delhi, and at IIMA as a Professor of behavioral sciences. He was also one of the founders of ISABS where he was both Dean and Executive Director. He is considered an NTL style T-Group trainer and OD consultant. He is known for his OD practice using a variety of T-group-based methodologies. He has changed the way of thinking in L&T, ECC, SAMTEL, and many other organizations. He worked very closely with Dr Udai Pareek, Rolf Lynton, etc., and was a student of Lee Cronbach at Stanford Univeristy. He co-authored books with Manohar Nadkarni and Udai Pareek on managing change.

Ganesh Chella, with over three decades of professional experience, has come to be respected and acknowledged for his contributions as a practitioner, consultant, coach, and thought leader in the field of organization development, human resources management, and executive coaching. He is the founder of totus consulting, a strategic HR consulting firm that addresses the OD needs of high growth organizations, and totus HR School, an institution that strives to make HR functions in organizations effective by promoting the professional development of its practitioners. He is also the Executive Vice Chairman of Coaching Foundation India, India's leading executive coaching institution.

Keith C. D'Souza, PGDIR&PM (XLRI), Fellow (IIMA), is currently a Professor, S. P. Jain Institute of Management Studies and Research, Mumbai. He has 36 years of experience, covering academic teaching and administration, research and consulting in management and human resource development, and management-level employment in the corporate sector, including 11 years as member of the faculty of XLRI Jamshedpur and 13 in the corporate sector in senior-level HR positions. Among the positions he held were: Dean (Academics) at XLRI, Executive Director at Academy of HRD, Associate VP (HR) at Ion Exchange India Limited, Director (Organizational Effectiveness) at Pfizer Limited, and Vice President (OD) at Wockhardt Limited. Dr D'Souza was a founding member of the NHRDN in India and the first full-time Executive Director of the Academy of HRD in Ahmedabad. He was a Visiting Professor at S. P. Jain Center of Management, Dubai and Singapore, for several years. He has published some books and a number of journal articles in the areas of OB and HRD. He consults with business and nonbusiness organizations in the areas of HR and OD and is a member of the governing board of the Academy of HRD.

Ravindra Dey, MHRDM (Mumbai University), MDSE, M.Com, is currently a Professor, Head of Organizational Behavior at Xavier Institute of Management & Research (XIMR), Mumbai. He has a specialization in OB, OD, HRD, and competency assessment. He is currently pursuing Fellow Program in Management from the Academy of HRD, Ahmedabad. He is actively engaged with consulting and training work in the corporate, education, and social development sectors. He has about 20 years of experience, spanning the corporate sector as well as the academic sector. In the corporate sector, he worked as Vice President (HR and Admin), EWDL, Indore; General Manager-HR, TCGRE, Mumbai; DGM-HR at Wockhardt; and Head of HR, OMCI, Mumbai. He has published research papers in reputed journals. He is an author of two books on perspective management and organizational theories structures and design.

Lalitha Iyer is an independent consultant, with a special interest in institutional change facilitation in social sector, education, and livelihoods and microfinance. A professional member of ISABS, she is currently a Dean (Professional Excellence). She is the Chairperson of Sathi, Bengaluru, an NGO working on child protection and the Director of Bharatiya Samruddhi Financial Services Ltd, NBFC of the Basix group. She is the author of *The Strategic Business Spiral* (2001) and co-edited *Institutions Consultants and Transformation: Cases Studies from the Development Sector* (2009) and *Learning Crucible* (2014) . She is a co-author of *Whose Sustainability Counts?* (2011) and *Rescuing Runaway Children* (2013). She holds an M.Sc. degree in statistics and the PhD degree in economics, and has senior-level experience in development banking with the State Bank of India. She headed Vidyaranya, a progressive school, as the Principal for three years (1998–2001) and has been interested in nonformal education for many years. She is a founder trustee of Plustrust (www.plustust.org) that offers fellowships on inclusive education and animal welfare.

Mary Jane "Maja" Balasi is an organization change consultant with Sullivan Transformation Agents. She is currently working on her PhD degree in social organizational psychology at the Ateneo de Manila University. Her experience as a facilitator in eight countries in transforming systems in different sectors such as youth, business, women, faith-based organizations, economically poor communities, and academe for more than 20 years helped her become a natural change agent. Maja is passionate in researching and mastering whole system transformation. The largest interactive group she facilitated was 30,000 participants.

Anil K. Gupta, currently a Professor at the IIMA, has been Executive Vice Chair, National Innovation Foundation; Founder, Honey Bee Network, SRISTI, NIF, and GIAN; Fellow, The World Academy of Art and Science, California 2001. Professor Gupta's *mission* is to expand the global as well as local space for grassroots and young innovators including children and youth to ensure recognition, respect, and reward for them, creating knowledge network at different levels for augmenting grassroots green inventions and innovations in informal and formal sectors. He supports social innovations in education, culture, institutions, and technology in public and private sectors to expand entrepreneurial opportunities for disadvantaged communities and social change agents.

Rajen K. Gupta is a Professor of Human Behavior and Organization Development at MDI Gurgaon. An electrical engineer from IIT-Kanpur and doctorate from IIMA, his rich experience includes working in corporate, research, teaching, authoring/editing books, and consultancy. He has four books and over 100 scientific publications to his credit. He is or has been an independent Director on the Board of Power Grid Corporation of India Ltd, a member of Customer Service Committee of Bank of Baroda, a member of Expert Group on Psychology of ICSSR, a Research Advisor to KIIT School of Management, and a member of faculty selection panels of IIM Indore, Kozhikode, and Rohtak.

Kritvi Kedia is currently working as a Project Leader with the Executive Education and Consulting Division at the School of Inspired Leadership (SOIL). She has been involved in multiple projects in the domain of leadership development and organizational transformation. With a strong research background, Kritvi has co-authored papers that have been presented at renowned national and international conferences. She is a post-graduate in organizational and social psychology from the London School of Economics and Political Sciences and completed her graduation in applied psychology from the University of Delhi where she was a University Gold Medallist.

Pradip Khandwalla was a faculty at IIMA, 1975–2002, in the OB Area. Before that, he taught at McGill University, Montreal, 1969–1975. He is a CA, MBA (Wharton School, USA), and PhD in industrial administration (Carnegie-Mellon University, USA). At IIMA, he held the L&T Chair in OB, 1985–1991, and was IIMA's Director, 1991–1996. He has received two Lifetime Achievement Awards and Seagram Research Award, Canada. His main contributions have been in organization design and theory, creativity, and management of government and turnarounds. He is also a poet and a translator. He has authored over 16 books and over 100 papers and articles in various journals. His book, *Design of Organizations*, was an international textbook used across various schools worldwide.

Inderjit Khanna had his schooling at Sherwood College, Nainital, and then did his MSc degree in mathematics from St. Stephen's College, New Delhi. He joined the Indian Administrative Service in 1966 in the Rajasthan Cadre and after initial years as SDM, Collector, Deputy Secretary, and Director Education, he spent fairly long periods of posting in the fields of rural development, education, planning, and finance. This included two tenures of over six years each with Government of India and a spell of two years as a Visiting Professor at the IIMA from 1981 to 1983. His last posting as Chief Secretary, Rajasthan, was for three years. After retirement in December 2002, he worked as the State Election Commissioner, Rajasthan, for 5 years. He is currently associated with the Gita Mittal Foundation that is running three centers for skill development of youth in Rajasthan. He is also on the committees and boards of seven institutions in the fields of health, education, and rural development.

P. Sethu Madhavan is an established HR professional who has the rare combination of experience as an HR practitioner, trainer, consultant, researcher, and teacher. Currently, he is working as an advisor with Tawazun in UAE. Earlier, he has worked with Center for Organization Development, Academy of Human Resource Development (AHRD), E&Y, ADCO (Abu Dhabi Company for Onshore Oil Operations), and L&T. He has rendered training and consulting services to leading companies in India and abroad. He has many publications, case studies, psychometric, and survey instruments to his credit. He has served as an editor of journals and as a member of boards. He has been speaking regularly at many HR events in Europe and Asia.

Gopal P. Mahapatra is Chief Learning Officer, RPG Enterprises, Mumbai. A post-graduate from Xavier Institute of Social Service, Ranchi, and Fellow from IIM-Bangalore, he has over 29 years of rich and diverse professional experience in MNCs (10 years as Senior Director HR at Oracle India), private and public sector units in SHRM, talent management, leadership development, OD, executive coaching (ICF and BCC), 360-degree feedback, performance management, assessment development centers, and TQM. He is the Associate Editor of *South Asian Journal of Human Resources Management* by SAGE. He was President, NHRDN, Bengaluru. He has co-edited 3 books and wrote over 20 journal papers. He serves in the Academic Council of B Schools and has been a visiting faculty at many leading B Schools in India. He has received Lifetime Achievement and Distinguished Alumni Award from Xavier Institute, Ranchi.

Sheba Mathew has completed her MPhil degree in interdisciplinary study from IIT Bombay. She has specialized in OB and industrial psychology in her bachelors and masters degrees from the Maharaja Sayajirao University of Baroda. She has received gold medal in masters as well as her research work in OB. She is currently working as a Research Associate at S. P. Jain Institute of Management and Research. Her area of interest includes research in organizational culture, multi-culture organizations, women in organization, psychology of consumer behavior, etc. She is trained in psychometric scale and test-making for industry and various other psychometric test administrations. She looks forward to engage in teaching and research as a career.

Naresh N. Mehta is an HR professional passionate in the field of strategy management and transformation. He has managed operations and organization transformation for over 35 years. He is a mechanical engineer with masters degrees in HR and finance. He holds Fellowship, AHRD, and PhD in management. He is a Fellow of Academy of Human Resource Development.

D. Nagabrahmam is a Fellow of IIMA and has about 40 years of experience in teaching, research, and academic administration. He had been involved for about 16 years with T. A. Pai Management Institute (TAPMI), Manipal, as the Director and known to be responsible for its development, growth, and reputation. Earlier, he was a faculty member in the Osmania University and at the IRMA. During his tenure at TAPMI, he was instrumental in ushering in many significant changes. These include introduction of the "thematic curriculum" for the first time in India, commencement of a weekend Executive PGDM program, being the first school in India seeking accreditation to the Association to Advance Collegiate Schools of Business, and introduction of the innovative course "Management In Practice" in management. Currently, he is the Honorable Director at Asian School of Business, Trivandrum.

Rajeshwari Narendran is PhD in management and MBA from Andrew Towl Scholar-Harvard Business School (GloColl and CWW), Boston, USA, and Faculty Development Program, IIMA. She has the distinction of being the first ever and youngest National President of Indian Society for Training and Development, New Delhi (2013–2014). She is a Professor of HR/OB at M. L. Sukhadia University, Udaipur, and Visiting Professor at IIMA, IIM Udaipur, and many universities across the globe. She is currently deputed as Director, Academy of HRD, Ahmedabad. She has to her credit over 140 articles and research publications, and recently, a book comprising 55 case studies titled *Innovations in HRD and Training*. She is the board member of NHRDN, and also a member of Developing Countries Concern Committee, (IFTDO). She was earlier a board member of IFTDO. She has represented in the UN and is working on Global Policy for Diversity and Women Empowerment at Global Level.

Indira J. Parikh is the Founder President of FLAME. She has been involved in creating the academic vision and shaping FLAME. She has conceptualized the Center for Organizational Growth and Excellence (COGE) with its focus on thresholds of life, lifelong learning, development, and growth. Professor Parikh has done MEd from the University of Rochester, New York, USA, and the Doctorate from the Gujarat University. She was a faculty at IIMA for over 30 years and the Dean from 2002 to 2005. She taught at INSEAD, Fontainebleau (France), and Texas A&M University. She has specialized in organization development and design, and institution building. She has designed and offered management and leadership development programs in public sector, private sector, and multinational organizations. She has been a consultant to various national and international organizations. Professor Parikh has been honored with several lifetime achievement awards both nationally and internationally. She has written numerous articles published in national and international journals and is the co-author/author of several books.

Dennyson Francis Pereira, MA (1952), PhD (1978), was the first to start and head the HRD function in L&T. He retired in 1989 from L&T after working for over two decades. Dr Pereira is trained in Tavistock Rorschach, NTL, and in power motivation by Dr David McCleland of Harvard University. He worked with Marguette Hertz and Dr Muriel James on TA and conducted many workshops. He co-authored *Recent Experiences in HRD*. He was a regular Visiting Professor at IIMA since its inception, NITIE, Bhavans Institute of Management, Narsee Munjee Institute of Management, Indian Society for Training and Development (ISTD), and National Productivity Council. Widely read and research-driven Pereira conducted several international workshops and seminars and was key-note speaker at United States of America Agency for International Development Conference and for several national conferences. He presented papers in various international forums including International Conference of Cross Cultural Psychology, Jerusalem (1968), Academy of Management Conference Louisiana, and National Conference of Counseling Psychology, Ottawa. He was awarded for Outstanding Contribution to HRD in 1989.

D. M. Pestonjee completed MA and PhD from Aligarh Muslim University, was a Professor (OB) at IIMA, and has been Chair Professor (GSPL) at Pandit Deendayal Petroleum University. He was awarded D.Litt (Honoris Causa) by Banaras Hindu University. He was associated with IIMA from 1979 to 2001. He has had the privilege of teaching, researching, and publishing with Dr Udai Pareek. He has several publications to his credit.

Kavil Ramachandran is the Executive Director at the Thomas Schmidheiny Centre for Family Enterprise at the Indian School of Business (ISB). He as a founding faculty had set up the Wadhwani Centre for Entrepreneurship Development at the ISB in 2001. Later, he was the Associate Dean (Academic Programs) before becoming the Thomas Schmidheiny Chair Professor of Family

Business and Wealth Management. He has specialized in family business, entrepreneurship, and strategy, and has three decades of combined experience as an academic at IIMA, and ISB. He did his PhD from the Cranfield University, UK. His latest book *The 10 Commandments for Family Business* has been published by SAGE in January 2015. He has been a pioneer academic entrepreneur, propagating the message of strengthening family business in India and outside.

S. Ramnarayan is an engineer, MBA, and a PhD in OB from the Case Western Reserve University, Cleveland, Ohio. He has worked as a manager, a consultant, and a professor at different times in his career. After being a Professor at IIMA, for 14 years, he is presently with ISB for close to a decade. He has carried out research funded by Ford Foundation, World Bank, Commonwealth Secretariat, Department for International Development, Human Capital Leadership Institute, and German Science Foundation. Apart from research papers, monographs, and case studies, Ram has co-authored nine books on topics of change management, organization development, and behavioral challenges in strategic organizations.

G. P. Rao is currently Management Advisory and Managing Partner, GPR HR Consulting LLP. He has worked in HRM for 37 years with SAIL, JK Organization, Birlas, and Reliance group in India and Malaysia. His last assignment was as chief of human resources and management services, Recron Malaysia, a Reliance group company. He did a degree in law; masters in commerce, public administration, social work, and business management; and certification in assessment tools of MBTI, OPQ, SHL, Human Edge, etc. He also received an honorary PhD. He has received a number of awards including the Outstanding Contribution Award from NHRDN, HR Leadership Excellence Award at Singapore, and Achievers Awards at Colombo. He is associated with NHRDN, National Institute of Personnel Management, ISTD, CII, Associated Chambers of Commerce of India (ASSOCHAM), and PHD Chamber of Commerce and Industry having held senior positions. He has a vision to create and nurture a large NGO in India to identify poor youngsters having passion to grow, provide them with skill sets, and make them employable. He is currently working as a management advisor in India and abroad, helping corporations, academic institutes, NGOs, and government. His focus areas include leadership development, coaching and mentoring, diversity management, action learning, and ER capability.

Anil Sachdev, founder and CEO of School of Inspired Leadership, is recognized as a thought leader in talent management, leadership development, and organizational transformation. He is a member of the World Compassion Council, serves on the Academic Committee of CEDEP, the Leadership Institute at Fontainebleau, is the Trustee of the Chinmaya Mission, a global nonprofit organization, and is part of the Advisory Board of Schneider Electric. He began his career with Tata Motors in 1975 after completing his MBA and joined Eicher in 1978 and worked in HR, operations, and TQM.

He is the founder and CEO of Eicher Consultancy Services and also of Grow Talent Company Limited, which was created in 2000 and became the leader in the strategic HR consulting industry by 2006.

Surabhi Sharma works as a Program Leader with the Executive Education and Consulting Vertical at the School of Inspired Leadership. Her diverse professional experience comes from over eight years of exposure in IT, public relations, and human resources domain. Her journey with SOIL has helped her in gaining valuable experience through various leadership development and organizational transformation interventions with large Indian and global organizations across sectors. She also had a short entrepreneurial stint with her PR consulting firm. She has done her post-graduation in HR from SOIL. She also holds BE degree in IT from the University of Rajasthan. She enjoys quizzing, traveling, and meditation.

Paul Siromani is one of the founding members of ISABS. He specializes in training for greater awareness and skills in personal, interpersonal, and group processes. He worked with different levels of managers and supervisors in BHEL, Neyvelli Lignite, and other corporations in the private sector, NGOs, services organizations, college principals, hospital staff, religious order priests and nuns, and staff of community development projects using T-group methodology and process work for transformation. He also conducted "open space" sessions, LSIP workshops, 360-degree feedback, and extension motivations labs along with Dr Udai Pareek and Dr TV Rao, and facilitated large-scale interactive process workshops for feedback and self-renewal of projects and communities in Orissa, Bihar, Madhya Pradesh, West Bengal, Manipur, and for church congregations in Bengaluru.

S. Y. Siddiqui is currently the Chief Mentor at Maruti Suzuki India Limited. He is a post-graduate in HRM with a career track of 35 years in HR and business roles of good Indian corporates and MNCs, including New Holland Tractors India (FIAT GROUP), DCM Toyota; DCM Daewoo Motors and DCM Benetton India; and Escorts Limited. Having widely traveled abroad, Siddiqui handled global and multiculture HR issues having a unique experience of working in multicultures and multinationalities. He has handled a broad management and leadership role at board level at Maruti since 2006 and from January 2008 operated as COO for five years as head of all central functions. He is passionate about cricket and played at the North Zone Inter Varsity level from 1974 to 1979 representing Jamia Millia Islamia, New Delhi.

Roland L. Sullivan is an original 100-Change Agent who has consulted in 44 countries with over a thousand organizations. He has taught OD in over 20 universities around the world. Roland is known for his exemplary work in change. He coined the phrase "whole system transformation" and has been the main focus of his transformation journey with client-organizations for more than 50 years.

K. K. Verma is the former Director of Academy of Human Resources Development. He is a fellow of All India Management Association and holds a post-graduate degree in social work from Delhi School of Social Work, the University of Delhi, and a post-graduate diploma in labor laws from Indian Law Institute, New Delhi. He has been associated with HRD functions for over 40 years in corporate sector as well as in academic bodies. He has worked as a Research Associate at IIMA, in early part of his career and held various positions subsequently including Corporate General Manager-HRM at Bank of Baroda (BOB), zonal manager of the smallest and largest networks of bank, Managing Director at BOB Capital Markets Ltd, and Principal of BOB Staff College. He also worked as corporate head-HR of DS Group of Industries, Noida. He has authored two books and published several papers and articles in reputed journals and newspapers.

P. Vijayakumar is the Chairperson of the Centre for Social and Organizational Leadership (C SOL) at Tata Institute of Social Sciences (TISS), Mumbai. C SOL offers a diploma program in OD and change for the working executives. His current research interests are focused on the reconceptualization of the contemporary organizational space with a special emphasis on (a) strategizing, (b) OD and change, and (c) leadership and innovation. He argues for a practice perspective with a focus on collaboration to foster the linkage between micro- and macro-organizational aspects.

Zeb O. Waturuocha is an honors graduate in economics, from Punjab University, gold medalist in MBA, doctorate in management science from the University of Mysore, masters in psychology from Anamali University, and PG diploma in HRM from IGNOU. Zeb is a professional member of ISABS and an accredited facilitator for future search, appreciative inquiry, emotional intelligence, and sensitivity training using the laboratory methodology. He is a certified MBTI practitioner and a certified coach. He has more than two decades experience of working among slum dwellers, tribal community, differently abled people, women and children, corporate consulting, OD, and training. He has authored two books—*Facilitating Self Awareness* and *Excuses Galore.* He is the Principal at the Center for Emotional Literacy and Leadership (CELL) and the founder head of aui consultants (the human process people).

Index